HISTORY OF TECHNOLOGY

HISTORY OF TECHNOLOGY

History of Technology

Volume Twenty-two, 2000

Edited by
Graham Hollister-Short

CONTINUUM
London and New York

First published 2001 by
Continuum
The Tower Building, 11 York Road, London SE1 7NX
370 Lexington Avenue, New York, NY 10017-6503

British Library Cataloguing-in-Publication Data

History of Technology – 22nd volume (2000)
1. Technology – History – Periodicals

ISBN 0–8264–5339–2
ISSN 0307–5451

Typeset by BookEns Limited, Royston, Herts.
Printed and bound in Great Britain by
The Cromwell Press, Trowbridge, Wilts

Contents

Editorial

To Euro-sceptics the miserable fate of ELDO (European Space Vehicle Launcher Development Programme) in the days when the 15 of the EU were still only the six of the EEC may well seem emblematic of much more than 'a failure to communicate'. The failure to communicate is the subject of Stephen Johnson's account of the ELDO project and of Europa I, Europe's response to the US Apollo programme. It makes absorbing, if melancholy, reading. When the ELDO project was finally wound up in 1974 after 12 years of 'negotiation, compromise and struggle' and six consecutive launch failures, it was clear to everyone involved that the project's fundamental flaws were social and political. For many Europeans 'Apollo highlighted European fragmentation and organisational ineptitude'.

The success of a variety of bodies in the Ruhr valley in restoring in large measure a landscape ruined by industrialization, the subject of Edmund Todd's paper, demonstrates by contrast that collaboration, albeit on a less grandiose scale than great European projects, can achieve much. The lesson here, perhaps, is that as a result of central direction being largely absent or weak, local interest groups could fill the vacuum and negotiate mutually satisfying outcomes. The storage lakes judged necessary for ecological and aesthetic reasons by the Ruhrverband (Ruhr Association) to assist in the cleaning up of the River Ruhr, for instance, could also be made to fit the needs of the RWE (Rhenish Westphalian Electricity Company). The storage reservoirs created the high heads needed to run the company's hydro-electric plants. County commissioners, mayors and industrialists co-operated in all these water undertakings to create 'a new way of life that was politically, socially, economically, aesthetically and technically viable'.

In 'The Craft of the Bell-Founder' Tatyana Shashkina puts flesh, so to speak, on the bones of her earlier joint paper (with Raisa Podmogil'naya) on bell-founding practice: 'Universals in Traditional Methods of Design in the Bell-Founding Arts' which appeared in Volume 3 of *ICON* in 1997. This last paper investigated the modular nature of constructional practices in England, France, Germany and Russia. That tradition ceased abruptly in Russia after the October Revolution of 1917 but the materials discovered by Dr Shashkina have permitted her to reconstruct something of the history of the Chyaryshnikov family, one of the leading firms of bell founders in the nineteenth century, as also to locate a number of the bells cast in their workshops.

John Farey's *Treatise on the Steam Engine*, Volume 1 of which appeared in 1827, has been described by G. von Tunzelmann as 'the finest book on technology published during the Industrial Revolution'. Farey's second volume on the post-Watt period, and projected on the same scale as the first, was never completed. Even the truncated version was published only in the century following his death. The greatly abridged version omitted, other than the work of Arthur Woolf, accounts of contemporary steam engineers that would rank now at first-class primary source material, for Farey, a consulting mechanical engineer, had an unrivalled knowledge of the high pressure steam engines then being built by some of the best machine constructors in England. A.P. Woolrich, in relating the vicissitudes that beset Volume 2, has also assembled what to all intents and purposes is a complete set of the illustrations planned by Farey for this volume.

Michael Fores shows in the exemplary case of Werner Siemens how divorced most historical writing on engineering and on its history has been from what happens on the ground. The reification of historical ideas has produced, as Fores says, 'the advent of conditions ... which have never been detected outside the written page ...' The world of difference between luciferous and lucriferous activity, or rather failure to understand the difference (the ultimate source of the confusion complained of by Fores, as it seems to me), beautifully enunciated by Francis Bacon in the *Novum Organon*, was understood at least by the men who brought the Royal Society into existence in 1660.

Yuji Jido examines the introduction of chemical engineering into Japan and the conceptual problems that had to be overcome before the true nature of chemical engineering operations was understood. Uncertainty about this was reflected in the fluidity of nomenclature adopted, illustrating once again how important it is to understand how technical vocabulary is generated.

Friedrich Toussaint mentioned Ludwig von Eschwege to me at the ICOHTEC conference in Belfort in 1999, and since there was nothing in the English language on this distinguished metallurgist and his role in transplanting new technology to Brazil in the early nineteenth century, he undertook to contribute a sketch for this volume of von Eschwege's work, in Portugal and Brazil, and as well a summary description of his magnum opus, *Pluto Brasiliensis*, on the mines of Brazil.

Readers will notice that this volume contains a new feature: a symposium on the current state of history of technology in Britain. The idea for this began to take shape when I read a paper some two years ago by Lutz Engelskirchen on the development of the discipline in Germany after 1945. The reforms made in higher education in the Bundesrepublik transformed the situation from one in which the subject was sustained by a handful of individuals (outside academe) to one in which some 20 university posts had been created. Why, I wondered, should the case have been so very different here? After talking the situation over several times with Ian Inkster, I decided to invite those to take part whose names appear in the symposium. With Ian Inkster's active help in the work of preparation, the symposium papers were finally assembled. I hope the thoughts and reflections there stated, sometimes with vigour, will stimulate further debate. The discussion may well continue into future volumes, but that will be for Ian Inkster, the next editor, to determine.

Graham Hollister-Short
London

The Contributors

Professor Hans-Joachim Braun
Universität der Bundeswehr
Hamburg
Holstenhofweg 85
D-22039 Hamburg
Germany

Professor R. Angus Buchanan
Centre for the History of
Technology
School of Social Sciences
University of Bath
Claverton Down
Bath, BA2 7AY
England

Professor David Edgerton
Centre for the History of Science,
Technology and Medicine
Sherfield Building
Imperial College
London, SW7 2AZ
England

Michael Fores
80 Lexham Gardens
London, W8 5JB
England

Dr Graeme Gooday
Department of Philosophy
University of Leeds
Leeds, LS2 9JT
England

Professor Rupert Hall
14 Ball Lane
Tackley, Oxfordshire
OX5 3AG
England

Dr Graham Hollister-Short
Centre for the History of Science,
Technology and Medicine
Sherfield Building
Imperial College
London, SW7 2AZ
England

Professor Ian Inkster
International Studies
Nottingham Trent University
Clifton Lane
Nottingham, NG11 8NS
England

Professor Yuji Jido
College of Asia Pacific
Management
Ritsumeikan Asia Pacific
University
1-1, Jumonjibaru, Beppu
Oita 874-8577
Japan

Dr Stephen B. Johnson
Department of Space Studies
University of North Dakota
School of Aerospace Sciences
4149 Campus Road, Clifford Hall
Grand Forks, ND 58202-9008
USA

Dr Tatyana Shashkina†
Pushkinskaya 58–5
65011 Odessa
Ukraine

Professor Edmund N. Todd
University of New Haven
300 Orange Avenue
West Haven, CT 06516
USA

Professor Hugh Torrens
Department of Geology
Keele University
Keele, Staffordshire, ST5 5BG
England

Dipl. Ing. Friedrich Toussaint
Geschichtsausschusses des VDEh
Haus Fischerkotten
Kuhlendahlerstr. 299
D-42553 Velbert-Neviges
Germany

A. P. Woolrich
Canal Side
Huntworth
Bridgewater
Somerset, TA7 0AJ
England

Notes for Contributors

Contributions are welcome and should be sent to the editor. They are considered on the understanding that they are previously unpublished in English and are not on offer to another journal. Papers in French and German will be considered for publication, but an English summary will be required. The editor will also consider publishing English translations of papers already published in languages other than English. Include an abstract of 150–200 words.

Authors who have passages originally in Cyrillic or oriental scripts should indicate the system of transliteration they have used. Be clear and consistent.

All papers should be rigorously documented, with references to primary and secondary sources typed separately from the text, double-line spaced and *numbered consecutively*. Cite as follows for:

BOOKS

1. David Gooding, *Experiment and the Making of Meaning: Human Agency in Scientific Observation and Experiment* (Dordrecht, 1990), 54–5.

Only name the publisher for good reason.

Reference to a previous note:

3. Gooding, *op. cit.* (1), 43.

Titles of standard works may be cited by abbreviation: *DNB*, *DBB*, etc.

THESES

Cite University Microfilm order number or at least Dissertation Abstract number.

ARTICLES

13. Andrew Nahum, 'The Rotary Aero Engine', *Hist. Tech.*, 1986, 11: 125–66, esp. 139.

Please note the following guidelines for the submission and presentation of all contributions:

1. Type your manuscript on good quality paper, on one side only and double-line spaced *throughout*. The text, including all endnotes, references and indented block quotes, should be in one typesize (if possible 12 pt).
2. In the first instance submit two copies only. Once the text has been agreed, then you need to submit three copies of the final version, one for the editor and two for the publishers. You should, of course, retain a copy for yourself.
3. Number the pages consecutively throughout (including endnotes and any figures/tables).
4. Spelling should conform to the latest edition of the *Concise Oxford English Dictionary*.
5. Quoted material of more than three lines should be indented, without quotation marks, and double-line spaced.
6. Use single quotes for shorter, non-indented, quotations. For quotes within quotes use double quotation marks.
7. The source of all extracts, illustrations, etc., should be cited and/or acknowledged.
8. Italic type should be indicated by underlining. Italics (i.e. underlining) should be used for foreign words and titles of books and journals. Articles in journals are not italicized but placed within single quotation marks.
9. **Figures.** Line drawings should be drawn boldly in black ink on stout white paper, feint-ruled paper or tracing paper. Photographs should be glossy prints of good contrast and well matched for tonal range. Each illustration must be numbered and have a caption. Xerox copies may be sent when the article is first submitted for consideration. Please do not send originals of photographs or transparencies but if possible have a good-quality copy made. While every care will be taken, the publishers cannot be held responsible for any loss or damage. Photographs or other illustrative material should be kept separate from the text. They should be keyed to your typescript with a note in the margin to indicate where they should appear.

   Provide a separate list of captions for the figures.
10. Notes should come at the end of the text as endnotes, double-line spaced.
11. It is the responsibility of the author to obtain copyright clearance for the use of previously published material and for photographs.

A Failure to Communicate: The Demise of ELDO*

STEPHEN B. JOHNSON

> The failure of F11 in November 1971 brought home to the member states – and this was indeed the only positive point it achieved – the necessity for a complete overhaul of the programme management methods.
>
> General Robert Aubinière, 1972[1]

In July 1969, the world looked on in amazement as the United States landed the first humans on the moon. Despite growing disenchantment with the war in Vietnam, student protests, assassinations and race riots, the *Apollo* lunar programme showcased the United States at its finest: a technological and organizational marvel for peaceful purposes. A number of observers noted that the real importance of *Apollo* was its demonstration of managerial prowess. For some Europeans, *Apollo* highlighted European fragmentation and organizational ineptitude, as they watched the fourth consecutive failure of their own grandiosely named rocket, *Europa I* on 3 July 1969.

Whereas *Apollo*'s mandate included a Presidential directive, national pride and an all-out competition with the Soviet Union for the loyalty of the world's peoples, *Europa I* began as a search to find a use for a cast-off ballistic missile. When British leaders decided to use American missile technology in the late 1950s, their own obsolescent rocket, *Blue Streak*, became expendable. The British decided to market it as the first stage of a European rocket, simultaneously salvaging their investment and signalling British willingness to co-operate with France, a gesture they hoped would lead to British acceptance into the Common Market. Complex negotiations ensued, as first Britain and France, and then West Germany, Italy, Belgium and the Netherlands warily decided to co-operate to build a European rocket. Each hoped to gain access to their neighbours' technologies and markets, while protecting their own as much as possible.

The European Space Vehicle Launcher Development Organisation (ELDO) reflected these national ambitions. Without the ability to let contracts or to direct the technical efforts, the Secretariat of ELDO tried

* For an explanation of this and other acronyms please see the list on pp. 19–20.

with growing dismay to integrate the vehicle, while its Member States minimized access to the data necessary for such integration. Not surprisingly, costs rose precipitously and schedules lengthened. After successful tests of the relatively mature British stage, integrated tests of the rocket's other stages between 1967 and 1972 failed miserably. Embarrassed engineers and managers struggled to identify and fix the technical problems, even as they recognized that ELDO's fundamental flaws were social and political.

The contrast between European failure and American success in July 1969 could not have been starker, with American astronauts returning to earth to lead a world publicity tour, while European managers and engineers defended themselves from criticism as they analysed the latest explosion. ELDO's record of failure continued for over four more years before frustrated European leaders dissolved the organization and started afresh. Ultimately, ELDO's technical failure resulted from the poor communication and conflicting goals of its member organizations. Its internal problems reflected the political and cultural divisions of Western Europe, resulting in a technical device with little internal consistency. The *Europa I* and *Europa II* launchers failed because European countries believed that their individual nationalist agendas overrode the need for European integration.

EUROPEAN ROCKETRY AND THE CREATION OF ELDO

Interest in and development of rockets in Europe began before World War II in a number of countries. By far the most important of them was the German programme leading to the *A-4* ballistic missile, better known as the *V-2*. Since rocketry was one of the few technologies of military import not restricted by the Versailles Treaty, German Army Ordnance began to support the amateurs of the Verein für Raumschiffahrt. Just prior to Hitler's ascension to power in January 1933, Army Ordnance coaxed young engineering student Wernher von Braun to lead their rocketry programme. The Nazi regime, after some delay, began to pour large sums into the programme, resulting in an operational missile that terrorized the populations of Antwerp and London beginning in 1944. Although its military impact was minimal, it caught the attention of military technologists around the world.[2]

After the war, each of the victorious powers acquired German rocket technology and experts. The United States acquired the lion's share of both, including von Braun, and parts for some 60 *A-4s*. German rocket engineers assisted the Soviet Union's early programme, and helped the British launch three *A-4s* in October 1945 from a site near Cuxhaven.[3] The French acquired the services of a few experienced Germans from Peenemünde, and began working at Vernon, near Paris, on an *A-4* derivative vehicle and a new rocket engine.[4]

In March 1949, the French Directorate for Armament Studies and Fabrication decided to build *Véronique*, a single-stage, liquid-fuelled sounding rocket. From 1951 to 1964, engineers extended the design, improving altitude capability from two to 315 kilometres. They initially tested the unguided rocket in southern France, later at Hammaguir in the Algerian desert under the direction of Colonel Robert Aubinière, and eventually in Kourou, French Guiana.[5]

After the 1956 decision by the French to build a nuclear force, missile development expanded rapidly. French engineers began development of rockets capable of placing small satellites in orbit. The French state rocket consortium, the Society for Study and Development of Ballistic Engines (Société pour l'Étude et la Réalisation d'Engins Balistiques – SEREB) studied the issue, and in 1960 ruled positively on the feasibility of the rocket eventually known as the *Diamant*. The newly created French space agency, Centre National d'Études Spatiales (CNES), funded the launcher project, while SEREB developed the stages and the military tested them. With some assistance from Colonel Edward Hall, the American Air Force officer who initially developed the *Minuteman* Intercontinental Ballistic Missile, these efforts came to successful fruition with France's launch of a small test satellite in November 1965.[6]

The British also developed rockets, and nuclear weapons to place on them. They tested their first fission bomb in 1952 off the coast of Australia, and a fusion weapon in May 1957 over Christmas Island. From the late 1940s on, they developed a variety of missiles, including air-to-surface, surface-to-air, air-to-air and ship-to-air weapons. When in 1954, American Secretary of Defense Charles Wilson offered to collaborate with Britain on an Intermediate Range Ballistic Missile, the British expressed interest. The Americans allowed the formation of agreements between the British companies de Havilland and Rolls-Royce, and American corporations Convair and North American Aviation. American experts evaluated the British design soon known as *Blue Streak*.[7]

However, American efforts quickly surpassed those of the British, and in 1956 the United States offered to place *Thor* missiles in Britain five years sooner than *Blue Streak* would be available. British officials accepted the offer in 1958, and ordered their engineers to increase *Blue Streak*'s range to 2500 miles with underground basing. In April 1960, the British military cancelled *Blue Streak* altogether, instead purchasing American air-launched *Skybolt* and sea-launched *Polaris* missiles.[8]

Blue Streak's cancellation led British officials to consider its potential as a satellite launcher. The British had sunk £60 million into the project, and conversion to a launcher would require £240 million more. Some argued that the technology would be obsolete by the time of completion. On the other hand, Britain would no longer be dependent on the United States to launch satellites. *Blue Streak* could also be used to forge closer relationships with France. Needing de Gaulle's support to join the Common Market, Prime Minister Macmillan hoped a joint launcher programme with France would smooth the path of Britain's application. Supported by the United States, which continued to promote integrated European endeavours, Macmillan decided to approach France in late 1960.[9]

The French reaction was cautiously optimistic. They wanted to make a joint approach to other European governments, but had two prerequisites: that the second stage should be French, and that the cost be carefully studied. They also expressed keen interest in gaining access to inertial guidance and nose cone re-entry technologies. Because the United States prohibited the export of those technologies to France, this would have been a fatal objection. Unexpectedly, de Gaulle threw himself behind the project, even without the military technologies. In a meeting with Macmillan in January 1961, he agreed to join the project.[10] They scheduled a conference the month following in Strasbourg to broach the subject to other governments.

Whereas Macmillan saw the launcher programme as a way to gain French support for British entry to the Common Market, de Gaulle saw it as a means of fulfilling French technological ambitions. de Gaulle used space and nuclear programmes to create a permanent 'technological revolution' to support a strong and independent France with 'grandeur'. He supported integrated European efforts insofar as they supported an independent France. France would lead European efforts, and benefit the most from them. French contributions to European space and nuclear efforts were a fraction of French expenditures on national programmes.[11]

After the Strasbourg Conference, the Germans accepted an offer to build the launcher's third stage. This gave them the opening to rocketry that they had been looking for. They would not build rockets alone because of the Nazi *V-2* heritage, but as part of a European effort they could develop rockets without antagonizing their neighbours. Desperate for an agreement by autumn 1961, the British put substantial diplomatic pressure on Italy to join as well. Already building sounding rockets under American licence, the Italians reluctantly agreed to build the satellite test vehicle. Negotiations produced the convention for the European Space Vehicle Launcher Development Organisation (ELDO) in March 1962.

The three-stage *Europa I* launch vehicle mirrored its political origins. Britain's determination to keep *Blue Streak* alive was the foundation of the organization, and of the first stage. The French seconded the British proposal, and developed the second stage. West Germany wanted to rebuild rocket capability without political repercussions, and so piggy-backed upon the British and French with the third stage. Italy would build the test satellite, almost an afterthought to the main work on the launch vehicle. The Netherlands and Belgium re-enacted their role as 'glue' between the larger countries, manufacturing the ground equipment. Australia provided the launch site at Woomera. Britain paid a heavy price for its desperation and was saddled with 38.79 per cent of contributions to the projected £70 million Initial Programme. They would pay this percentage of actual costs, even if they overran. The Initial Programme, scheduled for completion by the end of 1965, divided the remaining costs, with France, West Germany and Italy paying 23.93 per cent, 18.92 per cent, and 9.78 per cent, respectively, and Belgium and the Netherlands,

2.85 per cent and 2.64 per cent. The convention would come into force when Britain, France and West Germany ratified it. Until that time, the countries organized a 'Preparatory Group'.[12]

ELDO's structure emphasized national interests. Because they based their contributions on existing programmes, Britain and France insisted upon managing their stages through their national government organizations, according to their own procedures but at ELDO's expense. ELDO could place contracts only 'in agreement with the government of the Member State in whose territory the work is to be carried out'. In short, ELDO provided but could not control funding. Since member states contributed funding in fixed proportions, but expended the funding according to actual expenses, each country had a built-in incentive to increase costs to recoup its investment. For example, if Belgium overran its budget by 50 per cent, it would contribute only its 2.85 per cent share to that overrun! Another troubling factor was that the launcher had no definite mission. The Convention referred to users that *could* either be the member states or other organizations.[13]

Although created as an organization for technological co-operation, the Member States severely circumscribed ELDO's authority, rendering co-operation difficult at best. The job of the Secretariat required delicate negotiation skills, a fact recognized in the appointment of Italian ambassador R. Carrobio di Carrobio to the post of Secretary-General of ELDO, which came into official existence in February 1964.[14] Carrobio would need all of his negotiating skills for the problems of ELDO.

AGEING TECHNOLOGY AND CHANGING OBJECTIVES

The ELDO Preparatory Group established a Technical Committee in December 1961 that initially focused on launch facilities, the guidance system, the Italian satellite test vehicle and analyses of future design options. However, without any staff except for those supplied by the national governments, and technical problems more complex than originally thought, ELDO moved slowly in its first three years. Delayed for two years, the first and second *Blue Streak* launches in 1964 succeeded, an auspicious, albeit tardy, beginning. However, delays and technical difficulties led to substantial cost escalation, from the 198 MAU (million accounting units)[15] originally budgeted for (the equivalent of £70 million) to 400 MAU at the end of 1964.[16]

In the mid-1960s the prospect of commercial communication satellites became a reality as the United States tested its first commercial satellites, including *Telstar*, *Syncom* and *Early Bird*. Possessing the only launchers and communications satellites, the United States dictated the terms of the first international consortium for satellite telecommunications, INTELSAT. European leaders wanted to break the American stranglehold on the technology and its implementation, and thought ELDO might help. In 1964, the ELDO Secretariat reported that an upgraded *Europa I* launcher could place 20–40 kilogrammes (kg) of equipment in polar orbit, and that a more powerful *ELDO B* rocket would be able to place a 1000 kg communications satellite in geostationary orbit in the 1970s. The next year, the French delegation proposed that ELDO scrap the under-powered *Europa I*, and instead immediately begin work on *ELDO B*. The other delegates vetoed this as too risky, but agreed to reconsider it later.[17]

While Britain managed to stave off the massive new investment that ELDO B represented, her disproportionate funding of ELDO's huge overruns took its toll. In April 1966, British leaders reversed their strong support for ELDO. They believed that the rocket would be obsolete when completed and its commercial potential limited, and stated that they would neither participate in rocket upgrades, nor contribute beyond existing commitments. Under pressure from the other delegations, the British agreed in June 1966 to remain in ELDO, but only if the organization reduced Britain's contribution. In July, the delegates agreed to fund an equatorial base, inertial guidance improvements, and the *Europa II* rocket that could launch up to 150 kg into geostationary orbit. The new cost would be 626 MAU, over three times initial ELDO estimates. Believing that they could purchase American launches at costs far lower than ELDO could supply, Britain made 626 MAU a hard limit on ELDO's expenditures.

The new compromise was not enough. In February 1968, after two failures of the French second stage, the British invoked ELDO's new procedures for projected cost overruns. Two months later, the British announced that they would make no further contributions to ELDO. Italian delegates, angered by the refusal of France and Germany to include them in the Franco-German *Symphonie* communications satellite, and also by their inability to recoup ELDO contracts for their own space industry, refused to agree to a French-German 'austerity plan' that would have cut Italian portions of the programme. After yet another rocket failure in November 1968, this time of the German third stage, France, Germany, Belgium and the Netherlands agreed to make up the shortfall in British and Italian contributions to complete a scaled-back programme. Italy finally agreed to rejoin, but beyond supplying stages for two more flights, Britain withdrew. The remaining partners agreed to fund studies for a first stage replacement.[18]

NASA International Programs chief Arnold Frutkin noted the 'half-hearted and mutually suspicious character of participation by its [ELDO's] members'. With Europeans united only in their suspicions that the Americans intended to monopolize communication satellites, Frutkin believed that 'US offers in space and other fields of technology will continue to be regarded with extreme and often irrational suspicion until the comsat issue is resolved . . .'. European governments remained at least as interested in promoting their own national interests as in co-operation with each other, and held together only insofar as the United States resisted European commercial interests. American leaders sent mixed signals, both promoting and resisting European interests, and so helped keep ELDO alive.[19]

Changing technical goals contributed greatly to ELDO's problems. Without firm objectives, ELDO and French studies showed that ELDO needed a more powerful launch vehicle to place the new communications satellites into orbit. The French, who viewed ELDO as an essential part of their drive for independence from the United States, were willing to pay the price. So too were the Germans, who were willing to subsidize their re-entrance into rocketry. The Belgians and Dutch believed they needed to go along with their powerful neighbours. As long as ELDO guaranteed the Italians a technologically interesting task, Italian leaders would contribute. However, the British soon tired of funding other members' overruns, and had nothing of technological interest to gain from the programme since their first stage was operational. Convinced of America's willingness to launch European satellites, the British believed it more important to spend their money on applications satellites than on launchers. Perhaps these differences could have been overcome if ELDO's rockets had proved successful.

ORGANIZING FOR FAILURE

Like other large projects in Europe and the United States, ELDO distributed tasks to a number of widely dispersed organizations. ELDO funded the British Ministry of Aviation for the first stage. They in turn channelled funds to the Royal Aircraft Establishment, which contracted de Havilland Company for most tasks, and Rolls-Royce for the engines. De Havilland subcontracted to Sperry Gyroscope for the guidance package. ELDO funded the French Space Agency (Centre National d'Études Spatiales – CNES) for the *Coralie* second stage. They in turn contracted with the French Army Laboratory for Ballistic and Aerodynamic Research (Laboratoire de Récherche Balistique et Aérodynamique – LRBA) for second-stage development, and with the government's National Society for Study and Construction of Aviation Engines (Société Nationale d'Étude et de Construction de Moteurs d'Aviation – SNECMA) for the engines.[20]

West Germany, which had no space organization prior to the ELDO discussions, initially placed its space activities under the Ministry for Atomic Energy and Water Power (Bundesministerium für Atomkernenergie und Wasserwirtschaft). In August 1962, the Germans formed a government-owned corporation, the Space Research Company (Gesellschaft für Weltraumforschung) to study space activities, under the guidance of the German Commission for Space Research (Deutsche Kommission für Weltraumforschung), and the Ministry for Atomic Energy and Water Power expanded to become the Ministry for Scientific Research (Bundesministerium für wissenschaftliche Forschung) under the Ministry for Education and Science (Bundesministerium für Bildung und Wissenschaft – BMBW). ELDO delegated the BMBW as the national agency for the *Europa I* third stage, which in turn contracted with the newly created industrial consortium, Arbeitsgemeinschaft Satellitenträger

(ASAT). ASAT was an uneasy, and according to some, involuntary alliance between two aerospace firms, Messerschmitt-Bölkow-Blohm (MBB) and Entwicklungsring Nord Raumfahrttechnik (ERNO). Similar arrangements held for the Italian, Belgian and Dutch portions of ELDO.[21]

Complexity of the organizational structure was not in itself a problem. The difficulty was that the working groups of private companies and government–industry consortia reported to their national government organizations. These in turn reported to the ELDO Secretariat in Paris. Since ELDO gave its funding to the national governments, which distributed it according to their own procedures, industry took its orders from the national governments, not the ELDO Secretariat. This 'indirect contracting' structure interposed an extra layer of bureaucracy, and gave that layer final authority.[22]

Because contributions to ELDO were in fixed national proportions, and ELDO based expenditures on needs defined by the national governments, ELDO's structure encouraged each national programme to incur more expenses to maximize financial returns. NASA's International Programs head Frutkin noted that this structure was 'inherently and seriously faulty', with 'built-in motivation to escalate costs'.[23]

The Secretariat had no authority to force the governments or contractors to make changes; it could only make suggestions to the national governments. Nor could the Secretariat take legal action, both because contractual authority lay with the national governments, and because of vaguely worded contracts. As late as 1972, the *Europa II* Project Review Commission stated that 'there is no clear definition of responsibilities within the ELDO organization, nor between ELDO staff and ELDO contractors'.[24]

Uncertainty about roles and responsibilities led to 'a complete effacement of the Secretariat's role' in Britain and France, where governments were 'strongly structured'. When the national agencies were weak, as in the case of Germany's new organizations, this led to 'confusion in the minds of firms about the technical responsibilities of the Secretariat and those of the national agency'. In some cases the Secretariat 'did not respect the responsibilities of the national agencies', and undermined their authority.[25]

Unclear and changing requirements did not help. The 1972 review commission noted that '*Europa II* seems in a continuous state of research and development with major changes made from one launch to the next almost independently of whether the previous flight objectives have been achieved'. No single complete specification existed for the entire vehicle. Without clear specifications, engineers had no clear goals for defining telemetry measurements, for limiting the vehicle's weight, or for quality and redundancy across the project. The end result was 'a launch vehicle with little design coherence, and posing complicated integration and operational problems'.[26]

Because the British and French had designed their first and second stages before ELDO existed, they ensured that their own government organizations determined methods and standards for their own stages. ELDO had no authority to impose standards. This led to inconsistent and incomplete specifications, documentation, and quality standards and procedures. The Secretariat had no quality control organization until 1970, relying upon the national teams to enforce good manufacturing practices, use high-quality components and adhere to thorough testing procedures. At best, the result was variable quality components, processes and documentation. In practice, variable quality led to flight failures.[27]

The Secretariat's small engineering staff had limited ability to analyse problems. Before ELDO came into official existence, the Preparatory Group relied on engineers supplied by the national governments. After February 1964, the Secretariat built a small engineering staff in the Technical Directorate. Often the engineering staff 'endeavored to promote the solution of technical problems, but in some cases important solutions ... [were] ... refused on budgetary grounds'. Without access to the necessary information, the staff to perform substantial work, or the authority to make changes, the Technical Directorate performed little systems engineering. Unless contractors resolved interface problems among themselves, they remained unresolved.[28]

Problems remained unresolved because of poor communications throughout the project. No single location existed for project documentation. Nor were there standards for what documentation should be produced. Project reviewers noted that 'while certain documents were available, there was nothing systematic about this'.[29] British and French government organizations and contractors had launch vehicle experience and their own communication methods.

In communication across national boundaries, barriers of language, industrial competition and national factionalism took precedence. The most extreme case was with the German third stage contractor ASAT, which displayed

> total disinterest in the IGS [Inertial Guidance System – built by British contractor Marconi], a refusal to attend acceptance or bench integration tests, a lack of cooperation in defining strict working procedures, a total refusal of responsibilities.

The ELDO Secretariat failed to bridge the gap between ASAT and Marconi. Communications between manufacturing and testing groups were poor, as were communications between the launch teams and the engineering teams. In the case of the guidance systems, Marconi 'built a wall between users and manufacturers, a wall which was accepted, if not liked, by everybody and which ELDO, among others, did not make much effort to destroy'.[30]

The ELDO Secretariat had somewhat larger financial and schedule teams, but their problems were similar to those of their technical compatriots. ELDO created a Project Management Directorate that used tools such as the Program Evaluation and Research Technique (PERT) to track three levels of schedules: the contractors, national programmes and

ELDO Secretariat.[31] Unfortunately, the Secretariat had no authority to force timely or accurate reporting. Analysts lamented that 'the reports of the member states are always late ...'[32] Even when they acquired timely data, they could do little more than watch the schedule slip and remind offenders that they were deviating from the plan. Without authority to do anything about cost or schedule overruns, tools and organizations to implement them were of little use to the Secretariat other than reminding them of their powerlessness over the national governments and contractors.

By design, ELDO's member states created a weak organization. The ELDO Secretariat had few staff members and little authority to do anything but watch events happen, and to try to coordinate its unruly member states and contractors. When troubles came – and come they did – the Secretariat's staff tried to co-ordinate and plan around the problems. What it could not do was manage or control them.

'PARIS WE HAVE A PROBLEM ...' WITH INTERFACES

ELDO's technical troubles could be traced in most cases to problems with interfaces: component boundaries that were also organizational boundaries. Here ELDO's inability either to impose standards or to ensure communication among engineering groups produced its logical result: failure. ELDO engineers and managers soon recognized that they had a major problem with interfaces and communications. Through the Secretariat and national organizations they tried to make this point to the politicians who governed ELDO. Despite efforts to improve ELDO's communications and systems engineering capability, ELDO's basic flaw was a lack of authority that no piecemeal measures could repair. The symptoms became evident first in cost overruns and schedule slips, and then in flight failures.

The first test flights in 1964 and 1965 with Britain's *Blue Streak* were deceptively promising.[33] De Havilland's first stage design incorporated several years of design experience prior to the formation of ELDO, as well as American techniques from the *Atlas* programme, which itself had been developed for a number of years before Britain acquired some of its technologies. *Blue Streak*'s initial flights did not involve interfaces with any other stage. Managed by a single government organization that had prior rocket and missile experience, and built by firms experienced with these technologies and with each other, communication problems were not significant. *Blue Streak*'s successful example was not to be repeated.

Problems soon manifested themselves at the interfaces between launcher stages and the organizations responsible for them. Under the ELDO agreements of 1963, member states made the 'lower stage' contractor responsible for interfaces between two stages. Thus the British were responsible for the interface between the first and second, the French for the interface between the second and third stages, and the Germans for the third stage – test satellite interface. Meetings in 1965 further defined interface procedures, specifying that the 'Interface Design Authority' (the lower stage contractor) would 'freeze' the design, make the information available to all parties, and provide for hardware inspection. The Interface Design Authority would submit a 'Certificate of Design' to ELDO to certify the correctness of the interface. However, the system had a fatal flaw:

> it was not the intention that an Interface Authority should do again work already allocated and being performed by another Design Authority. The Interface Design Authority would therefore base his design declaration on statements, made by the other Design Authorities concerned, that the relevant specifications had been met.

The contractor responsible for the interface merely ensured that other organizations provided the appropriate documentation, but no single organization analysed both sides of the interface for discrepancies. Without anyone checking *both* sides, misunderstandings went unnoticed until the organizations tried to connect the stages together, or tested them in flight.[34]

Misunderstandings became painfully evident the moment contractors tried to connect their hardware. In an early test of the interface between the French second stage and the German third stage, the structure failed due to the wrong kinds of connecting bolts. When the ELDO Secretariat decided to make changes to the French second stage in response to this problem, the French complained, questioning who had the 'power to impose a solution'. In another case, the Germans developed a table to mimic the structural interface of the Italian test satellite 'before the Italian Authorities had completed their examination [of] the requirements'. This led to a mismatch between the assumed size of the connecting ring and the actual ring later designed by the Italians, and the Germans had to scrap their hardware and rebuild a new table. The Germans also could not control third stage development, because some of the modules, such as the British guidance system, were under the 'National Leadership of other Member States'.[35]

These complaints reached the ELDO Council through Member State delegations, leading to a study of ELDO's organization in early 1966. The Belgians, who had to collect data from all ELDO members to design the telemetry system, felt the interface problem in full. They suggested that the Secretariat be given substantially more authority, recommending a two-level management scheme. At the first level would be a study bureau to establish specifications for technical developments. It would be at the national level, but under the 'functional authority' of the ELDO Secretariat, and have authority to approve modifications and make technical decisions. Through control of the national bureaux, the Secretariat would impose consistent standards and processes. At a higher level in the organization, the Secretariat would have greater power, 'corresponding in the English sense to the word "control" (monitoring plus decision authority)'. This level would be staffed with 70 engineers, with seven 'inspector generals' from each national programme under the direction of a Management Director. These engineers would focus above all on integration problems and looking for future problems. The seven inspector generals and the seven 'national programme managers' would meet the Council to discuss problems at least every other month, supplemented with PERT-Cost networks.[36]

Countering the Belgian proposal, the French proposed an 'Industrial Integrating Group' that would exchange information among the governmental and industrial firms building *Europa I*. The French solution provided information, but did not give the Secretariat the power to enforce solutions. The Industrial Integrating Group would collect information for the Secretariat and then pass its recommendations to the Secretariat, which in turn could recommend changes to the national programmes and the ELDO Council. Perhaps not coincidentally, the Industrial Integrating Group would be led by SEREB, the French organization that coordinated the French rocket programme.[37]

In matters political, the French point of view carried more weight. German support helped the French, because they, like the French, did not want a strong project manager. With the British, French and Belgians all supporting stronger ELDO management, the Germans chose the weaker of the two proposals, and backed the French. Not wanting a project manager to have financial control, the Dutch supported the Germans.[38]

Along with appointing a project manager, the ELDO Council also requested that the Secretariat investigate programme management procedures, and agreed with 'the necessity of adopting a system for providing delegations and the Secretariat with continuous and full preventative information on the progress of the current programmes . . .' Secretary General Carrobio reported back, agreeing that a 'Corps of Inspectors' should review ELDO and make recommendations concerning ELDO's processes, structure and management. Carrobio proposed an 'integrating group set up by industry and subordinate to the Secretariat's authority'. Such a group would enhance ELDO's position.[39]

In December 1966, the Corps of Inspectors delivered their report, which described 'the problems of the interfaces' and 'the role played in this matter by the Secretariat'. L.T.D. Williams, the Chairman of the group, noted that

> the German Authorities and the German industrials repeatedly stressed the difficulties which have resulted from an incomplete solution of the interface problems, which they attribute to gaps in the methods of coordination.

Williams described a number of interface problems, concluding that the Secretariat should

> prepare for the Council a paper in which it set out: (a) what procedures and principles have been defined for the methods by which it may intervene in the solution of interface problems, and what deficiencies it sees in these; (b) what are its intentions

> concerning the further development of this question, and what systems it expects to set up eventually.[40]

Carrobio responded the next month, first to specific problems noted by Williams' team. The Secretariat had 'played a determining role in reconciling the viewpoints' of the French and Germans in making design changes after the failure of the 'attempted design qualification of F4 in February 1966'. However, in the case of problems with the German third stage, neither the German government, nor German contractors, nor ELDO could control non-German suppliers for third stage components. Even here, Carrobio noted that 'these were not so much a matter of principles as of practical difficulties due to slippages in the development programme of Germany's foreign suppliers'. Since German contractor Bölkow had no control, 'it is then up to the Secretariat to intervene and endeavour to find and win acceptance for the least harmful compromise, and this has been done on all occasions. Improvement in this area will come with greater experience in the initial forecasting of time-scales for development of components.' Carrobio believed that only minor adjustments to ELDO's organization were necessary, not a complete overhaul.[41]

The French proposal prevailed. In July 1966, the ELDO Council approved the *Europa II* vehicle,[42] and agreed to create an integration group, known as the Company for the Study and Integration of Space Systems (Société d'Étude et d'Intégration de Systèmes Spatiaux – SETIS), to strengthen the Secretariat. SETIS began as a division of SEREB, and then spun off into a separate organization.[43]

SETIS had only an advisory capacity, reporting to the ELDO Project Management Directorate, which the Council also created at this time. Under the new system, the Project Management Directorate assigned project managers to *Europa I* and *Europa II*. Each country also selected its own project manager, who reported to his national organization, and to the ELDO Project Manager, who distributed information to the member states through the Scientific and Technical Committee. SETIS worked only on *Europa II*, since *Europa I* was soon to begin integration testing. For *Europa II*, the Secretariat now had authority to place contracts directly with industry. ELDO's *Europa I* project manager remained as a co-ordinator, without authority over funding, and required to consult with national directors who controlled the funds.[44]

Secretary-General Carrobio expected that SETIS would strengthen ELDO's technical capabilities. Because its engineers came from *Europa II* contractors, SETIS would provide better contact between ELDO and the manufacturers 'by means of direct contacts'. The Secretariat directed SETIS, not the member states, and SETIS had the capability to direct contractors, 'especially concerning interfaces'. SETIS planned to hire 40 engineers for the *Europa II* programme, and 60 more after that, arranged in three divisions, the 'PAS Vehicle [*Europa II*] Development', 'Planning and Information', and 'System Integration'. Despite its apparently broad charter, SETIS' power was strictly limited since it could not amend contracts or change costs, schedules or technical performance except through the Secretariat. Because the Secretariat did not have this authority either, SETIS could only analyse information that the member states and contractors were willing to provide.[45]

The new processes and procedures took some time to implement. After an ELDO visit to NASA Goddard Space Flight Center to learn more about project management techniques, the ELDO Council finally ratified the new Programme Management procedures in September 1967. SETIS came into official existence as a separate company on 1 January 1968, aiding *Europa II*. Despite the changes, ELDO management had no authority to enforce decisions on either *Europa I* or *Europa II*.[46]

ELDO contracted with Hawker Siddeley Dynamics to develop an Electrical Mock-Up of the *Europa I* and *II* launchers at its facility in Stevenage. In conjunction with engineers from other contractors, the company assembled complete stage mock-ups (sometimes without engines) prior to the F5 through F11 tests and flights. Hawker Siddeley ran several kinds of tests, including injection of faulty signals, electromagnetic interference and flight sequence testing. These tests uncovered numerous design problems, which Hawker Siddeley then reported to the ELDO Secretariat and the participating companies.[47] Although Hawker Siddeley found numerous problems, it did not find enough of them.

As ELDO, the national governments, and the contractors started to build and integrate rockets, they found that communication and interface problems were a major stumbling-block. Complaints bubbled up from the contractors through the national delegations to the ELDO Council, leading to an enhancement of ELDO's project management capability. The new procedures gave the *Europa II* project manager the authority to make direct contracts, and gave him a staff that could monitor events much more closely. Even for *Europa II* the Secretariat still had only limited authority to modify contracts, costs, and schedules. However, ELDO's immediate future hinged on *Europa I* and its much weaker organization.

DISASTER AND DISSOLUTION

ELDO's next major flight test, originally scheduled for 1965, mated France's *Coralie* second stage with *Blue Streak*. After delays due to a failure of the interface structure between the second and third stage, and numerous problems in transportation and launch operations, the test flight F6/1 of *Coralie* with *Blue Streak* came in August 1967. The first eight launch attempts were delayed. ELDO called off three because of bad weather, three because of safety system problems, and two due to ground equipment failures. On the ninth attempt, the first stage operated properly, and the second stage successfully separated from the first. However, the second stage engines never fired, and it crashed prematurely to earth. Subsequent investigations showed that the problem stemmed from an electrical ground fault in the second stage, which de-energized a relay in the first stage, leading to a failure of the second stage sequencer, which then failed to issue commands for the second stage to fire. In short, the launch failed because of an electrical interface problem between the first and second stages.[48]

The next attempt with *Coralie* came in December 1967. It failed just like the first flight, with its engines 'failing to light up'. In this case, an electrical interface problem between the second stage and its 'umbilical' connection to the ground system on the launch pad caused the problem. In addition, electromagnetic interference hampered communication with the vehicle and its safety system, causing the loss of flight data and the potential for an inadvertent explosion.[49]

After the loss of the first *Coralie* in August, the French considered second-stage problems to be minor. However, with another failure of French equipment, French authorities this time reacted with urgency. In a major internal reorganization, the French contracted with SEREB to manage the second stage programme. SEREB had been involved with the programme up until 1963, performing feasibility studies and initial designs. After that time, authority rested with the military Laboratory for Ballistic and Aerodynamic Research (Laboratoire de Récherche Balistique et Aérodynamique – LRBA) and the Bureau Permanent Nord Vernon. The French Space Agency interposed SEREB between itself and LRBA to keep closer surveillance and management of the programme.[50]

SEREB quickly analysed the second stage's problems, and proposed a major vehicle redesign to improve reliability. Their proposal involved performing more qualification tests on *Coralie* components, and the replacement of several major components with others that SEREB used in its *Diamant* rocket design. ELDO's technical group rejected the French proposals on the grounds of their very high costs, and more importantly because they would disrupt the entire programme, including designs for the first and third stages.[51]

ELDO engineers supported the Laboratory for Ballistic and Aerodynamic Research against SEREB, noting that the detailed qualification tests SEREB recommended did not exist when the programme started. While agreeing with further qualification tests, ELDO engineers rejected SEREB's proposal to replace components. They stated that 'the approach adopted by the French authorities is mainly due to the formation of a new technical direction team which is naturally anxious to use equipment with which it is familiar while being less familiar with the equipment it proposes to replace'. ELDO engineers chided SEREB, stating that they

> will have to make great efforts to be as familiar with the programme as the present team – which has 'lived with' the *EUROPA I* launcher for five years and is at present very experienced – not only in order to develop its equipment but also to take account of the specific contingencies of ELDO.

The other Member States rejected SEREB's expensive proposal in favour of ELDO's proposal to upgrade and test existing second stage components.[52]

While the French regrouped, ELDO management assessed the impact of their project management reforms. The Secretariat divided its project management directorates into two divisions, 'one responsible for technical and timescale aspects and the other for financial and contract aspects', while other directorates provided support to them. Secretary-General Carrobio noted that 'the principle is now acknowledged, inside and outside the Secretariat, that additional work or modifications to approved work require the prior agreement of the project management directors'. The co-ordination between Member States and the Secretariat was improving, but still presented problems in schedule reporting because the Member States delivered PERT reports late, and in cost control because Member States did not thoroughly check contractor proposals. On *Europa II*, ELDO's system of monthly progress reports and meetings was working smoothly.[53]

In the next flight in December 1968, ELDO engineers felt vindicated in their earlier resistance to SEREB's proposals, because both *Blue Streak* and *Coralie* worked perfectly. Even though the German *Astris* exploded shortly after separation, ELDO and German engineers believed that they would isolate and repair the propulsion system problems they thought responsible for the failure.[54]

Their confidence was unfounded, for on the next attempt in July 1969, *Astris* failed precisely as before, with an explosion within a second of separation from the second stage. Just as the French realized two years before, the Germans found they had major problems. The Germans formed four committees: a government committee to investigate the failure, another to investigate the rest of the design, an internal committee of the contractor Arbeitsgemeinschaft Satellitenträger (ASAT), and a fourth committee to oversee and co-ordinate the other three committees. Contrary to expectations, the investigators found that the explosions resulted not from a third-stage propulsion problem, but rather an electrical failure in the interface between the third stage and the Italian test satellite, in which an electrical failure ignited the safety self-destruct system. The Italians had already noticed sensitivity in these German circuits during their tests, but neither they nor the Germans recognized the importance of the finding.[55]

The Germans fixed the electrical problem, but the third stage showed new problems in the last flight of *Europa I* in June 1970. This time, the resulting investigation showed two third-stage failures. First, an electrical connector disconnected prematurely, preventing separation of the Italian satellite test vehicle. Engineers traced this mechanical failure to the pressure between trapped air in the mechanism and the vacuum of space. Second, the third stage propulsion feed system failed, probably due to contaminants, which kept a helium pressure valve open when it should have closed. These failures led the ELDO Council to create a Quality Assurance organization in 1970, but due to a lack of staff, it could not cover all of the sites and processes.[56]

In November 1971 *Europa II* flew for the first and last time. This vehicle, which included the first three stages plus a new 'Perigee-Apogee' stage, blew up two and a half minutes into the flight due to an electrical malfunction caused by a failure of the third-stage guidance computer. This too was an 'interface problem', because the British contractors manufactured the computer, and delivered it to German contractor ASAT, which took no responsibility for proper integration.[57] At last, all ELDO Member States reacted strongly. As stated by General Aubinière, ELDO's new Secretary-General, 'the failure of F11 in November 1971 brought home to the member states – and this was indeed the only positive point it achieved – the necessity for a complete overhaul of the programme management methods'.[58]

The ELDO Council appointed an investigation committee to review every aspect of the programme, with senior engineers and executives from government and industry in the United States and Europe.[59] The commission's report was a devastating indictment of ELDO's organization and management, finding numerous technical problems that resulted from the lack of authority and poor communications. Poor electrical integration in the third stage was the primary problem, having caused the failures of flights F7 through F11. This was a result both of the poor management of the German company ASAT and the contracting policy of the Secretariat, which did not assign integration authority to any third-stage contractor. In other words, failures resulted from interface problems with components delivered by foreign contractors to Arbeitsgemeinschaft Satellitenträger (ASAT), and between ASAT's two partners: Messerschmitt-Bölkow-Blohm (MBB), and Entwicklungsring Nord (ERNO).

Communications between ASAT and its two 'partners', MBB and ERNO, were extremely poor. ASAT was a very small organization created by the German government solely to co-ordinate MBB and ERNO on the ELDO third stage. Communications between ASAT and other firms supplying third-stage components were even worse, and led to a design that 'obeys none of the most elementary rules concerning separation of high and low level signals, separation of signals and electrical power supply, screening, earthing, bonding, etc.'. This poor electrical interface design made the British guidance computer and sequencer extraordinarily sensitive to noise and minor voltage variations, which in turn caused it to fail in the F11 flight.[60]

By far the most significant commission recommendations were to abolish indirect contracting, and to define clear responsibilities for all stage and inter-stage interfaces. The Member States at last gave the Secretariat the authority to place contracts directly. Desperate for solutions, ELDO adopted a number of American techniques, including the full adoption of phased planning, work breakdown contractual structures, and preliminary and final design reviews for the *Europa III* programme, ELDO's hoped-for improvement to *Europa II*.[61]

General Aubinière, ELDO Secretary-General and the former director of the French Space Agency, hoped that stronger project methods would turn ELDO around. Unfortunately, ELDO never got another chance. After the British withdrew financial support in 1971, the remaining partners had either to develop a new first stage or purchase *Blue Streak* stages. Disillusioned after the F11 failure, the Germans threatened to withdraw. With continuing political disagreements over launch vehicles, and over co-operation with the United States, ELDO's support evaporated. European governments eliminated the *Europa II* programme while the F12 launch vehicle was on its way to Kourou in April 1973. ELDO had hoped for a part in the American Shuttle programme, but when the Americans withdrew the Space Tug as a co-operative project, ELDO's time was up. The member states dissolved it in February 1974.[62] Twelve years of negotiation, compromises and struggle came to an end.

CONCLUSION

Political and industrial interests drove the formation of ELDO, Europe's largest co-operative space project of the 1960s. Each country preserved its national interests through the system of indirect contracting, by which the national governments maintained authority. Even after being strengthened in 1966 by adding a Programme Management Directorate and an integrating organization, the Secretariat remained limited to collecting information, and distributing it through the Technical Directorate to the ELDO Council. Member States compounded the Secretariat's weakness by changing objectives, as British support weakened and France pushed to create a more powerful rocket that could boost communications satellites into geostationary orbit.

The result of ELDO's weakness was a series of failures caused by interface problems between and internal to the launcher stages. Britain and France came to the project with existing rockets, and both insisted on using their own standards and procedures. The ELDO Secretariat could neither create nor enforce consistent documentation, processes or quality. Nor could they force contractors to communicate with each other. These problems resulted in badly designed electrical circuitry between the British, French and German stages, and also internal to the German third stage due to the poor management, organization and attitude of German contractors and British subcontractors. Six consecutive failures were the end result, all but one due to interface failures directly traceable to poor communications between the member countries and their contractors.

Controlled by politicians representing strong national military and industrial interests, and headed by an Italian ambassador, ELDO did not have a single strong central authority to enforce changes and communication. Although populated with engineers and managers, neither of these groups controlled the organization. ELDO's engineering teams could not overcome industrial and political barriers such as those that typified German contractor ASAT and its poor relationships with both its German sponsors and its foreign suppliers. Without authority, neither management hierarchies nor engineering teams could make ELDO function. In a final effort to fix ELDO, the Council appointed the military officer, French General Robert Aubinière as the Secretary-General, and gave him the authority necessary to unite ELDO's fractious members. Unfortunately, the stronger methods he and the Member States promoted came too late, for the Member States gave up before ELDO's next launch attempt.

ELDO combined many of the worst management ideas into a single, pitiful organization. Its engineers, managers and directors struggled against a fatally flawed management structure. ELDO's project managers were virtually powerless. ELDO initially ignored interfaces, and later simply trusted the contractors to do their job. Component quality, assured through inspections, testing and documentation, were randomly present in ELDO. ELDO's hapless record and defective structure served as a warning to European leaders that co-operative technology development required more than a weak organization left helpless in the face of national interests. Europeans would begin launcher development again. However, the next time, learning the lessons of ELDO, the European Space Agency (ESA) ensured strong project management that could ensure both communication and control.

The results would include *Ariane*, which became an icon of successful European co-operation, and *Spacelab*, ESA's portion of the United States' shuttle programme. On *Ariane*, ELDO veterans who had developed their own testing methods, supplemented by significant American management techniques imported for *Europa III*, created a reliable space launcher that would compete with American expendables and the new shuttle. Other European countries, led by West Germany, would work directly with the United States on *Spacelab* to learn NASA's managerial methods. The result of both programmes would be that ESA's methods would essentially mimic those of NASA, in a highly successful transfer of managerial expertise. ESA's success would owe a great deal to ELDO, whose negative example kept European nationalists from weakening ESA.[63]

ACRONYMS USED IN THIS PAPER

ASAT	Arbeitsgemeinschaft Satellitenträger
BMBW	Bundesministerium für Bildung und Wissenschaft
CNES	Centre National d'Études Spatiales
ELDO	European Space Vehicle Launcher Development Organisation
ERNO	Entwicklungsring Nord Raumfarhttechnik
ESA	European Space Agency
IGS	Inertial Guidance System
INTELSAT	International Telecommunications Satellite Organization
LRBA	Laboratoire de Récherche Balistique et Aérodynamique
MAU	Million Accounting Units
MBB	Messerschmitt-Bölkow-Blohm
NASA	National Aeronautics and Space Administration
PAS	Perigree-Apogee System (an upper stage for *Europa II*)
PERT	Program Evaluation and Research Technique
SEREB	Société pour l'Étude et la Réalisation d'Engins Balistiques

SETIS	Société d'Étude et d'Intégration de Systèmes Spatiaux
SNECA	Société Nationale d'Étude et de Construction de Moteurs d'Aviation

ACKNOWLEDGEMENTS

I would like to thank John Krige and the ESA history project for their early support of the research that went into this article, Gherardo Bonini at the Historical Archives of the European University Institute for his outstanding support in locating documents, and the anonymous reviewers of *History of Technology* for their comments.[64]

Notes and References

1. R. Aubinière, 'Tenth Anniversary of the Establishment of ELDO', *ESRO/ELDO Bulletin*, March 1974, 24: 13.

2. Michael J. Neufeld, *The Rocket and the Reich: Peenemünde and the Coming of the Ballistic Missile Era* (New York, 1995).

3. Harrie Massey and M.O. Robins, *History of British Space Science* (Cambridge, 1986), 11–12.

4. Jean Corbeau, 'A History of the French Sounding Rocket Veronique', in Kristan R. Lattu (ed.), *History of Rocketry and Astronautics*, AAS History Series Volume 8 (San Diego, 1989), 147.

5. Corbeau, *op. cit.* (4), 147–167. Carlier, Claude, and Marcel Gilli, *The First Thirty Years at CNES, The French Space Agency, 1962–1992* (Paris, 1994), 6.

6. Guy Collins, *Europe in Space* (New York, 1990), 13–14. Jean-Pierre Causse, 'Les lanceurs européens avant Ariane', in Emmanuel Chadeau (ed.), *L'Ambition Technologique: Naissance d'Ariane* (Paris, 1995), 15–17. Berry Sanders, 'The French *Daimant* Rockets', *Quest*, 1999, 7: 1, 18–22. Colonel Edward N. Hall, Oral History Interview, 11 July 1989, with Jack Neufeld, AFHRA K239.0512–1820, pp. 1–2, 26–29. Hall's involvement shows that American policy was not consistent. On the one hand the US government blocked direct technology transfers to France, but on the other, it is likely that Hall gave them more capability through his expertise. Secretary of the Air Force Donald Quarles sponsored Hall's work in Europe. I do not know if Hall's mission was known to others in the US governement.

7. See John Krige, *The Launch of ELDO* (Noordwijk, The Netherlands, 1993), 2–4. Stephen Robert Twigge, *The Early Development of Guided Weapons in the United Kingdom, 1940–1960* (Chur, Switzerland, 1993), 23–24, 40–43, 188–191. John Krige and Arturo Russo, *Europe in Space 1960–1973* (Noordwijk, The Netherlands, 1994), 29.

8. Krige, *op. cit.* (7), 4–6. Twigge, *op. cit.* (7), 193–202.

9. Krige, *op. cit.* (7), 6–10.

10. Krige, *op. cit.* (7), 11–17.

11. Walter A. McDougall, 'Space-Age Europe: Gaullism, Euro-Gaullism, and the American Dilemma', *Technology and Culture*, 1985, 26: 179–203. See also Robert Gilpin, *France in the Age of the Scientific State* (Princeton, 1968).

12. Krige, *op. cit.* (7), 11–17. Michelangelo De Maria, *Europe in Space: Edoardo Amaldi and the Inception of ESRO* (Noordwijk, The Netherlands, 1993), 28–34. Michelangelo De Maria, *The History of ELDO Part 1: 1961–1964* (Noordwijk, The Netherlands, 1993), 7–12, 17. *CECLES – ELDO 1960–1965*, Report to the Council of Europe, Paris, 23 December 1965, 7–16.

13. *ELDO Convention*, Arts 6(1), 6(2), 9, 11, 16(1), and 16(2). Arnold Frutkin, memorandum for Robert F. Packard, Director Outer Space Affairs, SCI, Department of State, Subject: Embassy Brussels observations on ELDO, July 5 1966, NASA History Office.

14. *CECLES – ELDO 1960–1965*, *op. cit.* (12), organigramme, final page.

15. MAU = million accounting units, a unit calculated from a mix of European currencies. In the mid-1960s one AU approximately equalled one dollar.

16. De Maria, *The History of ELDO Part 1*, *op. cit.* (12), 17–19, 23, 24, 27, 30. Krige and Russo, *op. cit.* (7), 71. *CECLES – ELDO 1960–1965*, *op. cit.* (12), 15–17.

17. *CECLES – ELDO 1960–1965, op. cit.* (12), 18–20. Andrew J. Butrica (ed.), *Beyond the Ionosphere: Fifty Years of Satellite Communication* (Washington, DC, 1997).

18. Krige and Russo, *op. cit.* (7), 71–82. Michelangelo De Maria and John Krige, 'Early European Attempts in Launcher Technology: Original Sins in ELDO's Sad Parable', *History and Technology*, 1992, 9: 109–137.

19. 'Intra-European Cooperation on Space Programs', author probably Frutkin, estimated date, 1968, NASA History Office. Lorenza Sebesta, *The Availability of American Launchers and Europe's Decision 'To Go It Alone'* (Noordwijk, The Netherlands, 1996).

20. *World-Wide Space Activities*, Report Prepared for the Subcommittee on Space Science and Applications of the Committee on Science and Technology, US House of Representatives, Ninety-Fifth Congress, 1st Session, by the Science Policy Research Division of the Congressional Research Service, September 1977, 263, 266.

21. Werner Büdeler, *Raumfahrt in Deutschland: Forschung, Entwicklung, Ziele* (Frankfurt, 1978), pp. 31–37, 141–143. Peter Fischer, *The Origins of the Federal Republic of Germany's Space Policy 1959–1965: European and National Dimensions* (Noordwijk, The Netherlands, 1994): 35–37, 41–45, Table II.

22. Groupe ad hoc de la Gestion du Controle, 'Note de la Délégation belge', ELDO/C(66)26, 3 mars 1966, Paris, HAEUI. *Organisation Européenne pour la Mise au Point et la Construction de Lanceurs d'Engins Spatiaux, Structure et Table des Matières du Rapport de Synthèse de la Commission de Revue de Projet, 1ère partie, Rapport de Synthèse de la Commission de Revue de Projet*, Neuilly, 19 April 1972, ELDO/CRP(72)38, Historical Archives of the European University Institute Fond (HAEUI) 2885, p. 8.

23. Arnold Frutkin, memorandum for Robert F. Packard, Director Outer Space Affairs, SCI, Department of State, Subject: Embassy Brussels observations on ELDO, July 5 1966, NASA History Office.

24. *European Space Vehicle Launcher Development Organisation, Europa II Project Review Commission, Group No. 5, Sequencing, Separation and Safety Systems Final Report*, Neuilly, 18 April 1972, ELDO/CRP(72)40, HAEUI Fond 2887, Section 0.4.7.1. ELDO *Rapport de Synthèse, 1ère partie, op. cit.* (22), 6–7.

25. *Final Report of the Project Review Commission*, ELDO/C(72)18, Neuilly, France 19 May 1972, HAEUI, p. 14.

26. ELDO, *Europa II Project Review Commission, Group No. 5, op. cit.* (24), Sections 0.4.1.1., 0.4.2.3.

27. *Final Report of the Project Review Commission, op. cit.* (25), 13, 14, 17.

28. *Final Report of the Project Review Commission, op. cit.* (25), 12–13. ELDO *Rapport de Synthèse, 1ère partie, op. cit.* (22), 8.

29. ELDO, *Europa II Project Review Commission, Group No. 5, op. cit.* (24), Section 0.4.2.1., *Final Report of the Project Review Commission, op. cit.* (25), 13.

30. *European Space Vehicle Launcher Development Organisation, Europa II Project Review Commission, Group 3 Final Report*, Neuilly, 18 April 1972, ELDO/CRP(72)40, HAEUI Fond 2884, pp. 15–18. ELDO, *Europa II Project Review Commission, Group 5, op. cit.* (24), Sec. 0.4.6.2.

31. J. Nouaille, 'The ELDO PAS Programme and its Management', *ESRO/ELDO Bulletin*, May 1968, 1: 10. I. Stevenson, 'L'Analyse de Réseau, contribution à la gestion du CECLES', *ESRO/ELDO Bulletin*, August 1968, 2: 12–14.

32. Stevenson stated 'malheureusement, les rapports des Etats membres sont souvent retardés . . .', Stevenson, *op. cit.* (31), 15.

33. *CECLES – ELDO 1960–1965, op. cit.* (12), pp. 64–65.

34. L.T.D. Williams, Chairman, Corps of Inspectors, Letter to Secretary-General, ELDO/CECLES, ELDO/T(67)5, London, December 17 1966, HAEUI ELDO fond 2958. See Annex B, which includes ELDO/PG(63)T/27: 'Report on ELDO Initial Programme, October 1963 – Part III – Division of Responsibilities', and 'Technical Acceptance Procedures for Application to F4 Vehicle'.

35. Williams, *op. cit.* (34).

36. Groupe ad hoc de la Gestion du Controle, 'Note de la Délégation belge', ELDO/C(66)26, 3 mars 1966, Paris, HAEUI.

37. Nouaille, *op. cit.* (31), 10.

38. '14. Session of ELDO Council', Paris, 03/03/66, Discussion of 'Structure of Secretariat', pp. 15–19, HAEUI ELDO 1140. '15 Session of ELDO Council', 30/03/66,

Paris, Section on 'Management and Control of ELDO programmes', pp. 7, 8, HAEUI ELDO 1141. 'Decision of the Council on the ELDO Structure 1966', European Space Vehicle Launcher Development Organisation, ELDO/C(66)27, Paris, 12 April 1966, HAEUI ELDO 1177.

39. 'Resolution of the Council on the Management and Control Procedures for the Programmes', European Space Vehicle Launcher Development Organisation, ELDO/C(66)27, Paris, 12 April 1966, HAEUI ELDO 1177.

40. Williams, *op. cit.* (34).

41. 'Comments by the Secretariat on interface problems following the first report of the Corps of Inspectors', Note by the Secretariat, ELDO, Paris, 7 February 1967, ELDO/T(67)5, HAEUI Fond 9458.

42. *Europa II* was also known as the *ELDO-PAS* system, for Perigee-Apogee System.

43. 'Creation of an ELDO Integrating Group', December 2 1966, ELDO Fond 1210, HAEUI. '21 Session of ELDO Council', Paris, 15/12 and 16/12/66, section on 'Interim Report by the Secretariat on the constitution of the Industrial Integrating Group' (EDLO/C(66)64), pp. 16–18, HAEUI ELDO 1147. 'Industrial Integrating Group', Note by the Secretariat, European Space Vehicle Launcher Development Organisation, ELDO/T/(67)1, Paris, 24 January 1967, HAEUI ELDO Fond 2954.

44. Nouaille, *op. cit.* (31), 10–11. *World-Wide Space Activities*, *op. cit.* (20), 271–272. '17 Session of ELDO Council', Paris, 08/07/66, section on 'Modifications to Structure', pp. 2, 3, HAEUI ELDO 1143. 'Draft Resolution – Integrating Group', ELDO/C(66)WP/16, Paris, 29 September 1966, HAEUI ELDO 1229. 'ELDO Integrating Group Operating Proposal in SEREB International Division', November 28 1966, HAEUI ELDO 1210. Krige and Russo, *op. cit.* (7), 76.

45. 'Industrial Integrating Group', Note by the Secretariat, European Space Vehicle Launcher Development Organisation, ELDO/T/(67)1, Paris, 24 January 1967, HAEUI ELDO Fond 2954.

46. Lorenza Sebesta, *op. cit.* (19), 20–21. *Control and Management of ELDO Programmes, Report by the Secretary General*, ELDO/C(67)47, 9 October 1967, ELDO Fond 1288, HAEUI.

47. T.A. Graham and E. Dombrowski, 'Electrical Interface Problems in Multi-Stage Launchers as Illustrated in the ELDO Europa Project', *ESRO/ELDO Bulletin*, November 1970, 12: 16–23.

48. 'Report on Launch of F6/1 Vehicle', Note by the Secretariat, ELDO/T(67)24, Paris, 22 September 1967, HAEUI ELDO Fond 2977.

49. 'Report on Launch of F6/2 Vehicle', Note by the Secretariat, ELDO/T(68)1, Neuilly, 8 February 1968, HAEUI ELDO Fond 2992.

50. 'Consequences of F6/2 on the Programme', Note by the Secretariat, ELDO/T(68)2, Neuilly, 20 February 1968, HAEUI ELDO Fond 2993.

51. 'Operation Coralie Development Plan', Annex A to ELDO/T(68)2, Courbevoie, 31 January 1968, SEREB Technical Directorate, DTA/C – no. 30/9497, HAEUI ELDO Fond 2993.

52. 'Secretariat comments on the French authorities' proposal', Annex B to ELDO/T(68)2, 20 February 1968, HAEUI ELDO Fond 2993.

53. 'Functioning of the Project Management Directorates created in the Secretariat', Note by the Secretariat, Neuilly, 9 May 1968, ELDO/C(68)12, HAEUI. Minutes of the 29th Session of the Council held in Neuilly on 24 and 25 June 1968, ELDO/C(68)12 and corrigendum, HAEUI 1308.

54. P. Rochefort, 'The Flight Test of the Europa I Satellite Launcher', *ESRO/ELDO Bulletin*, January 1969, 4: 14–19. 'F7 Flight Test, Note by the Scientific and Technical Committee', ELDO/T(68)20, Neuilly, 11 December 1968, HAEUI ELDO Fond 3010.

55. 'Preliminary Report on the Launch of the F8 Vehicle', Note by the Secretariat, ELDO/T(69) 8, Neuilly, 17 September 1969, HAEUI ELDO Fond 3026.

56. 'F9 Assessment – Progress Report', Note by the Secretariat, ELDO/T(70)10, Neuilly, 9 October 1970, HAEUI Fond 3052. *Final Report of the Project Review Commission*, *op. cit.* (25), 16.

57. *Final Report of the Project Review Commission*, *op. cit.* (25), 29–30.

58. R. Aubinière, 'Tenth Anniversary of the Establishment of ELDO', *ESRO/ELDO Bulletin*, March 1974, 24: 13.

59. *Final Report of the Project Review Commission, op. cit.* (25), Annex 2.

60. *Final Report of the Project Review Commission, op. cit.* (25), 15, 20–21. United States Department of Commerce, Bureau of International Commerce, *Report of the Space Industry Trade Mission on the European Aerospace Industry*, January 1966, 'U.S.–Europe 1965–1972' folder 014548, LEK 7/9/4, NASA History Office.

61. *Preliminary Report of the Europa II Project Review Commission to the Council of ELDO*, Annex V, pp. 9–10, in *Final Report of the Project Review Commission, op. cit.* (25). 'Europa III, Preparatory Phase', *ESRO/ELDO Bulletin* (May 1972), 18: 10–14. ELDO adopted these American methods for *Europa III* planning even before the F11 failure. They could do this because it was only in the planning stages, not as an officially approved programme subject to the national governments.

62. *World-Wide Space Activities, op. cit.* (20), 282–287.

63. Stephen B. Johnson, 'Building an American Bridge over the "Management Gap": The Adoption of Systems Management in ESRO and ESA', *History and Technology*, 1999, 16: 1–32.

64. This article is adapted from Chapter 6 of Stephen B. Johnson, *Secrets of Apollo: Systems Management in American and European Space Programs* (Baltimore, forthcoming), title tentative, Johns Hopkins University Press.

Building a Hybrid Landscape to Purify the Ruhr Region, 1890–1935

EDMUND N. TODD

ABSTRACT

Established in 1913, the Ruhrverband (Ruhr Association) combined cities, counties and industrial corporations in a hybrid organizational form that began restructuring the landscape of the Ruhr river to provide for water purification. During 1920s the Ruhrverband built the first of three new lakes, Lake Hengstey. Two others were completed during the 1930s. In Lake Hengstey, the Ruhrverband combined functions in a hybrid landscape to build a constituency for water purification. The dam creating the lake generated electricity, while the lake allowed particles to settle out. Beaches allowed for bathing, and a new bridge improved transport. In addition, Lake Hengstey supplied water to a pumped-storage station that enabled a power company to help balance load on its thermal plants.

INTRODUCTION

The Ruhr region changed dramatically during the 1800s. Draining 4500 square kilometres, the Ruhr river flows 219 kilometres from near Winterberg in the Sauerland to Duisburg and Ruhrort on the Rhine.[1] In 1800 several duchies, an imperial city, and various religious territories divided the last 60 kilometres of the river.[2] South of the river, small iron-working facilities used Ruhr tributaries for power. Fishing, agriculture, small towns and cottage industry characterized the area north of the river, as well as the Emscher valley and the southern part of the middle Lippe valley. Towards the end of the century, this northern area became known as the Ruhr region because of its heavy industrial development (Figure 1). By 1900, coal-mines, steel mills and cities were polluting the Ruhr region and, by threatening the traditional order, promoting 'cultural despair'.[3] These large-scale forms of pollution required responses.

But Berlin did not respond to the situation with solutions. Instead, in the Ruhr, it was left to urban, rural, state and industrial representatives to seek viable ways of constructing an infrastructure that would make the

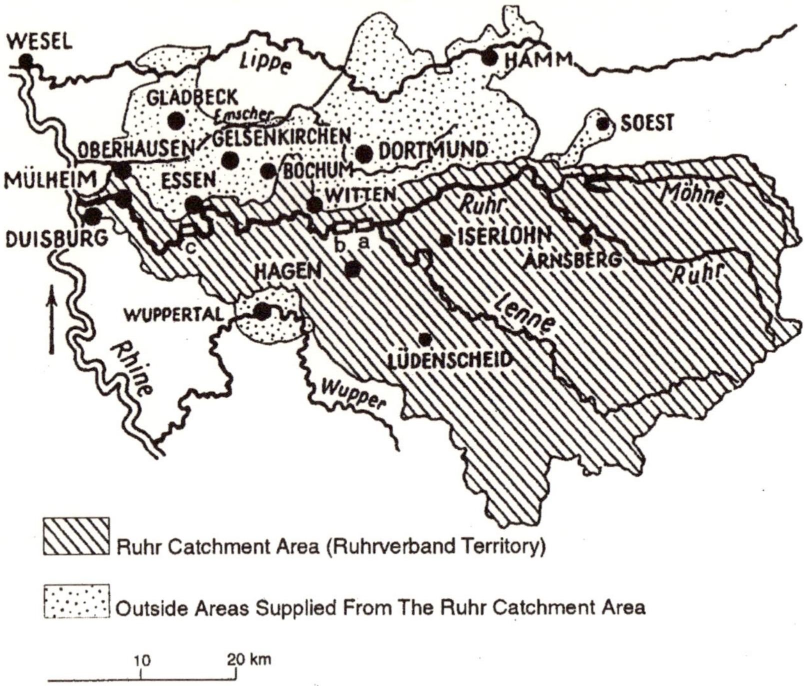

Figure 1 Water supply in the Ruhr region: the Ruhr industrial region stretched from Dortmund to Duisburg. The Ruhr river's catchment area lay to the south, but it supplied outside areas. Lakes Hengstey (a), Harkort (b), and Baldeney (c) created hybrid landscapes.
Source: *Gesundheits-Ingenieur*, 1934, 57: 200.
Artist: P. Falcone

new urban and industrial way of life tolerable. Members of these contending subcultures struggled to put into practice different conceptions of development. They resolved problems locally without giving up either local or national autonomy.[4] For instance, in electrification, they fought one another to a standstill, then expanded their contending power systems to include territory outside the Ruhr.[5] In questions relating to water, however, they collaborated in forming water associations to provide fresh water and to remove sewage. Combining different organizations and interests, these hybrid associations resolved political, social and economic problems not in general, but those dealing with water. They built multi-functional or hybrid landscapes that stabilized local social arrangements as they strove to resolve problems of urban and industrial pollution.[6]

One such hybrid landscape is Lake Hengstey, built by the Ruhrverband (Ruhr Association), which had been established in 1913 to purify the Ruhr river. In the 1920s, Ruhrverband officials were planning a series of eight lakes along the river. These would purify and store water, raise the level of ground water, and allow 'artificial flushing of river sections', while catching 'flushed sludge in the next reservoir'. By adding other functions in building hybrid landscapes, the Ruhrverband created a constituency and reduced building and operating costs paid by members. Ruhrverband officials promised that the eight lakes would turn the Ruhr river into a vacation wonderland. By 1934, they had built three: Lakes Hengstey, Harkort and Baldeney.[7] Lake Hengstey, the first of the lakes planned and built, will be the focus of this essay.

'INDUSTRIAL PRESERVE' AS 'TRADING ZONE'

Since it was neither unified nor uniform, the Ruhr does not fit well into that version of German history seen as controlled from Berlin. Nevertheless, belief in top-down control from Berlin is widespread among social, political and urban historians.[8] Some environmental historians follow suit. For instance, a leading historian of the Ruhr's pollution, Franz-Josef Brüggemeier, argues that

> the laws, administrative procedures, etc., dealing with pollution came from Berlin; the Prussian government had the last say whenever a conflict arose, and the system established in Prussia served as a model for the rest of Germany.[9]

Displaying a prescriptive intent, perhaps associated with belief in the reality of top-down control, Brüggemeier argues that state officials and industrialists were aware of problems but did not respond as they should have. Industry was more important for Prussian officials in Berlin. They believed what they were told by industrial representatives who stressed problems of costs, and, as a result, rather than resolving problems as they should have, allowed the Ruhr to become an 'industrial preserve (*Industrieschutzzone*)', within which industrial pollution was meant to be contained.[10]

The sources for German history, however, support alternative interpretations that stress local development rather than control from Berlin. Geoff Eley and David Blackbourn argue that Berlin was much less in control than other historians believe.[11] Some urban historians emphasize the importance of local elites.[12] In his short history of environmental policy, Klaus-Georg Wey emphasizes local origins of laws for local implementation of solutions.[13] Although some environmental historians portray nature as 'an uncontested and transcendent category', others present it as a contested terrain. Thus, members of different groups promote different conceptions of nature and must work out their differences to create policies that then help to shape the environment.[14] These approaches support a commonly encountered view in the history of technology: solutions to problems must be politically, economically and technically feasible. Resolving problems involves more than technical fixes and requires implementing solutions in contexts that are always local. In addition, as Donald MacKenzie argues, artefacts and technologies stabilize social relations and 'make larger, more complex societies possible'. Thus, we should see arguments about policy, law, pollution, technology and costs as arguments about how to organize social systems. In Germany, as elsewhere, gaining political backing to resolve environmental problems was quite difficult.[15]

Fragmented, the Ruhr required local solutions for local problems. Steeply rising terrain provides a narrow catchment area north of the river. The heavy-industrial region, mainly north of the Ruhr river, and the Ruhr basin, mainly south of the river, included members of contending subcultures, each of which had different stakes in existing landscapes.[16] Gary Herrigel characterizes the region north of the river as containing a large-scale, centralized, autarkic industrial order, in which companies integrated production horizontally and vertically within their own managerial hierarchies. The region south of the river contained small to medium-sized firms in a decentralized industrial order. Here small firms and local governments developed a variety of regional institutions to provide training, financial support and other kinds of assistance.[17] For instance, the owners of the small, metal-working companies around the city of Hagen felt squeezed between the heavy-industrial producers of their raw materials and the heavy-industrial consumers of their products. Electric power companies based in the heavy-industrial region threatened to expand south. As a result, in 1906, municipal and county (*Landkreis*) leaders in the south, led by Mayor Willi Cuno of Hagen, established their own power company, which then expanded further to the south and east.[18]

The Ruhr heavy-industrial region and basin was a 'trading zone',[19] in which members of different subcultures sought solutions to the problems of urban and industrial development without giving up their 'abstract principles' and commitments to industrial capitalism or 'rural romanticism'. They did not resolve all their problems in the same way. County commissioners (*Landräte*), mayors and industrialists formed competing electric power companies but co-operated in water undertakings. Mayor Erich Zweigert in Essen, Mayor Cuno of Hagen, and *Landrat* Karl Gerstein in Bochum promoted both independent electric power systems and unified water systems. The mining company Gelsenkirchener Bergwerks-AG promoted a power company and a large waterworks in the Ruhr, as did the Thyssen steel corporation.[20] Some sought better water without resolving their 'cultural despair' over the transition from an agrarian to an urban, industrial nation. H. Breme, the Provincial Meadow Master-Builder (*Provinzialwiesenbaumeister*) in Münster, doubted in 1901 that anything could be done to correct the industrial and urban 'abuses' along the Emscher River:

> according to [his] experience, the watercourses in the industrial region would only become clean again if all businesses, factory towns and cities were demolished and the population, again limited, returned to its former cultivation of beets.

Nevertheless, he suggested several steps to improve the situation.[21]

Members of subcultures could not find overall solutions to the problems of industrialization. Neither national nor state governments developed a legal basis for regulating electric supply, and both eventually built their own power systems.[22] Because municipal and county governments owned streets and controlled rights to cross those streets, they became integral parts of electric power systems. The Rhenish Westphalian Electrical Company (RWE) provides an example. Established in Essen in 1898, the RWE became a 'mixed' corporation soon after two industrialists, Hugo Stinnes and August Thyssen, gained control of it in 1902. During a short losing battle to expand eastwards into Westphalia, the RWE began including city and county shareholders to create the mixed corporate form that would allow its expansion in the Rhineland. Two Westphalian corporations opposed to the RWE also became mixed. Finally, in 1908, an agreement among leaders of the competing companies excluded the RWE from the Westphalian section of the Ruhr. Combining private, municipal and county ownership, mixed corporations constructed regional electric power systems that divided the Ruhr.[23]

Water, however, was a different matter. Drawing their water from the Ruhr and its tributaries, representatives of agrarian, autarkic and decentralized orders worked out their differences in respect of water. They could do so in part because the various German states, in this case Prussia, controlled navigable rivers. Intended to consolidate Prussian law, the Prussian Civil Code (*Preussisches Allgemeines Landrecht*) of 1794 provided the basis for a general regulation of water law in Prussia for most of the nineteenth century. Although a law in 1879 provided special conditions for establishing water associations, these latter would have limited control over members and membership. For instance, the associations could force membership only for agricultural reasons. In 1891, a Prussian law removed the agricultural constraint for the Ruhr's tributaries, and a further law in 1900 lifted the constraint for the Ruhr river itself. In 1913, Prussia established a new water law.[24] However, special laws to deal with the pollution of the Ruhr were still needed.

FORMS OF POLLUTION

'Pollution' has several meanings Members of institutions use accepted concerns about pollution to demarcate an 'inside' that must be protected from an 'outside'. People who identify new sources of pollution must make those sources credible in order to create new responses by reorganized institutions operating within modified boundaries. Sometimes new institutions must be created to provide new boundaries and responses. The responses must be made 'technically and politically feasible'. For instance, the Ruhr river brought dangers resulting from geography and weather, as well as from industrial and urban development. Flooding and drought periodically endangered the social and economic order, old or new. Industrial and urban growth threatened the 'old order' and its agrarian way of life. Members of the 'old order', who identified industrial and urban development as sources of cultural and biological pollution, promoted a 'politics of cultural despair'. And, of course, industrial and urban waste produced another kind of pollution.[25]

Developing adequate responses required complex negotiations between representatives of the heavy industrial firms and state institutions. After the Napoleonic period two Prussian Provinces, three government districts (*Regierungsbezirke*), and numerous cities and counties (*Landkreise*) divided the Ruhr. During the nineteenth century, this institutional arrangement provided the agrarian order with mechanisms for extending control over some issues down to the local level. As locations for creating new social, cultural and economic arrangements, rapidly developing towns threatened the agrarian order, especially when they became cities independent of counties. Developing new industrial and urban ways of life through institutional arrangements created by representatives of an agrarian order proved to be difficult. However, cities and counties were more than administrative units applying Prussian law created in Berlin. Their legislative councils could deal with local problems by creating economic organizations to provide water, gas, streetcars and electricity.[26] Those building the infrastructure had also to deal with representatives of other counties, cities and corporations.

Responses were necessary. Coal-mines pumped increasing amounts of waste water into the Emscher, and communities along the Emscher and its tributaries used them to carry away untreated sewage. By the 1880s, mines and communities discharging material into the Emscher and its tributaries were already threatening agriculture, fishing, and water-powered craft production. Foul water killed fish, turf and trees, and gave livestock diarrhoea. Built without adequate culverts, railroad embankments curtailed drainage. Drainage problems along the Emscher also increased, because mining caused the ground to settle. Settlement of three metres was common, but near the mine Hibernia on the outskirts of Gelsenkirchen the ground had sunk up to five metres. The depressions exacerbated swampy conditions along the Emscher. A report of 1901 described the Emscher as 'a thick, black substance sluggishly [moving] forwards'. Towns were built 'in the middle of the swamps', and it seemed that builders 'often even sought out the most dangerous locations', where 'the most dangerous sanitary conditions' could be found.[27]

The slow-moving Emscher slowed even more as it neared the Rhine. For its part, the Ruhr river carried away discharges from coal, iron, paper, beer, leather and textile plants. A report of 1902 stated that between Witten and Essen, river water could 'no longer be used for drinking'. 'Never odourless', the water contained 'excessive amounts of ammonia, chlorine, nitrous acid, nitric acid, live worms and parasites; its taste [was] stale, when warm nauseating'. During dry spells, sewage and industrial wastes could be seen coating the river bottom near its mouth. Upriver, the coating frequently prevented water from passing into wells through the sand and gravel beds lining the river's banks. To circumvent that problem, many waterworks built filter basins fed by river water to increase artificially the flow of water into wells. In severe droughts, cities sometimes pumped water directly from the river into city water mains. In 1901 that practice caused a typhoid epidemic that killed 350 people in Gelsenkirchen.[28]

In addition, the Ruhr river experienced water flow problems. It flooded periodically, while in time of drought people could walk across the riverbed near the Rhine. Industrial development and urban growth increased demand dramatically. Waterworks pumped 90 million cubic metres (m^3) out of the Ruhr in 1893 and 500 million m^3 per year in the late 1920s. Industry consumed 85 per cent of the water, the rest providing drinking water. Water usage created two problems with water flow. First, during droughts waterworks and pumping stations between Arnsberg and the Rhine pumped more water out of the Ruhr than flowed down it at those times. Second, they pumped water into gravity tanks on the watersheds between the Ruhr and neighbouring tributaries of the Rhine. From there, 75 per cent of the water in 1893 flowed out of the Ruhr catchment area altogether to supply areas external to it and then north into the Emscher and Lippe, or south into the Wupper. That problem continued: very little of the water taken from the Ruhr flowed back into it.[29]

Members of contending institutions resolved their differences over Ruhr water by forming new hybrid organizations. After negotiating for several years, in 1899 representatives of private and communal waterworks joined owners of water-power facilities to establish the Ruhr Reservoir Association (*Ruhrtalsperrenverein*) under private law. The Association began building impounding reservoirs on tributaries of the Ruhr.[30] More common in northern Germany than in water-rich southern Germany, impounding reservoirs collected water during wet periods in order to regularize flow and to provide water during dry periods.[31] After 1904, the Emscher Association (*Emschergenossenschaft*) by contrast turned the Emscher and its tributaries into sewage canals carrying industrial and urban waste as quickly as possible to the Rhine. Membership comprised municipal, county and industrial polluters.[32]

The Ruhr Reservoir and Emscher Associations provided models and personnel for the Ruhrverband. Inadequate for dealing with the problems along the Emscher and Ruhr rivers, the new Prussian water law of 1913, left the special law for the Emscher unaltered. However, in 1913, in consultation with local interests, the Prussian state produced two special laws for the Ruhr: one established the Ruhrverband, and the other re-established the Ruhr Reservoir Association. Law for public bodies then governed both associations. Uniting cities, counties and industrial firms, both associations gained legal powers necessary to carry out their tasks. The three associations could force membership on organizations using the rivers and could tax members to pay for facilities.[33] Although exercising a form of sovereignty over their water systems, these organizations still had to deal with other parts of the state, local governments and industry. Within the new water associations, representatives of different institutions struggled to promote their own visions for development. These hybrid organizations created new landscapes for their rivers.

THE RUHRVERBAND AS AN ORGANIZATIONAL HYBRID

The Ruhrverband became increasingly active after World War I. To resolve problems on the lower Ruhr, the Ruhrverband built a 12-kilometre long intercepting sewer to direct sewage from Mülheim, Oberhausen and Duisburg away from the Ruhr river. Additionally, filters and traps for tar and oil cleaned the water before it reached the Rhine. By 1926, 45 purification facilities were treating municipal and industrial waste. Thirty-one of them were below Hagen. Eight years later, in 1934, the Ruhrverband was operating 75 purification facilities, 200 km of waste water collectors, and 15 waste water pumping stations. Nevertheless, significant sources of pollution remained.[34]

The Lenne flowing into the Ruhr north of Hagen produced pollution downstream that threatened water supplies. Flowing north and west past Hagen to the Ruhr, the Lenne became increasingly polluted after the 1870s. More than 100 small iron businesses operated in the mid-1920s along the Lenne river, into which they discharged sulphuric and hydrochloric acid and iron salts.[35] The acidic Lenne flowed into the alkaline Ruhr, which carried waste water discharged into it further east by paper mills and wood pulp industries. With a relative density of 1.07 to 1.15, the resulting flocs created a flaky sludge that clogged the river bed's sand and gravel, thereby hindering the flow of water into wells. Especially during the dry years, waterworks pumped water into filter basins near wells to increase their yield. For instance, in 1918, the Hagen waterworks pumped 49,600 m^3 of water during eight days into its filter basins, and in 1921, it pumped 400,000 m^3 during 161 days.[36]

Ruhrverband officials wanted an inexpensive solution, because its members had to pay for construction and operations. Purification in the Lenne valley itself appeared unfeasible. In 1928, the Ruhrverband's assistant manager, Oskar Spetzler, argued that the valley was often too narrow for individual purification facilities, which, even if they could have been built, would have been too expensive for the economically weak businesses to support. In addition, he claimed that even after purification, some iron in solution would remain, producing sludge. Upgrading mechanical processes biologically in individual facilities would increase costs significantly. Spetzler argued that building a settling basin would provide the 'fastest, most effective and most economical' method for dealing with iron sludge. Built between the mouth of the Lenne and Hagen's waterworks, a settling basin would allow harmful materials to settle out for eventual dredging. The first two plans failed to gain financial and political support in the early 1920s. The first plan no longer appeared viable after the period of inflation ended, and the mayor of Hagen opposed the second as a threat to Hagen's waterworks. Then, to reduce construction and operating costs as much as possible, Ruhrverband officials allied themselves with the Rhenish Westphalian Electrical Company – the RWE. In turn, the RWE gained new hydroelectric facilities built together with the settling basin.[37]

During the 1920s, the RWE expanded dramatically. However, RWE officials hoped to avoid adding new steam capacity simply to cover peak demand and sought ways of better utilizing already installed capacity.[38] That could, of course, be done by finding customers who would provide varied load so that peaks and troughs of consumption would even more closely approximate to a steady generation of electricity. Power companies in Germany were beginning to seek domestic load to help balance industrial load.[39]

Another possibility was to find ways of storing energy during off-peak periods and using it to cover peak periods of consumption. Storing energy was not a new idea. Fly-wheels, Ilgner systems and batteries were already used to store energy to regularize load.[40] Reservoirs already stored river flow, but in the 1920s power companies introduced pumped-storage stations with high heads. Water pumped up into storage basins during off-peak hours could be released to drive turbines during peak periods of consumption. Pumped-storage stations could also improve system reliability by allowing thermal units to operate steadily and by providing instant reserves in case of disturbances in supply. By 1924, several south German cities had pumped-storage stations, as did the Badenwerk, the state power company in Baden.[41]

The RWE explored several locations for its own pumped-storage station. In 1924, RWE officials began investigating a storage facility near Coblenz at Lake Laacher, five kilometres west of the Rhine and 220 metres higher. The owner of Lake Laacher, the Abbey Maria-Laach, rejected the RWE's proposal in order to preserve the 'lake's landscape'. The RWE also sought to build a pumped-storage facility on the Our, on the border between the Rhine Province and Luxembourg. However, negotiations dragged on through 1926 and 1927. The RWE also considered sites at Lay on the Mosel and at Niederhausen on the Nahe. In southern Germany, it collaborated with other power companies to build several pumped-storage stations. Without abandoning negotiations concerning the Our, the RWE negotiated with the Ruhrverband. Building a pumped-storage station on the Ruhr river attracted the RWE because of its potential scale and its proximity to Rhenish-Westphalian consumers. By 1930, the Ruhrverband and the RWE were planning a series of pumped-storage stations and reservoirs from Herdecke down to the Rhine.[42] Collaboration would improve electric generation and water purification.

The alliance between the Ruhrverband and the RWE allowed both organizations to shift their operational boundaries. With RWE financial support, the Ruhrverband could build a larger settling basin with a larger surface area to increase biological purification. The RWE could control the small hydro station at the dam and could build a pumped-storage station in territory from which it had been excluded in 1908.[43] By building its storage station under the Ruhrverband's rights to modify the Ruhr, the RWE avoided authorization procedures controlling the taking of river water to which it would otherwise be subject. In addition, the RWE would not need to gain permission for each modification of its facility and could avoid demands for modifying the storage station to suit the changing goals of state officials.[44] The RWE also would thereby circumvent local control of building permits. In 1927, the Ruhrverband transferred property to the RWE for its pump-storage station.[45]

Ruhrverband officials dealt with other boundaries. Some opponents objected to the Ruhrverband becoming a productive facility – perceived as a form of social and economic pollution. In 1928, assistant manager Spetzler, countering such opposition, argued that the head created by the dam should be used to generate electricity. Allowing the water simply to fall over the dam would be 'economic nonsense'. He asserted that building and operating a power station was one of 'the most practical means' of making treatment of industrial waste water 'useful for the economy' and thereby creating 'the most economical purification method'. Since the main goal was purification, selling electricity was no different from selling sludge from the reservoir for 'mine filling or fertilizer', or 'selling methane produced by purification facilities'. Spetzler insisted that the Ruhrverband was not seeking to use 'public funds cloaked by a tax privilege' to move into the electrical, fertilizer or gas businesses.[46]

BUILDING A VIGOROUS HYBRID SYSTEM

Ruhrverband officials planned and built a hybrid project serving multiple constituencies and functions. Agricultural experts believed that the mud-flat meadows, or foreland, along the southern shore of the river had little agricultural value, at least compared to the increased purification that would result from enclosing the area in a reservoir. In order to prepare for an eventual increase in the size of the reservoir, the Ruhrverband had purchased almost all the remaining foreland. The settling basin, called Lake Hengstey, was four kilometres long and at its widest 500 m. To achieve the 2.8 million m^3 planned capacity some 500,000 m^3 of soil and rock were removed in the upper section (Figure 2). Downstream, the dam created a head of 4.5 m that a hydroelectric station could use to provide the RWE with 3900 kilowatts (kW). In addition, the lake was to supply the RWE with some 1.2 million m^3 of water daily for its pumped-storage station.[47]

Beginning just below the mouth of the Lenne, the Ruhrverband built a dike and small island stretching almost one kilometre along the right bank of the Ruhr. The narrow channel served to mix the waters of the Lenne and Ruhr. In the lake, the water slowed to allow flocs to settle out. The 50-metre-wide island provided the base for a bridge. A section of 69 m spanned the mixing channel. Resting on two piers, the other section crossed the remaining 180 m of the lake. The road improved transportation between the two sides of the river. On the island, the Ruhrverband built a restaurant, which generated income.[48]

The Ruhrverband rearranged the left bank of the Ruhr to fit the lake into the neighbouring communities. Although streams at the upper, eastern end continued to drain into the lake, the Ruhrverband adjusted

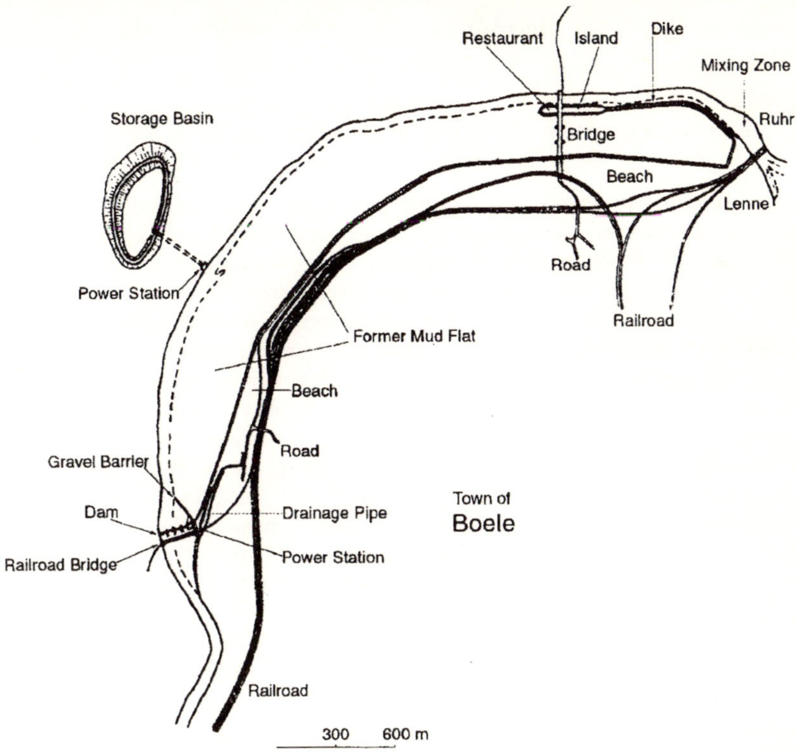

Figure 2 Lake Hengstey: the dam generated electricity and created a settling basin south of the old river bed (shown by a dotted line) with two beaches. The dike allowed water from the Ruhr and the Lenne to mix before entering into the settling basin, which provided water for a pumped-storage station.
Source: *Wasserkraft und Wasserwirtschaft*, 1928, 23: 335.
Artist: P. Falcone

the lower, western end of the lake. The town of Boele's drainage channel, which occasionally carried tar and oil, was re-routed to the Lenne after passing through a special purification facility. A concrete pipe carried drainage further west to the other side of the dam. Higher than flood levels, the new embankment was covered with stone on the water face and with grass on the air face. The Ruhrverband used dredged material to level land adjoining the lake for sports. It also constructed beaches near both ends of the lake's left bank.[49] Sunny summer Sundays during the 1930s found up to 100,000 visitors bathing along the Ruhr.[50]

Completed in March 1928, Lake Hengstey was meant to appeal aesthetically. Ruhrverband officials sought to fit the bridge and restaurant into what Spetzler referred to as 'one of the most beautiful landscapes of the Ruhr valley'. 'Aesthetic demands' led to designs suiting the 'charm of the landscape' and the 'beautiful contrast' between the steep, wooded incline on the river's right bank and the flat area on the left. Spetzler noted that

> by forgoing plaster and almost all form of ornament, the modest and austere concrete construction stood in expressive association with the technical content of the station as well as with the landscape.

He argued that the dam and other structures demonstrated that 'such facilities [could fit] into even the most beautiful natural settings'. In addition, the RWE and Ruhrverband were working 'hand in hand to beautify the scene with new plants and green areas'.[51]

The Ruhrverband added other functions. With local communities, it established the Lake Hengstey Association to create 'a first-rate recreational facility'. Following the directives of the Chief Fish Warden (*Oberfischmeister*) and Prussian law, the Ruhrverband protected fish by installing grates of appropriate size on the turbine intakes. Stocking Lake Hengstey with fish promoted biological purification and returned fauna that had all but disappeared from the river prior to World War I. After five years of success with fish stocks, the Ruhrverband leased fishing rights to a fishing association. By the mid-1930s, Lake Hengstey was providing opportunities for fishing to 'many workers, miners, craftsmen and tradesmen'. However, the uselessness of the sludge, for which the lake had been created, disappointed officials. It contained up to 40 per cent iron, but the Dortmunder Union, a steel firm connected to the RWE, found that the iron could not be recovered. The high iron content made the sludge unusable for fertilizer or for making bricks. It could only be used as landfill.[52]

Other problems of the hybrid landscape remained. From the beginning, officials from the Ruhrverband planned one use of the lake, while representatives of the RWE planned another. In 1928, Spetzler, the Ruhrverband's assistant manager, believed that storage-facility operations would improve purification, but tests found no discernible effect.[53] A more important problem arose from water usage. RWE operations seemed simple and straightforward. According to the contract of 1926, during 14 night-time hours, electricity from the RWE's system would power pumps, drawing water down to 70 cm below the lake's maximum surface level. The agreed-upon 1.2 million m^3 could then be used to drive turbines during six to ten daytime hours to cover peak demand for electricity. However, in 1928, Spetzler thought that the RWE would rarely use all of the water that it was allowed to take.[54]

RWE officials, however, planned a larger facility. In 1927, they described a facility with a storage basin of 1.5 million m^3 supporting a peak capacity of 140,000 kW and generating about 550,000 kW-h per day.[55] They built a basin for 1.6 million m^3.[56] Using 400,000 m^3 more than the contract specified, the expanded system would require changes in the Hengstey reservoir. To provide such additional water, the lake surface would have to be raised by 25 cm or the RWE would have to take water from the flow of the Ruhr itself, which would require the next reservoir – not yet built – to balance the water system.[57]

In practice, the Ruhrverband and the RWE adjusted their systems to make them work together. Legally, the Ruhrverband could not raise the level of the lake. Perhaps because it was experiencing difficulties making its station economically viable, the RWE did not set up a planned fourth pump.[58] Nevertheless, it still had to take water from the flow of the Ruhr, because Ruhrverband officials had miscalculated the surface area of the lake. They had thought the area would be 176 hectares, but discovered that it was only 136 hectares. Contractually the RWE had a claim on the top 70 cm of the lake, which from 136 hectares could only provide 952,000 m^3. A solution was at hand. In 1931, the Ruhrverband opened Lake Harkort, which was just below Lake Hengstey, and in 1933, Lake Baldeney, south of Essen. Lake Harkort allowed the Ruhrverband to adjust its operations to fit contractual and legal obligations. As suggested at the Second World Power Conference in 1930, the Ruhrverband operated Lake Hengstey, Lake Harkort, and the pumped-storage station together to provide the RWE with water *and* to ensure a steady flow of river water to waterworks. While the pumped-storage station drew water from Lake Hengstey and directly from the Ruhr river to fill its storage basin, Lake Harkort provided water for the continued flow of the river. As the pumped-storage station used stored water to drive its turbines and generate electricity, that water and the flow of the river refilled first Harkort then Hengstey.[59]

CONCLUSION

The Ruhrverband's lakes worked well, if not ideally, their dams and reservoirs serving many purposes. They helped regulate and clean the Ruhr, made it possible to use the river for shipping, and opened it up for recreational purposes. Along their stretches of the river, Lakes Hengstey, Harkort and Baldeney increased the flow-through period fivefold, thus permitting particulate matter to settle in the reservoirs. The enlarged surface area increased purification by natural aeration, the effect of the sun, fish, bacteria and plants helping also. It was also thought that the expanded water surface would help clean the air by absorbing industrial gases.[60]

Tests from 1928 to 1932 demonstrated the benefits of Lake Hengstey. Slow passage through the lake reduced the amount of plankton by 94.5 per cent from 45 to 2.5 cm^3 per m^3. Clarity of the water increased significantly. From the mouth of the Lenne to the dam, the lake reduced iron content by 63 per cent from 8.9 mg per litre to 3.3 mg per litre. Mixing the waters of the Ruhr and the Lenne also reduced the pH of the Ruhr, but biological action then slightly increased it. About 57 per cent of the biological material and 54 per cent of the phenol compounds carried into the lake decomposed there. The lake reduced the number of bacteria (*Gesamtkeimzahl*) by 50 per cent from an average count of 45,470 per m^3 after the confluence of Ruhr and Lenne to 23,100 at the dam. E. coli (*Colikeime*) were reduced almost 62 per cent from 55 to 21 per m^3.[61]

A hybrid organization, the Ruhrverband built a vigorous hybrid landscape. Members of local subcultures – cities, counties and industries, large and small – co-operated sufficiently to provide enough local agreement in 1913 for the Prussian state to create a new law, which, together with Prussian water law, armed the Ruhrverband with the legal powers to force membership on water users and to tax them. While the Emscher Association converted the Emscher river into a working sewer system, the Ruhrverband turned the Ruhr river into a viable source of clean water. To build an acceptable entity serving its many constituencies, however, Ruhrverband officials promoted multi-functional facilities like Lake Hengstey. The RWE funded a larger settling basin. Building Lake Hengstey improved purification, electrification, transportation, bathing and fishing, while the Ruhrverband levelled land for sporting facilities and built a restaurant. Such multiple uses and functions produced multiple constituencies and reduced the costs that members had to pay. In this way, the hybrid landscape helped create a new way of life that was politically, socially, economically, aesthetically and technically viable and provided local continuity in a century of political turmoil.

ACKNOWLEDGEMENTS

I would like to thank Hans-Joachim Braun, Helmut Maier, Jean Nocito-Gobel, and Graham Hollister-Short for their suggestions. A version of this paper was presented at the ICOHTEC meeting in Belfort, France in 1999.

Notes and References

1. K. Imhoff, *Der Ruhrverband* (Berlin, 1928), 5, has the figure of 4,500 km^2; Helmut Wolter, 'Die Ruhrstauseen,' *Der Naturforscher*, 1935/36, 12: 384–387, esp. p. 384, provides the figure of 235 km, but a sign at the source states 219 km. Helmut Maier provided me with a copy of this article; *Meyers Enzyklopädisches Lexikon* (Mannheim, 1977), Vol. 20: 438, provides figures of 4,444 km^2 and 218 km.

2. Wolfhard Weber, *Industrialisierung: Das Ruhrgebiet – ein Fallbeispiel* (Brunswick, 1982), 7.

3. For the Ruhr's development see: Wolfgang Köllmann, *et al.* (eds.), *Das Ruhrgebiet im Industriezeitalter: Geschichte und Entwicklung*, 2 vols. (Düsseldorf, 1990). Wolfgang Köllmann, 'Beginn der Industrialisierung', *ibid.*, Vol. 1: 11–79, esp. 12, identifies the core of the Ruhr as the area from just east of Dortmund to the Rhine and from the Lippe to the high ground south of the Ruhr valley. For discussions of the Ruhr's environmental pollution see Thomas Rommelspacher, 'Das natürliche Recht auf Wasserverschmutzung', in Franz-Josef Brüggemeier and Thomas Rommelspacher (eds.), *Besiegte Natur: Geschichte der Umwelt im 19. und 20. Jahrhundert* (Munich, 1989), 42–63; Franz-Josef Brüggemeier and Thomas Rommelspacher, *Blauer Himmel über der Ruhr: Geschichte der Umwelt im Ruhrgebiet, 1840–1990* (Essen, 1992); and F.-J. Brüggemeier, 'The Ruhr Basin 1850–1980: A Case of Large-Scale Environmental Pollution', in P. Brimblecombe and C. Pfister (eds.), *The Silent Countdown: Essays in European Environmental History* (Berlin, 1990), 210–227. For 'cultural despair' see Fritz Stern, *The Politics of Cultural Despair: A Study in the Rise of the Germanic Ideology* (Berkeley, 1961).

4. The Ruhr region, as noted below, was similar to a 'trading zone', in which members of subcultures met and worked together without resolving their global differences; see Peter Galison, 'Trading Zone: Coordinating Action and Belief', in Mario Biagioli (ed.), *The Science Studies Reader* (New York, 1999), 137–160; and Peter Galison, *Image and Logic: A Material Culture of Microphysics* (Chicago, 1997). See also James Clifford, 'Museums as Contact Zones', in his *Routes: Travel and Translation in the Late Twentieth Century* (Cambridge, MA, 1997), 188–219.

5. Edmund N. Todd, 'Industry, State, and Electrical Technology in the Ruhr circa 1900', *OSIRIS*, 1989, 5: 243–259.

6. Paul David, 'The Landscape and the Machine: Technical Interrelatedness, Land Tenure and the Mechanization of the Corn Harvest in Victorian Britain', in his *Technical Choice, Innovation and Economic Growth: Essays on American and British Experience in the Nineteenth Century* (Cambridge, 1975), 233–288, describes the landscapes as representing ossified decisions concerning how to work the land and organize land tenure. In order to introduce reapers, those decisions, embedded in the landscape, had to be changed. See also Edmund N. Todd, 'Electric Ploughs in Wilhelmine Germany: Failure of an Agricultural System', *Social Studies of Science*, 1992, 22: 263–281, esp. 272–275.

7. Wolter, *op. cit.* (1), 384; Karl Imhoff, 'Water Supply and Sewage Disposal in the Ruhr Valley', *Engineering News-Record*, 1925, 94: 104–106, esp. 105; F. Sierp and Hayo Bruns, 'Die Bedeutung der Stauseen für die Reinhaltung der Ruhr', *Gesundheits-Ingenieur*, 1934, 57: 199–206, esp. 199, 201. The quotations are from 'Der Ruhrverband 1929', *Wasserkraft und Wasserwirtschaft*, 1930, 25: 78–80, esp. 80. Imhoff was the Ruhrverband's chief engineer. Sierp worked for the Ruhrverband, and Bruns for the Institut für Hygiene und Bakteriologie in Gelsenkirchen.

8. Hans-Ulrich Wehler, *Deutsche Gesellschaftsgeschichte* (Munich, 1995), Vol. 3: *Von der 'Deutschen Doppelrevolution' bis zum Beginn des Ersten Weltkrieges, 1849–1914*; and Wolfgang Hardtwig, 'Großstadt und Bürgerlichkeit in der politischen Ordnung des Kaiserreichs', in Lother Gall (ed.), *Stadt und Bürgertum im 19. Jahrhundert* (Munich, 1990), 19–64.

9. Brüggemeier, 'Ruhr Basin.' *op. cit.* (3), 218.

10. Brüggemeier, *Blauer Himmel*, *op. cit.* (3), 47–49.

11. Geoff Eley, *Reshaping the German Right* (London, 1980); David Blackbourn and Geoff Eley, *The Peculiarities of German History: Bourgeois Society and Politics in Nineteenth-Century Germany* (Oxford, 1984); and David Blackbourn, *The Long Nineteenth Century: A History of Germany, 1780–1918* (Oxford, 1998).

12. Friedrich Lenger, 'Bürgertum und Stadtverwaltung in rheinischen Großstädten des 19. Jahrhunderts. Zu einem vernachlässigten Aspekt bürgerlicher Herrschaft', in *Stadt und Bürgertum im 19. Jahrhundert*, *op. cit.* (8), 97–169, presents cities as centres of local bourgeois control.

13. Klaus-Georg Wey, *Umweltpolitik in Deutschland: Kurze Geschichte des Umweltschutzes in Deutschland seit 1900* (Opladen, 1982), 36–104. See also Michael Wettengel, 'Staat und Naturschutz 1906–1945: Zur Geschichte der Staatlichen Stelle für Naturdenkmalpflege in Preußen und der Reichsstelle für Naturschutz', *Historische Zeitschrift*, 1993, 257: 355–399; and Michael Kloepfer, *Zur Geschichte des deutschen Umweltrechts* (Berlin, 1994).

14. William Cronon, 'Introduction: In Search of Nature', in William Cronon (ed.), *Uncommon Ground: Rethinking the Human Place in Nature* (New York, 1995), 23–56. Several articles in this volume are grouped under the heading 'contested terrains'; Joel A. Tarr, *The Search for the Ultimate Sink: Urban Pollution in Historical Perspective* (Akron, 1996), 123–128, 159–178, notes that, in the United States, sanitary engineers thought the cost of facilities too high to combine purification and sewage treatment. Instead, sewage would be dumped into rivers, and downstream water users would then purify water for their use. A different professional group, medical doctors in public health agencies, wanted prohibitively expensive combined systems.

15. Donald MacKenzie, *Knowing Machines: Essays on Technical Change* (Cambridge, MA, 1996), 14; Mary Douglas, 'Environments at Risk', in Barry Barnes and David Edge (eds.), *Science in Context: Readings in the Sociology of Science* (Cambridge, MA, 1982), 260–275; Mary Douglas, *How Institutions Think* (Syracuse, 1986); and Wiebe E. Bijker and John Law (eds.), *Shaping Technology/Building Society: Studies in Sociotechnical Change* (Cambridge, MA, 1992). For a discussion of organization efforts, see Raymond H. Dominick III, *The Environmental Movement in Germany: Prophets and Pioneers, 1871–1971* (Bloomington, 1992).

16. David, *op. cit.* (6), describes landscapes as embodying earlier decisions. Brüggemeier, 'Ruhr Basin', *op. cit.* (3), applies the term 'industrial preserve' to the Ruhr heavy-industrial region, which was north of the Ruhr River in the Emscher basin.

17. Gary Herrigel, *Industrial Constructions: The Sources of German Industrial Power* (Cambridge, 1996), specifically lists Württemberg, Baden, the Kingdom of Saxony, Bergisches Land, Siegerland, and the left bank of the Rhine as areas in which decentralized orders developed.

18. For the development of the region see Ludwig Beutin, *Geschichte der südwestfälischen Industrie- und Handelskammer zu Hagen und ihrer Wirtschaftslandschaft* (Hagen, 1956), 60–141. For electrification in that region see Edmund Neville Todd III, 'Technology and Interest Group Politics: Electrification of the Ruhr, 1886–1930' (PhD Dissertation, University of Pennsylvania, 1984), 107, 133–137; and Fritz Bürger, *Das Kommunale Elektrizitätswerk Mark (Hagen)* (Essen, 1937).

19. Galison, *op. cit.* (4).

20. Todd, *op. cit.* (5); Oskar Spetzler, 'Stausee und Pumpenspeicher Hengstey', *Wasserkraft und Wasserwirtschaft*, 1928, 23: 327–347, esp. p. 327, states that in the 1920s the Wasserwerk für das nördliche westfälische Kohlenrevier supplied an area of 1,300 km^2 and 1,280,000 people. See also Martin Weyer-von Schoultz, *Stadt und Gesundheit im Ruhrgebiet 1850–1929: Verstädterung und kommunale Gesundheitspolitik am Beispiel der jungen Industriestadt Gelsenkirchen* (Essen, 1994), 119.

21. H. Breme, 'Gutachten über die sanitären Verhältnisse in der Emscherniederung in dem Distrikte, welcher von der Wanne-Cranger und Essen-Horster Straße begrenzt wird', in H. Emmerich and F. Wolter (eds.), *Die Entstehungsgeschichte der Gelsenkirchener Thyphusepidemie von 1901* (Munich, 1906), 111–116, as reproduced in Brüggemeier, *Blauer Himmel*, *op. cit.* (3), 142–147, esp. 146–147.

22. Neither the Reich in 1911 nor Prussia in 1926 had the legal authority to control electrification: see Helga Nussbaum, 'Versuche zur Reichsgesetzlichen Regelung der deutschen Elektrizitätswirtschaft und zu ihrer Überführung in Reichseigentum, 1909 bis 1914', *Jahrbuch für Wirtschaftsgeschichte* (1968), pt. 2, 117–203, esp. 154; and Jaques, 'Preußen und das Reich in der deutschen Elektrizitätswirtschaft', *Elektrotechnische Zeitschrift*, 1926, 47: 1160–1163.

23. Todd, *op. cit.* (5). For recent discussions of the RWE's development see Helmut Maier (ed.), *Elektrizitätswirtschaft zwischen Umwelt, Technik und Politik: Aspekte aus 100 Jahren RWE-Geschichte, 1898–1998* (Freiberg, 1999).

24. Ruhrverband und Ruhrtalsperrenverein, *75 Jahre im Dienst für die Ruhr, 1913–1988* (Essen: Ruhrverband und Ruhrtalsperrenverein, 1988), 35–38.

25. See: Stern, *op. cit.* (3); Douglas, *op. cit.* (15); and Cronon, *op. cit.* (14). Robert W. Smith, 'The Biggest Kind of Big Science: Astronomers and the Space Telescope', in Peter Galison and Bruce Hevly (eds.), *Big Science: The Growth of Large-Scale Research* (Stanford, 1992), 184–211, describes the process of making the space telescope 'technically and politically feasible'.

26. See Heinz Günther Steinberg, 'Zur Verwaltungsgeschichte des Ruhrgebietes', in Walter Först (ed.), *Beiträge zur Neueren Landesgeschichte des Rheinlands und Westfalens*, Vol. 3: *Politik und Landschaft* (Cologne, 1969), 177–215; Wolfgang R. Krabbe, *Die deutsche Stadt im 19. und 20. Jahrhundert: Eine Einführung* (Göttingen, 1989), 15–16, 31–38; Edmund N. Todd, 'Prussian *Landräte* and Modern Technology: Electricity as a Source of Power in the Ruhr, 1900–1915', *ICON*, 1996, 2: 83–107; and Todd, *op. cit.* (5).

27. Brüggemeier, *Blauer Himmel*, *op. cit.* (3), 94–98; Hermann Korte, 'Die Entfaltung der Infrastruktur,' in *Das Ruhrgebiet*, *op. cit.* (3), Vol. 1: 583; Breme, *op. cit.* (21), 143, 146.

28. Rubner, 'Das Obergutachten der wissenschaftlichen Deputation für das Medizinalwesen', December 1, 1902, as quoted in E. Grahn, 'Die Gerichtsverhandlungen über die Gelsenkirchener Typhusepidemie im Jahre 1901', *Journal für Gasbeleuchtung und Wasserversorgung*, May 27, 1905, 48: 447–457, 475–502, 516–546, esp. 449; Ruhrverband, *op. cit.* (24), 164; Sierp and Bruns, *op. cit.* (7), 200; Brüggemeier, *Blauer Himmel*, *op. cit.* (3), 92–93; Imhoff, *op. cit.* (7), 104–106; and Weyer-von Schoultz, *op. cit.* (20), 112–121.

29. Ruhrverband, *op. cit.* (24), 45; Imhoff, *op. cit.* (7), 104–106; and Spetzler, *op. cit.* (2), 327. Spetzler was a graduate engineer (Dipl.-Ing.) and assistant manager (stellv. Geschäftsführer) of the Ruhrverband. Sierp and Bruns, *op. cit.* (7), 199–200, notes that with a population some 16 times larger, Germany used 2.5 billion m^3 per year.

30. Ruhrverband, *op. cit.* (24), 35, 45, and Brüggemeier, *Blauer Himmel*, *op. cit.* (3), 94, provide 1899 as the founding date of the Ruhrtalsperrenverein.

31. W. Soldan, 'Zur zweiten Weltkraftkonferenz. 15. bis 25. Juni 1930', *Wasserkraft und Wasserwirtschaft*, 15 June 1930, 25: 133.

32. Brüggemeier, 'Ruhr Basin', *op. cit.* (3), 213, and Imhoff, *op. cit.* (7), 104–106; Brüggemeier, *Blauer Himmel*, *op. cit.* (3), 98, states that the law creating the Emscher-

gennossenschaft protected its members from high demands and allowed them to 'operate at the lower boundary of the technically and economically possible' solutions.

33. Ruhrverband, *op. cit.* (24), 35–38, 48, 51–55.

34. See: Imhoff, *op. cit.* (7), 104–6; Spetzler, *op. cit.* (20), 327; and Sierp and Bruns, *op. cit.* (7), 200.

35. Wolter, *op. cit.* (1), 386.

36. Spetzler, *op. cit.* (20), 328.

37. *Ibid.*, 328–33. Sierp and Bruns, *op. cit.* (7), 201, note that a settling basin would also intensify the self-purification processes found in rivers. Although Brüggemeier, *Blauer Himmel*, *op. cit.* (3), 99–100, states that industrial concerns for costs prevented appropriate responses to pollution, Tarr, *op. cit.* (14), argues that in the United States costs represented a real issue for engineers. In evaluating debates concerning risk, Douglas, 'Environments at Risk', *op. cit.* (15), argues that money is one of four trump cards – 'time, money, God, and nature' – that people use as they struggle to organize their social systems and seek to make each other behave 'properly'.

38. See the RWE's business report of 1926/1927 as quoted in Ernst Henke, *Das RWE nach seinen Geschäftsberichten, 1898–1948*, (Essen, 1948), 44.

39. 'Germans Use Few Appliances But Increasing', *Electrical World*, 1929, 93: 448, shows that Cleveland was far ahead of Berlin in getting domestic electric appliances into homes. For strategies leading to overcapacities and efforts to overcome those problems see Norbert Gilson, 'Der Irrtum als Basis des Erfolgs. Das RWE und die Durchsetzung des ökonomischen Kalküls der Verbundwirtschaft bis in die 1930er Jahre', in Maier, *op. cit.* (23), 51–88.

40. Patented in 1901 by Carl Ilgner, Ilgner systems used large flywheels to regularize the load on electric motors driving hoisting equipment in mines. For their development and use see W. Philippi, *Electrische Fördermaschinen* (Leipzig, 1921).

41. A. Maas, 'Hydraulische Hochspeicherkraftwerke', *Verein Deutscher Ingenieure Zeitschrift* (hereafter, *ZVDI*), 1924, 68: 1161–67, 1195–99, esp. 1162–64; A. Maas, 'Die Wasserkraft-Speicheranlagen Tübingen und Reutlingen', *ZVDI*, 1924, 68: 1344–1345; and Spetzler, *op. cit.* (20), 332–333.

42. For these developments see Camilio Asriel, *Das R.W.E. Rheinisch-Westfälisches Elektrizitätswerk A.-G. Essen a.d. Ruhr: Ein Beitrag zur Erforschung der modernen Elektrizitätswirtschaft* (Zurich, 1930), 49–53; Henke, *op. cit.* (38), 44; A. Koepchen, 'Das RWE in der deutschen Elektrizitätswirtschaft (Vortrag, gehalten im Haus der Technik in Essen am 28. März 1930, von A. Koepchen, Essen)', in *RWE Geschäftsbericht* (1929/30), 3–12, esp. 7; and Spetzler, *op. cit.* (20), 332–335. The phrase 'lake's landscape' is from 'Rundschau: Elektrizitätswerk am Laacher-See', *Die Wasserkraft: Zeitschrift für die gesamte Wasserwirtschaft und Wasserkrafttechnik*, 1 October 1925, 20: 319. See also 'Das Pumpspeicherwerk Herdecke an der Ruhr des Rhein.-Westf. Elektrizitätswerkes, A.-G., Essen, unter Berücksichtigung der maschinellen Anlagen', *Wasserkraft und Wasserwirtschaft*, 1930, 25: 146–154.

43. For the RWE's earlier exclusion from Westphalia see: Todd, *op. cit.* (5).

44. For the Ruhrverband's rights see Abschrift: Verleiungs-Urkunde. B.A. II W.190/22, 114–115. Bezirksausschuss, Abteilung II. Arnsberg, gez. Schweiger, 23 Juli 1926. The Ruhrverband sent a copy to the city of Herdecke, Ruhrverband (Der Geschäftsführer I.V.) to Bürgermeister in Herdecke, 4 June 1927, Akte Herdecke.neu XI G5 (Bauakte Speicherwerk Stausee Hengstey), Stadtarchiv Herdecke. For an evaluation of the RWE's beneficial relationship to the Ruhrverband see: 'Die Unternehmungen des Ruhrverbandes und des Rheinisch-Westfälischen Elektrizitätswerkes an der Ruhr bei Hengstey. Rechtsgutachten des Senatspräsident a.D. Wirklichen Geheimen Oberregierungsrat Kisker zu Charlottenburg', 9 December 1927. The director of the Ruhrverband's Stausee division provided me with of copy of this.

45. For materials demonstrating Herdecke's inability to oversee the RWE adequately see Akte Herdecke.neu XI G5, *op. cit.* (44). For a form notarized on 30 July 1927 that transferred property in Herdecke from the Ruhrverband to the RWE, see Schlüter to Bürgermeisteramt Herdecke, 12 September 1927, *ibid.*

46. Spetzler, *op. cit.* (20), 328–329.

47. *Ibid.*, 333–339. Spetzler states '2.8 million m^3 contents' on p. 334 and 'about 3 million m^3 water contents' on p. 336. The lake would have a minimum bottom of 95.1 metres above sea level. See also Asriel, *op. cit.* (42), 87; and Henke, *op. cit.* (38), 44.

48. Spetzler, *op. cit.* (20), 335–336.

49. *Ibid.*, 334–339.

50. Sierp and Bruns, *op. cit.* (7), 200. For a celebration of Lake Hengstey see: Oskar Spetzler and Hans Strobel (eds.), *Der Hengsteysee im neugestalteten Ruhrtal als Erholungsstätte und Kraftquelle* (Essen: Seegesellschaft m.b.H. Hengstey, 1930).

51. Spetzler, *op. cit.* (20), 336–337, 340, 345. Spetzler's aesthetic concerns were similar to those expressed by members of völkisch groups; see Wettengel, *op. cit.* (13), 373.

52. Spetzler, *op. cit.* (20), 334, 341, 345, 346; 'Rundschau: Wasserkraftanlagen. Die Lichtweite der Rechenstäbe bei Turbinengerinnen', *Die Wasserkraft*, 1 October 1925, 20: 319; Sierp and Bruns, *op. cit.* (7), 201, 205–206.

53. Spetzler, *op. cit.* (20), 334–335; and Sierp and Bruns, *op. cit.* (7), 204.

54. Spetzler, *op. cit.* (20), 333, 335; Henke, *op. cit.* (38), 44.

55. Henke, *op. cit.* (38), 44.

56. Each pump could pump 14.6 m^3/s at the beginning of its operations to a height of 146.4 m and 12.3 m^3/s to a height of 165.2 m at the end of operations. Three could fill the basin in ten hours. A fourth pump was planned. Each of the four turbines could use water at a rate of 25.8 m^3/s. Together they were to provide maximum capacity for 4.2 hours, which would use 1,560,384 m^3 of water. See Deutscher Wasserwirtschafts- und Wasserkraftverband E.V. (ed.), *Die Wasserkraftwirtschaft Deutschlands: Festschrift zur Tagung der II. Weltkraftkonferenz, Berlin 1930* (Berlin: Deutscher Wasserwirtschafts- und Wasserkraftverband E.V., 1930), 141; and 'Das Pumpspeicherwerk Herdecke', *op. cit.* (42), 149, 154.

57. See Deutscher Wasserwirtschafts- und Wasserkraftverband E.V., *op. cit.* (56), 141.

58. For discussions of the RWE's economic problems see: Gilson, *op. cit.* (39); and Helmut Maier, 'Arthur Koepchen (1878–1954)', in Wolfhard Weber (ed.), *Ingenieure im Ruhrgebiet* (Münster, 1999), 195–202.

59. Ruhrverband, 'Vergleichende Untersuchung der Betriebszuständen zur Wiederinbetriebnahme des Alten Koepchenwerkes', November 26, 1991, 2–4. The director of the Ruhrverband's Stauseen provided me with a copy of this document.

60. Wolter, *op. cit.* (1), 385, 387.

61. The Ruhrverband's laboratory (Laboratorium des Ruhrverbandes) performed chemical tests; the Institut für Hygiene und Bakteriologie in Gelsenkirchen carried out bacteriological tests; and das Fischerei-biologische Institut in Münster investigated fishing and biological conditions. See Sierp and Bruns, *op. cit.* (7), 202–205.

The Craft of the Bell Founder: The Chyaryshnikovs, a Russian Bell-founding Family

TATYANA SHASHKINA†

The research on which this article is based was carried out during the period 1980 to 1999, supplemented by several visits to the old Russian towns of Yaroslavl, Nizhny Novgorod and Balakhna. Its main topic concerns the late stage of the Russian bell-founding art between the end of the eighteenth century and the beginning of the twentieth, when bell founding in Russia was undergoing a transition from traditional craft industry to a basis of capitalist production. This development was, however, to come to an end, as the activities of bell-founding firms were reduced to the minimum by the First World War and then fully stopped by the atheistic campaign launched after the 1917 Revolution. By the mid-1920s the profession of bell founder in Russia ceased to exist. As a result, the hitherto flourishing, century-long tradition came to an abrupt halt.[1]

Yet it is at this late stage, from *c*1900 onwards, that Russian bell founding received an important stimulus for its growth owing to the expansion of the Russian Orthodox Church. There were numbers of new churches being built and bell founders acquired a stable internal and international market, participating in important exhibitions and becoming famous for outstanding works. A definite trend towards monumentality which Russian bell art had revealed earlier, in the sixteenth and seventeenth centuries, persisted through the eighteenth into the nineteenth century, and culminated with two giants of the Moscow Kremlin – Tsar-Kolokol (*c*200 tons), cast in 1730–1735, and the Great Uspensky Bell (*c*65 tons), cast in 1817–1819.[2] This singular feature of Russian bell founding – monumentality – was more than once noted by foreign observers. Thus, *à propos* the former of these very large bells, both Russian national symbols, an English traveller Hanway reported in 1743:

> The most remarkable thing I saw is the great bell, which is indeed stupendous, and surprises equally on account of its size, and the folly of those who caused it to be made; but the Russians from time immemorial have had a strange ambition of this kind.[3]

while in connection with the recasting of the latter Robert Lyall, a member of several scientific societies in Great Britain and Russia, concluded in 1823:

> After examining the description of these bells, the reader will be ready to exclaim that the Great Bells of Britain are mere pigmies; and in Russia, so far from claiming particular notice, except on account of their tones, would only be deemed fit for country churches.[4]

It appears that the very specialization and complexity of bell founding have been the cause of its dynamism during this period, when new enterprises emerged as easily as they closed down, but, on average, the trade was represented by about 20 factories in the first half of the nineteenth century,[5] 25–30 factories in its second half[6] and 15–20 factories in the period between 1900 and 1914.[7] The major part of production was shared, however, by two centres – Moscow and Yaroslavl, where, by their scale of activity, the companies of the Samgin, Finlyandsky, Chyaryshnikov and Olovyanishnikov families rose to the forefront of the bell-founding trade.[8] It must be mostly these companies that ensured the state of progress of Russian bell founding on which N.F. Labzin, Professor Emeritus at the St Petersburg Institute of Technology, commented in 1896:

> Due to a long tradition of manufacturing and to a great demand bell founding in Russia could improve continually and has reached by now a very high degree of perfection to which foreign manufacturers of bells can hardly be compared, especially in casting very large bells of more than 500 to 1000 puds[9] which are a rarity abroad.[10]

Such assessments arose, evidently, from regular practice as can be found also in other, official, publications:

> this trade has grown up as a purely national affair and also closely associated with the religious attitudes of the popular masses. As a consequence, Russia possesses such works of bell-founding art which in other countries are rarities ... the industry itself has a unique, sacred significance.[11]

As a church-dependent, liturgical art and as an element of culture Russian bell ringing – namely, *zvon* – has a highly distinctive national character. Thus the monuments of this art which were created in its late, mature period reflect the crystallization of such principles of bell music which, similar to the fundamental Orthodox tradition of choral singing, appealed to Russian musical tastes. In this respect the economic history of nineteenth-century Russian bell founding is characterized by the fact that it reflects the latter's gradual industrialization within the circle of other metal-working industries, despite the unique specialization of this craft as a very complex occupation at the interface between monumental bronze casting and the building of musical instruments. Thus, for example, in the first half of the nineteenth century officially compiled statistics considered bell founding as a branch of the copper industry.[12] However, starting from the 1880s and especially with a series of statistical enquiries then being undertaken by the Department of Trade and Manufacturing of the Ministry of Finance, the relative significance of different crafts became clear.[13] As a consequence, bell founding acquired a position of its own in lists of the metal industries, placed after the casting of copper and bronze objects,[14] to become by 1913 one of 38 industrial occupations under the rubric 'Processing of metals, manufacturing of machines, apparatus and tools of craft trades' in the general classification of Russia's pre-revolutionary industry.[15] These changes show that in the course of the nineteenth century bell founding in Russia had been gradually moving from a closed, self-contained occupation towards a fully fledged branch of the metal-working industry.

* * *

The purpose of this article is to present some of the factual material which I managed to collect for the reconstruction of the historical heritage of one of the four bell-founding families mentioned above, namely, the Chyaryshnikov family. They merit being assigned a special place in the history of Russian bell founding on account of the unique workshop manual preserved in manuscript form in the State Library of Russia under the title *Kniga kolokol'nogo masterstva na raznye vesa kolokolov delannye i perevedennye iz tetrati nabelo mashtapy Ivanom Chyaryshnikovym* ('Book of bell art for bells of different weights with scales drawn and written out fair from the notebook by Ivan Chyaryshnikov in August 1808').[16] If we take into account the fact that not only the productive period, but the tradition itself of bell founding had been broken for a period of nearly seven decades as a result of the 1917 Revolution in Russia, it would seem almost a miracle that a highly important and authentic document should have come down to us as the only testimony of its kind to deal with a technical culture gone for ever.

* * *

Who then were these bell founders, the Chyaryshnikovs? When I started work on their 'Book of Bell Art' in 1980[17] I found myself in a situation with absolutely no contextual data, since the name of the Chyaryshnikovs was not mentioned in any of the pre-revolutionary publications on Russian bell history. The only indirect data could be found in the inscriptions on the bell 'Golodar' ('Hungry') of Rostov Veliky;[18] still more strange seemed the absence of references to the Chyaryshnikovs in either of the histories of 1906 and 1912 by N. Olovyanishnikov[19] who presented a picture of the contemporary state of the art in such a way as to suggest that the family name Chyaryshnikov was not known to him or as if in the history of bell founding inYaroslavl there had not been a period connected with the bell foundry owned by the Chyaryshnikovs.[20] Thus, in search of any historical sources I had to shift my attention to external evidence, and mostly to the materials on industrial statistics. These materials, despite their inexactness before 1870–1880 when the principles of data collection were yet to be elaborated, have substantially broadened the factual basis of my search; besides, they could be correlated with the accounts of fairs and art exhibitions as well as with the literature produced by industrial reviewers, inscriptions on bells and archival data discovered later; also such a rare piece of luck as meeting the living heirs of the family in 1989 deserves a special mention. The information all these documents and exchanges provided proved to be more than sufficient to establish that the Chyaryshnikov bell-founding tradition could be traced forward from the last third of the eighteenth century to 1922 (the death of the last active master-founder), that they were located first in Yaroslavl and then in Balakhna, and that they included several generations of bell founders. Another useful outcome of a broad approach to the historical evidence eventually collected was the possibility of deriving general trends in the Russian metal industry of relevant periods: this has been dictated by the necessity of relating particular facts of the family biography to the process as a whole.

* * *

Three distinct periods in the involvment of the family in the craft and then in the industry can be reconstructed. The first (1772–1830) is to be connected with the name of Ivan Grigorievich Chyaryshnikov who set up a workshop in Yaroslavl in 1802,[21] although he must have been casting bells long before that, because in their own documents the Chyaryshnikovs referred to 1772 as the year of the foundation of their company. At that time the foundry must have been a small private business which called itself a 'zavod' (workshop) and produced bell castings both in Yaroslavl and in various localities in Central Russia. It is, however, only in 1832 that the factory appeared in official statistics,[22] – earlier statistical accounts indicate, without detail, the existence in Yaroslavl of a bell-founding shop and several workshops of copper objects including the shop owned by the Olovyanishnikovs. There is no other external evidence about the Chyaryshnikovs' foundry in this early period, except that the predecessor of the presently existing bell 'Golodar' ('Hungry'), fourth in size, of the famous belfry of Rostov Veliky, was cast 'in 1807, on 20 December, during the reign of the Devoutest Sovereign Emperor Alexander Pavlovich, with the benediction of the Reverend Archbishop Antony, under the Archpriest Gavriil and Churchwarden Michail Ivanovich Smolin, in Yaroslavl, in the shop of master Ivan Chyaryshnikov, from the old bell cast under the metropolitan Iona of Rostov and Yaroslavl in 1654'.[23]

The second period in the activity of the family company covers 1830 to 1880 and is connected with the name of Semjen Dmitrievich Chyaryshnikov, the grandson of Ivan Grigorievich. Under Semjen Dmitrievich the foundry was turned into a manufactory with about 20 workers and an annual turnover (by 1879) of 84,000 roubles, sold its products at fairs in Nizhny Novgorod, Rostov Veliky and Yaroslavl, and in scale of production came to compete with the largest bell-casting factories, namely, the Olovyanishnikovs of Yaroslavl and the Samgins and Finlyandsky of Moscow.[24] An outstanding example of the work of the

Chyaryshnikovs of this period is the bell 'Blagovestnik' ('Ringing Good News'), cast in 1856 for the Solovetsky monastery on the White Sea by the order of the emperor Alexander II in commemoration of the events of 6–7 July, 1854 in the Crimean War, when the monastery was allegedly saved from destruction in the bombardment by the British Navy through a miracle performed by the patron saints of the monastery.[25] The story attached to this bell (Fig. 1a, b and c) is built on a popular belief in the supernatural powers of bells and icons to disperse harmful forces, transformed in this case into a historico-religious legend.[26] A fragment of the inscription on this bell, executed in Church Slavonic, reads as follows:

> GLORIOUS IS GOD IN HIS SAINTS
>
> In the summer of 1854 on July 6, in the time of the Father Superior archimandrite Alexander, two English 60-cannon steam frigates 'Brisk' and 'Miranda' came up to the Solovetsky cloister and one of them attacked the monastery with several shots of balls, but the two tiny 3-pounder cannons of the cloister responded so effectively that they damaged the frigate and made the enemy retreat. The next day, on July 7, after the monastery refused to surrender, both frigates bombarded the monastery for nine hours running with bombs, grenades, canister-shots and even 3-pud (*c*108 lb) red-hot balls. But despite all this, through the protection of God's saints, the Solovetsky cloister remained safe. Before the beginning of the bombardment and all the time during it a service was going on, and a procession went along the walls of the cloister, with all religious solemnity. But when the procession mounted the walls, singing 'Dare you, dare God's servants', the enemy's shooting intensified so that the monastery walls started to tremble and the wooden roof cracked as fiery balls burst through it and flew above the procession with terrible noise; whereafter the balls either fell on the ground after striking against the walls of the brothers' cells, or flew through the cells destroying everything on their way. Death was at a hair's distance from everybody, but – what a miracle! – during the bombardment not a single man was killed, nor even wounded, and none of the seagull chicks in the monasterial court was injured. To crown all this, a ball of the last enemy fire flew through the wall of the Cathedral above the western gates carrying with it the Image of the Sign of God's Mother who agreed to suffer this wound for the cloister, as Her Son has done for the whole world. After this all shooting stopped, and on the next day the enemy retired with shame. By the testimony of the enemy themselves, the number of shells was so great that not just a small, unfortified cloister, but six large towns as well could have been destroyed, so that even they acknowledged this as a manifestation of God's miraculous protection.[27]

Between 1881 and 1884 the Chyaryshnikovs' shop appears to have closed for business, but in 1885 or 1886 it reopened, although not in Yaroslavl, but in the provincial town of Balakhna, near Nizhny Novgorod,

Figure 1 The bell 'Blagovestnik' ('Ringing Good News') cast in memory of the Crimean war for the Solovetsky monastery by Semjen Dmitrievich Chyaryshnikov (courtesy of the State Solovetsky museum. Photo by Ju. B. Gendlin). a) General view.

b) The notice-board at the base of the bell: 'The Bell 'Blagovestnik', 1856. Donated to the Solovetsky monastery by Tsar Alexander II in memory of the events of 6–7 July 1854. Cast in 1856 at the Chyaryshnikov plant in Yaroslavl. Weight – 75 puds.'

c) Fragment of the decoration depicting the scene of bombardment of the monastery by the enemy ships.

where in 1890 a small bell foundry headed by the master, Sergej Semjenovich Chyaryshnikov, the son of Semjen Dmitrievich, was registered.[28] Sergej was succeeded in 1893 by his wife, Efrosinia Dmitrievna, and in 1905 their elder son, Alexander Sergeevich, took charge of the family business under the corporate name 'E.D. Chyaryshnikova and Sons'.[29] At this later period the foundry appears, however, to have been operating on a much smaller scale than in the mid-nineteenth century (two workers in 1890, seven workers in 1903, eight workers in 1910, with figures for the annual turnover being 2000, 2800 and 14,714 roubles respectively), evidently yielding in position to the Olovyanishnikov, Samgin and Finlyandsky factories.[30] The last owner of the foundry, Alexander Sergeevich, ran the business until 1921, but after 1919 there were practically no orders, and in 1922 the foundry premises were taken over by a carton-making factory; in the same year A.S. Chyaryshnikov died from asthma at the age of 46.[31] Fig. 2 is a photograph from this last period, and Fig. 3 shows the building of the former bell foundry as I found it on my visit to Balakhna in 1991.

* * *

Fourteen Chyaryshnikov bells have been established so far. The most distinguished of these are 'Lebed' ('Swan') cast in 1884 by Sergej Semjenovich for the Kirillo-Belozersky Monastery (443 puds or 7.256 kg),

Figure 2 Exterior view of the Balakhna foundry with the workforce in front, 1896. Alexander Sergeevich Chyaryshnikov is fourth on the left (Courtesy of the Chyaryshnikov family, private archive).

Figure 3 The former bell foundry at Balakhna, 1991 (Courtesy of Balakhna museum).

the above-mentioned bell 'Blagovestnik' ('Ringing Good News') of the Solovetsky Monastery (75 puds or 1.228 kg) (see Fig.1), 'Golodar' ('Hungry') of Rostov Veliky cast in 1856 (171 pud or 2.801 kg) and an undated, unnamed bell brought in 1988 from some provincial church to the newly restored belfry of the Danilov Monastery in Moscow (135 puds or 2.211 kg), all three latter bells cast by Semjen Dmitrievich. This bell founder (he was probably the most famous and successful of the family) is also known

Figure 4 A bell cast at the Balakhna foundry, *c*1910 (Courtesy of the Danilov monastery).

to have cast a very large bell of nearly 21 tons for the Irkutsk Cathedral in Siberia,[32] but this bell has not survived. Fig. 4 shows an example of Chyaryshnikov work from the latest, Balakhnian, period, decorated with the portrait of Emperor Nicholas II, the State Emblem of the Russian Empire (reverse side), two bands of ornament and the inscription 'From the Factory of the Trading House of E.D. Chyaryshnikova and Sons in Balakhna, province of Nizhny Novgorod'.[33]

* * *

The manuscript entitled *Kniga kolokol'nogo masterstva na raznye vesa kolokolov delannye i perevedennye iz tetrati nabelo mashtapy Ivanom Chyaryshnikovym* ('Book of bell art for bells of different weights with scales drawn and written out fair from the notebook by Ivan Chyaryshnikov in August 1808') (folio I, Fig. 5) is a collection of 38 loose folios measuring 32 by 20.7cm, which includes paper sheets covered with technical notes, sheets with drawings pasted on thick cardboard, several blank sheets and some ragged sheets with draft texts of inscriptions. Another title – *Sia kniga masterskich del Ivana Chyaryshnikova* ('This is the workshop book of Ivan Chyaryshnikov') – is

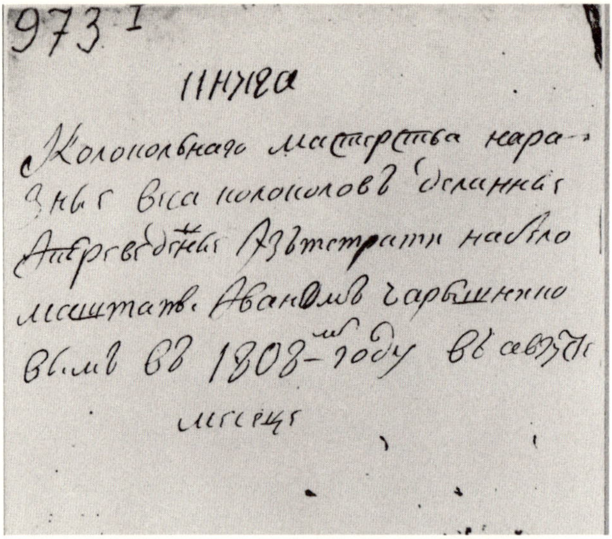
973 I

Книга

Figure 5 folio I. Title page of the 'Book of Bell Art' written by hand on the inner side of the leather cover (Figs. 5 to 8 courtesy of the Manuscript Department of Russia's State Library).

engraved on the leather cover and may not be contemporary with the first title under which the manuscript is catalogued.[34] The manuscript has been in the possession of Russia's State Library since 1940 or 1941, but nothing is known about how or from whom it got there.[35]

This is not the place to go into a detailed discussion of the manuscript, and only a brief general description will be attempted. The manuscript is, most probably, just one representative of a particular kind of craft literature originating directly from oral tradition.[36] Technical notes are in several hands; many notes are made in a hasty way and cannot be read easily; some notes are crossed out; the orthography is mostly archaic with little punctuation and many distorted or inaccurate spellings; some words are contracted or not completed; parts of the notes are in pencil. There is no overall system in the organization of notes; the notebook seems to have grown organically out of the needs of everyday practice.

An important feature is also that the year indicated in the title (1808) cannot be taken to date the manuscript, since it contains much earlier as well as much later entries; similarly, the name given in the title does not permit one to regard Ivan Chyaryshnikov as the only author of the 'Book of Bell Art'. Judging by the character of the notes and handwriting, the manuscript was composed by several persons over a long period of time, with the first dated entry (1804) (folio 19v) made by Ivan Chyaryshnikov, the main contributor to the 'Book of Bell Art', and the last dated entry (1889) by a certain Kuvshinov Alexander Dmitrievich (folio 21). Written as a textbook for the instruction of apprentices into the techniques of the art (such as it was conceived by Ivan Chyaryshnikov whose notes continue until 1822), it turned later into a working company notebook. The manuscript is thus a typical workshop manual based on a literary record of their craft by the bell founders themselves.

The entries in the manuscript are of the following types: estimates of materials, basic (copper and tin) and auxiliary (hemp, fat, bricks etc.) (folios 1, 1v, 3v, 6, 8, 8v); notes marking the completion of orders (folio 2); inventories of raw materials and equipment (folios 6, 7, 9); records of the number of bells cast during the year with the calculation of payments, overall and per pud weight (folio 1v); measurements and sketches of furnaces (folios 2v, 11v, 26); measurements of bell dimensions (folios 3v, 12, 19); notes of pre-estimated weights of bells as cast (folios 10, 10v); designs of inscriptions with fragments of texts from the Bible and Church Service books (folios 28, 29, 31, 32, 33), fragments of inscriptions in verse (folios 9, 9v); incidental notes as reminders for day-to-day work (folios 5v, 6, 17, 25v); and, the most voluminous part of the notebook, sheets of tables and drawings, with, related to them, rules and instructions (folios 3, 3v, 4, 4v, 5, 11, 11v, 12, 12v, 13, 13v, 15, 16v, 19, 19v, 20, 21, 34, 36, 37, 37v).

* * *

What follows below is intended to illustrate the contents of the 'Book of Bell Art' by means of several representative extracts chosen to reflect the salient features of bell founders' vernacular with its basis in the popular language and its self-created terminology; another aim is to show the variety of computation and design procedures used by Russian bell founders in the nineteenth century. In order to convey the individual style of the manuscript, quotations from the original text are transcribed with the punctuation and most of the orthographic peculiarities preserved. Specific terms of bell founders' vocabulary are translated literally; the terms for which no English equivalents have been found (e.g.,'perechen', 'otkos') are left in Russian (in square brackets). The transcription and translation are mine.

folio 1 (Fig. 6)
(Heading of the table setting out how to estimate the quantity of materials for moulds of bells of different sizes)

1811-goda marta dnya podrobnoe opisanie chto znanie nadobno dlya delania for[m] na litje raznych kolokol, a imyanno kirpich glina provoloka zhelezo poloski pen'ka pryadi vosk kalifon salo suslo drova i prochie ...

Translation
Of the year 1811 of the day of March detailed description what knowledge is necessary to make moulds for casting different bells, namely brick clay wire iron strips hemp hair wax rosin fat must wood and others...

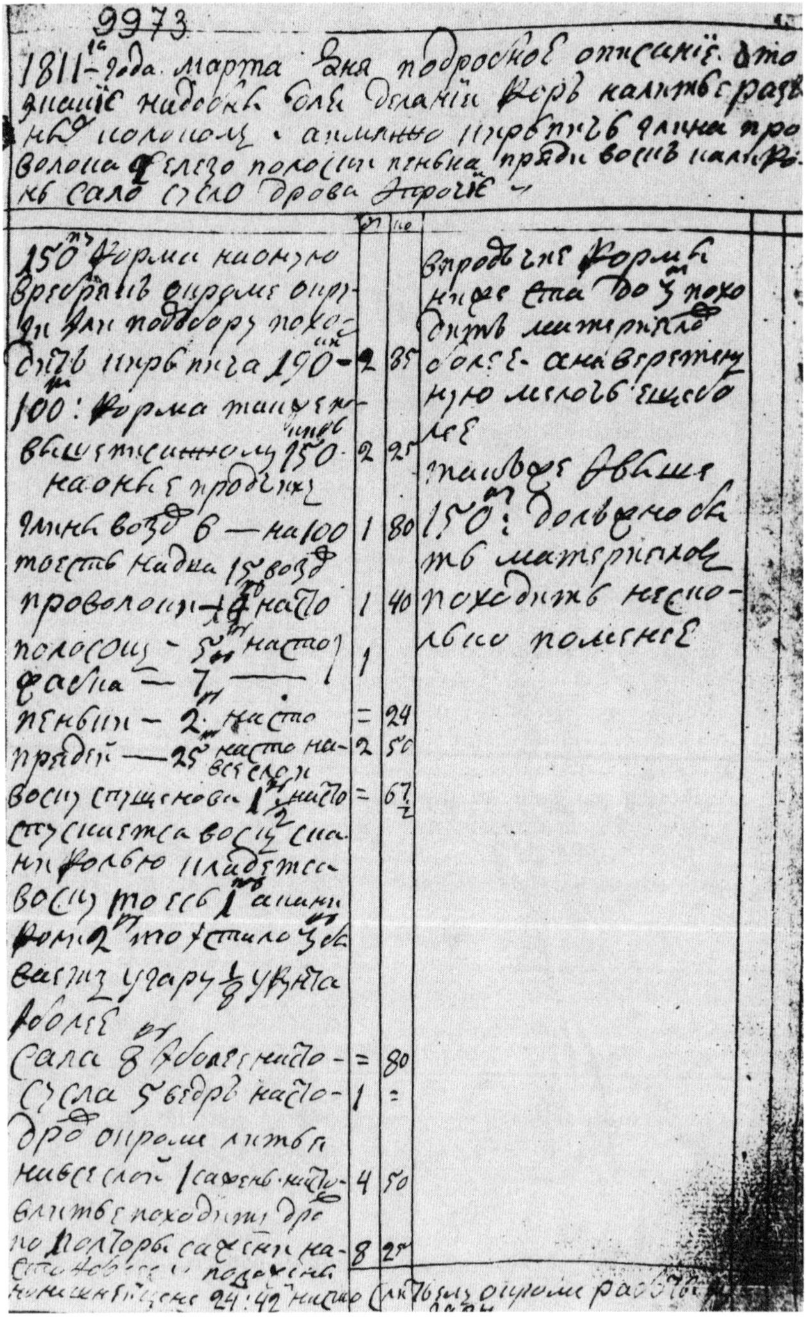

Figure 6 folio 1. Table with instructions on calculations of materials for making moulds.

folio 1v

(A record of the number of bells cast in 1810 and calculation of the payment obtained)

1810-goda s pervogo genvarya po 1811-god po 1-genvarya skolko v onom i gde gody vylito raznych i v raznye mesta kolokol imenno

v Yaroslavle	–	1720 pu[37]	:	1½ fu[38]
v Vologde	–	293	:	2
u Makaria[39]	–	966	:	26½
vsego		2979 pu	:	30 fu

Za onye vysheoznachennye kolokola za rabotu polucheno deneg 3981 r[40] : 93 k[41] vykhodit po slozhnosti za rabotu po 1 r 33½ k s puda...

Translation

From the 1st January of the year 1810 till the 1st January of the year 1811 how many different bells have been cast and where, namely

in Yaroslavl	–	1720 pu[37]	:	1½ fu[38]
in Vologda	–	293	:	2
near Makary[39]	–	966	:	26½
total		2979 pu	:	30 fu

For the above indicated bells for the making 3981 r[40] 93 k[41] have been received which gives for average work 1 r 33½ k for a pud...

folio 3v

(A note on the measurements of the bell of the church of St Nicholas in the village Melenki)

U Nikoly na Melenkakh[42] kolokol 190 pu s nego snyat i sdelan risunok vyshina po pryamoj cherte...[43] vershkov[44] shirina v perechne shirina 20½ vershkov iz krau v kraj – 42⅛ vershkov otkos – 3⅞ vershkov plecho – 3⅞ vershkov.

Translation

From a bell of 190 pu of Nicola in Melenki[42] a drawing is taken and made height along the straight line is...[43] vershkov[44] width in [perechen] is 20½ vershkov from edge to edge – 42⅛ vershkov [otkos] – 3⅞ vershkov [shoulder] – 3⅞ vershkov.

folios 4 and 4v

(Table with rules of proportion for designing bells of different dimensions)

folio 4 TABLITSIYA

kak rezat doski i kak syskivat razmery imenno pod sim

'1'

Delat na kazhduyu dosku mashtapy iz chastej syskat prezhde vyshinu po pryamoj cherte ot krayu do plecha kolokola i onuyu chertu razdelit na desyat chastej i odnu desyatinu razmeryat eshche na desyat zhe chastej s chastyami iz onych chastej i umnozhat delenia v razmere a kak syskat vyshinu pervago kolokola o tom budet skazano budet v '3' statje . . .

'3'

Ob umnozhenii velichiny kolokolov
Ot meryat verno chetyre vershka 3 osmuchi sdelat pryamuyu chertu onaya cherta ot krayu do plecha budet znachit mera 20 fu kolokola a kak delenia delat skazano v '1' statje.
1. Vyshepisannuyu chertu 4 vershka i $\frac{3}{8}$ razdelit tsirkulem v tochnosti na sem chastej i k onoj pribavit vyshiny osmuyu to budet kolokol v 30 funtov//

folio 4v
a jesli 4 vershka i tri osmuchi chertu razmeryat verno tsirkulem na chetyre chasti i k onym chetyrem chastyam pribavit pyatuyu to budet kolokol vesom 1 pud. Tak zhe i prodchie raznych sortov umnozhat bolshie.
2. Umenshat takzhe ili s kakogo snyat risunok. Jezheli umnozhat to najdi vyshiny pryamuyu liniyu i razmeryaj na chetyre chasti pribav pyatuyu to vesu bydet vdvoe = a jezheli tuzhe vysherechennyuy chertu razmeryaesh na sem chastej i pribavish osmuyu to vesu budet v poltora = tak zhe do unichtozhenia = (zri na oborotke pod sim '5')[45] kakova libo vesu kolokol najdi s nego meru vyshiny ot krayu do plecha smeryaj pryamuyu chertu i razmeryaj verno tsirkulem na pyat chastej iz onoj mery ubav pyatuyu ostav chetyre to primerno iz puda budet polpuda.
Tak zhe syshchi chertu pryamuyu i razmer′ verno tsirkulem na vosem′ chastej i vosmuyu ubav′ primerno budet iz polutora puda odin pud po semu primeru i prodchie imej znat′.

Translation
folio 4 TABLE

how to cut the boards and determine dimensions namely below

'1'

Make for each board scales from parts first find height along the straight line from edge to shoulder of the bell and divide this line into ten parts and one tenth part again into ten parts with parts from these parts and enlarge the divisions in size and how to find the height of the first bell will be said in the '3rd' article . . .

'3'

On the enlargement of bells
Measure correctly four vershkovs three eights make straight line and this line from edge to shoulder will be a measure for a bell of 20 fu and how to make divisions is said in the '1st' article.
1. Divide with the compasses the above said line of 4 vershkovs and $\frac{3}{8}$ exactly into seven parts add one part will be a bell of 30 funts//

folio 4v
but if you divide the line of 4 vershkovs and three eights with the compasses correctly into four parts and add one part the bell will have the weight of 1 pud. In this way also others of different sorts to be enlarged. 2. In the same way diminish or when you take a drawing from the bell. If you want to enlarge then find the straight line of height measure into four parts add one part then the weight will be two times more = but if you divide the said line into seven parts add one the weight will be one and a half = in the same way act before destroying the bell (see on the reverse side '5')[45] of any weight take the height from edge to shoulder measure the straight line and divide exactly by the compasses into five parts take one part away leave four then of the bell of one pud will be about half a pud. In the same way measure the straight line and divide correctly with the compasses into eight parts take away one then of one and a half puds will be one pud by this example learn to know about the others.

folio 19 (Fig. 7)
(A diagram and a note on the construction of the 'shoulder' of the bell profile)

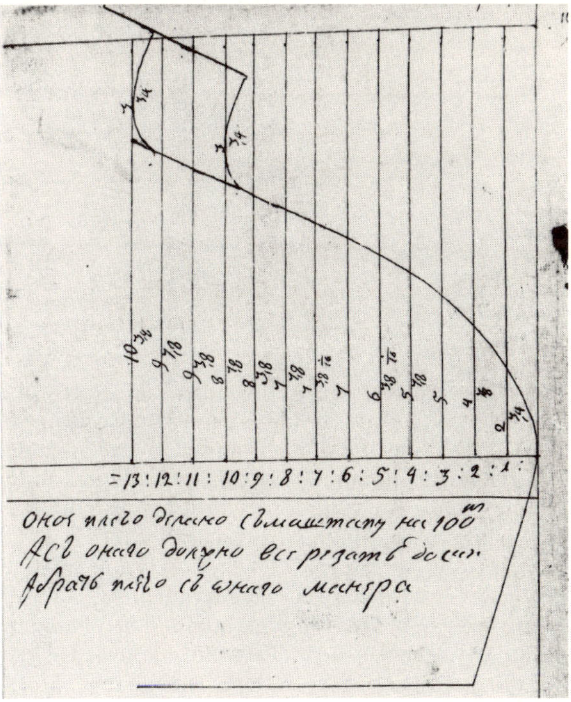

Figure 7 folio 19. A diagram and a note on the construction of the bell 'shoulder'.

Onoe plecho delano s mashtapu na 100 pu i s onago dolzhno vse rezat doski i brat plecho s onago manera.

Translation
This shoulder was made with the scale for 100 pu and from this cut all boards and make the shoulder by this pattern.

folio 20 (Fig. 8)
(Scales (‘*mashtaps*’) for bells of 1000, 900, 800, 700, 600, 500 and 400 puds; on the left margin is a note on the construction of the shoulder)

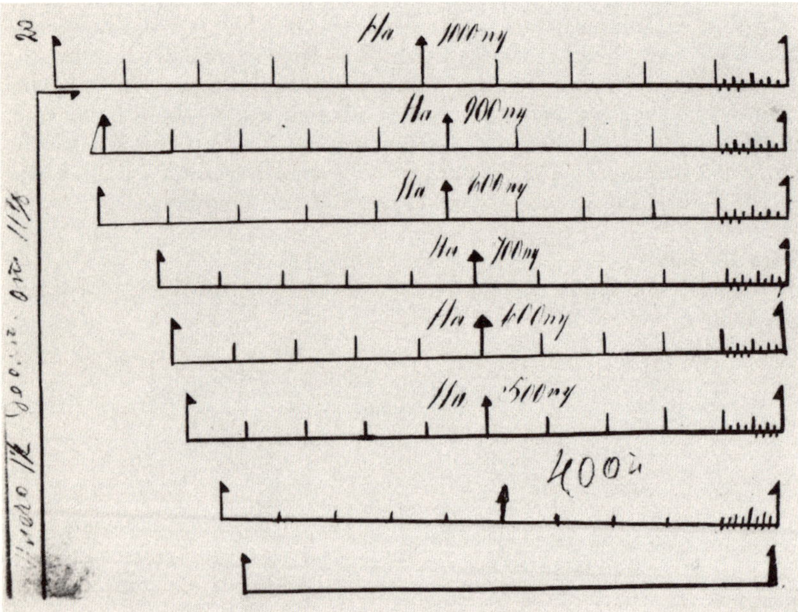

Figure 8 folio 20. Scales for bells of different dimensions.

* * *

The manuscript ‘Book of Bell Art’ is a historical document of exceptional importance for Russian culture, because an original system of bell design related to the national principles of bell music has been handed down by tradition and fixed in the manuscript in the form adapted to the technical thought of simple craftsmen. It took this form because the Russian art of bell founding in the nineteenth century was still a purely folk creation, a technology empirically developed by working people with little or no literacy. Hence the significance of the ‘Book of Bell Art’ as a repository of the mental constructions created by generations of bell founders and as an illustration of the indigenous system of technical knowledge crystallized in the depth of the national culture. Even though by the close of the nineteenth century Russian bell founding had evolved to become indistinguishable from other metal-working industries in terms of its social organization, intellectually, however, it was still medieval in character. Parallels in other national traditions are in Italian the ‘Zibaldone’ by

Buonaccorso Ghiberti,[46] in German the 'Buch der Buchsenmeisterei' by Christoph Sesselschreiber,[47] in Czech the 'Mathesis Bohemica' by Vavřince Křičky[48] and in several other fifteenth- to eighteenth-century manuscripts.[49] All these documents offer invaluable material for inquiry into traditional craftsmanship.

* * *

It may be worth noting that a scientific approach to the study of the Russian historical tradition in bell founding appears at present to be closely connected with the practical side, as in the past ten years or so the Russian tradition of the craft of bell-founding art has been undergoing a rapid revival due to the return of religious freedom in Russia, Ukraine and other post-Soviet countries. This has brought about an ever-growing demand for bells and international interest on the part of bell-founding companies. The craft of the bell founder is once again a living, ongoing process.

ACKNOWLEDGEMENTS

It is the duty of the author to express her gratitude to the heirs of the Chyaryshnikov family who generously communicated to the author their recollections and provided Fig. 2; the State Solovetsky Museum for Fig. 1, a to c; the Danilov Belfry for Fig. 3; the Balakhna Museum for Fig. 4; the MS Department of the Russian State Library for Figs. 5 to 8. My special gratitude is due also to Dr Graham Hollister-Short for his editorial handling of my manuscript.

Notes and References

1. See E.V. Williams, *The Bells of Russia: History and Technology* (Princeton, NJ,1985), 63–66.

2. Williams, *op. cit.* (1), 148–168. See also I.D. Kostina, 'K istorii sozdania Tsar-Kolokola', *Kolokola. Istoria i Sovremennost* ('Towards a History of Casting the Tsar-Kolokol', *Bells. Past and Present*) (Moscow, 1993), 119–126.

3. Quoted in Robert Lyall, *The Character of the Russians and a Detailed History of Moscow* (London, 1823), 203.

4. Lyall, *op. cit.* (3), 11.

5. *Spisok Fabrikantam i Zavodchikam Rossijskoj Imperii 1832 goda (A List of Manufacturers and Workshop Owners of the Russian Empire of the Year 1832)*. Pt. 2. (St Petersburg, 1833).

6. P.A. Orlov, *Ukazatel Fabrik i Zavodov Evropejskoj Rossii (A Directory of Factories and Workshops of European Russia)*, 1st ed. (St Petersburg, 1881); P.A. Orlov, *Ukazatel Fabrik i Zavodov Evropejskoj Rossii (A Directory of Factories and Workshops of European Russia)*, 2nd ed. (St Petersburg, 1884); P.A. Orlov and S.G. Budagov, *Ukazatel Fabrik i Zavodov Evropejskoj Rossii (A Directory of Factories and Workshops of European Russia)*, 3rd ed. (St Petersburg, 1894).

7. *Spisok Fabrik i Zavodov Evropejskoj Rossii (A List of Factories and Workshops of European Russia)* (St Petersburg, 1903); *Fabrichno-Zavodskaya Promyshlennost Rossii v 1910–1912 godach (Factory-and-Workshop Industry of Russia in 1910–1912)* (Petrograd,1914); *Spisok Fabrik i Zavodov Rossijskoj Imperii (A List of Factories and Workshops of the Russian Empire)* (St Petersburg,1912).

8. *Ibid.*

9. c. 8 to 16 tons (1 pud = 16.38 kg). See E.I. Kamentseva and N.V. Ustyugov, *Russkaya Metrologia (Russian Metrology)* (Moscow, 1975).

10. *Fabrichno-Zavodskaya Promyshlennost i Torgovlya Rossii (Factory-and-Workshop Industry and Trade of Russia)* (St Petersburg, 1896), 75 (Translation from the Russian by the author).

11. *Istoriko-Statistichesky Obzor Promyshlennosti Rossii (Historico-Statistical Review of Russian Industry)*. V.2 (St Petersburg,1886), 31 (Translation from the Russian by the author).

12. *Vedomost o Manufakturakh v Rossii za 1813 i 1814 gody (A Record of the Manufactories of Russia for 1813 and 1814)* (St Petersburg, 1816); Spisok . . . *op. cit.* (5).

13. *Jezhegodnik Ministerstva Finansov (Yearbook of the Ministry of Finance)*, N. XII (St Petersburg, 1882).

14. *Fabrichno-Zavodskaya Promyshlennost Rossii. Perechen Fabrik i Zavodov (Factory-and-Workshop Industry of Russia. A List of Factories and Workshops)* (St Petersburg, 1897).

15. *Svod Otchjetov Fabrichnych Inspektorov za 1913 god (A Book of Accounts of the Factory Inspectors for 1913)* ((Petrograd, 1914).

16. Russian State Library (Moscow), Manuscript Department, f. 178, N 9973.

17. It was Ju.G. Malkov, art historian of the Institute of Restoration Studies in Moscow who drew my attention to this unpublished manuscript.

18. A.A. Izrailev, *Rostovskije Kolokola i Zvony (Rostov Bells and Chimes)* (St Petersburg, 1884), 7–8.

19. N. Olovyanishnikov, *Istoria Kolokolov. Kolokololitejnoe Delo na Zavode Tovarishchestva P.I. Olovyanishnikova Synovia (History of Bells and Bell Founding at the Plant of the Company 'P.I. Olovyanishnikov's Sons')* (Yaroslavl, 1906); N. Olovyanishnikov, *Istoria Kolokolov i Kolokololitejnoe Iskusstvo (History of Bells and Bell Founding)*, 2nd ed. (Moscow, 1912). The author of these books, Nikolaj Ivanovich Olovyanishnikov (1875–1919), was one of the owners and directors of the 'P.I. Olovyanishnikov's Sons' company in Yaroslavl, the largest bell-founding firm in Russia in 1905–1914.

20. Olovyanishnikov, *op. cit.* (19, 1912), 409–413.

21. 'Yaroslavl v 1802 godu', *Yaroslavskie Gubernskie Vedomosti* ('Yaroslavl in 1802', *Records of Yaroslavl Province)*, non-official part, 1853, 16: p. 153.

22. *Spisok. . .op. cit.* (5), 843.

23. This is part of the inscription on the 1807 bell 'Golodar' ('Hungry') (recast in 1856), as quoted in the local Yaroslavl newspaper. See 'Kolokol'ny zvon pri Rostovskom Uspenskom Sobore', *Yaroslavskie Gubernskie Vedomosti* ('Bell Ringing at Rostov Uspensky Cathedral', *Records of the Yaroslavl Province*), non-official part, 1852, 12: p. 102 (translation from Russian into English by the author).

24. *Mejshen, Podrobnoe Statisticheskoe Obozrenie Fabrik i Zavodov Yaroslavskoj Gubernii (A Detailed Statistical Review of the Factories and Workshops of Yaroslavl Province* (St Petersburg, 1857), 8; Orlov, *op. cit.* (6, 1st ed.), 393*; Istoriko-Statistichesky Obzor. . . (Historical-Statistical Review. . .) op. cit.* (11), 33.

25. *Melety, Istoricheskoe Opisanie Stavropigialnogo Pervoklassnogo Solovetskogo Monastyrya (Historical Description of the Stavropigial First-Class Solovetsky Monastery)* (Moscow, 1881), 69–73.

26. See A.N. Davydov, 'Kolokola i Kolokol'nye Zvony v Narodnoj Kulture', *Kolokola. Istoria i Sovremennost* ('Bells and Bell Ringing in Popular Culture', *Bells, Past and Present*) (Moscow,1985): 7–17, pp. 7–9.

27. Translation from the Church Slavonic into Russian is in Davydow, *op. cit.* (26), p. 8. Translation from Russian into English is the author's.

28. Orlov i Budagov, *op. cit.* (6, 3rd ed.), 447.

29. Spisok. . . *op. cit.* (7, 1903), 296; Spisok. . . *op. cit.* (7, 1912), 237.

30. *Ibid.*

31. Personal communication of Valentina Alexsandrovna Chyaryshnikova (1910–1987), daughter of the last bell founder Chyaryshnikov.

32. I.I. Kozlov, *Kolokola ne umolkaut (Bells Do Not Stop)* (Irkutsk, 1979), 92.

33. At the time of photographing (July 1989) this bell was kept at the storehouse of the Danilov monastery, to which it had been brought in 1988.

34. Russian State Library, Manuscript Department. f. 178. Museum Collection, Russian and Slavic Part, Acquisitions of 1936–1947. Entry 4489, N 9973.

35. See T.B. Shashkina, 'Kniga kolokol'nogo masterstva', *Pamyatniki Kultury. Novye Otkrytia* ('Book of Bell Art', *Monuments of Culture. New Discoveries*), Yearbook 1985 (Moscow, 1987): 431–437, p. 437.

36. Shashkina, *op. cit.* (35), pp. 431, 436.

37. Contraction of pud (1 pud equals 16.38 kg).

38. Contraction of funt (1 funt equals 0.453 kg).

39. Makarievsky Monastery near Nizhny Novgorod.
40. Contraction of rouble.
41. Contraction of copeck.
42. The church of St Nicholas in the village Melenki.
43. Washed out.
44. 1 vershok equals 4.44 cm (see Kamentseva and Ustyugov, *op. cit.* (9)).
45. 'zri na oborotke pod sim' is written with thinner ink in the same hand.
46. Codex BR 228, Biblioteca Nazionale, Florence.
47. Codex germ. 973, Bayerische Staatsbibliothek, Munich.
48. N XVII B 17, University Library, Prague.
49. For an introduction to this area of study see Gustina Scaglia, 'A Miscellany of Bronze Works and Texts in the Zibaldone of Buonaccorso Ghiberti', *Proceedings of the American Philosophical Society*, 1976, 120: 485–511; Margarete Schilling, *Glocken: Gestalt, Klang und Zier* (Dresden,1988), 55–107; see also an introductory chapter by Prof.-Ing. Fr. Pišek in Pišek Fr., Sochor S., Stránský B. (eds.), *Vavřince Křičky z Bítyšky Návod k lití a připravě děl, kulí, hmoždiřu, zvonců, konvi ke zvedáni vody, k vodotryskům a p. četnými kresbami opatřený* (Mathesis Bohemica) (Prague, 1947).

John Farey and His *Treatise on the Steam Engine* of 1827

A. P. WOOLRICH

PART 1

John Farey's *Treatise on the Steam Engine* (1827) has been described as 'the finest book on technology published during the Industrial Revolution'.[1] It was originally planned to be a two-volume work, the first dealing with the history up to the death of Watt in 1819, and the second describing more modern patterns of engine and the scientific basis of steam technology. Problems beset the author and publisher so that only the first volume was ever published in the author's lifetime. The second volume was written in a greatly cut-down version, and was only finally published in 1971, reproducing the unique copy of the page proofs which by then had reached the British Library. None of the printed plates for this volume had survived.

No record survives in the Longmans' archive at Reading University of Farey's early dealings with the firm about the project, but their accounts and a number of letters to Farey present in some detail the printing and publishing history of the book. In addition, Farey's manuscript synopsis for the second volume, his manuscript list of the plates and woodcuts for the entire work, and his manuscript corrections to part of the first volume all survive in the British Library.[2] From various sources it has been possible to assemble virtually a complete set of the illustrations planned for the second volume, which are reproduced here for the first time.

John Farey, jr. (1791–1851) was a consulting mechanical engineer and a prolific writer about technology.[3] Following the appearance of his authoritative article on the steam engine in Rees's *Cyclopaedia* in 1816[4] his fellow engineers urged him to produce a full-length work about the steam engine. Longmans, the publisher with whom Farey had had dealings before, during the publication of Nicholson's *British Encyclopaedia* (1808), Smeaton's Reports (1812), and Rees's *Cyclopaedia* (1802–1819), was the natural choice to publish the proposed work.

Farey set out his aims for the book in the introduction to the first volume of the *Treatise*:

> The object of the present Treatise is to give a Historical, Practical and Descriptive account of the Steam Engine, and its application to useful purposes. The Historical part is intended to form a complete account of the invention, from its first origin, to the present state of perfection ... The Practical and Descriptive part is intended to form a course of instruction for professional students in the practice and principles of making steam engines ... The same part is intended to form a Manual, which, by means of tables and theorems for calculations will facilitate the practice of experienced engineers ... The principles of mechanical and chemical action upon which the operation of the Steam Engine depends ... will also be embodied in this part.
>
> Another part will contain a record and brief explanation of all speculative projects which have been proposed for the improvement of the Steam Engine....[5]

The synopsis and list of plates, printed in appendices 1 and 2, enable Farey's original intentions for the book to be examined in detail. The synopsis for the first volume does not survive, but the second volume was to begin with a highly detailed description of the engines then being made by some of the best engine makers in the country. This part was to occupy approximately 180 pages of text, and was to be illustrated by 35 plates and an unknown number of engravings. It was to be followed by a discussion of the scientific basis of steam technology, occupying a further 130 pages of text and seemingly not illustrated. The book was to conclude with a 200-page appendix of projects and patents on steam engines which fell outside the main flow of development. The second volume was thus to have been about 200 pages shorter than the first one but would have had ten more plates.

The engines which Farey chose to describe in the second volume were the products of well-known firms such as Maudslay, Son and Field of Lambeth, London, and Fenton, Murray and Wood of Holbeck, Leeds. He also included examples of engines made by Arthur Woolf of London, Thwaites, Hick and Rothwell of Bolton, and the London firm of Jukes, Coulson and Co.

At the time Farey was writing his book, the commonest pattern of engine was the beam engine, and this was of two varieties, house-built and portable. A house-built engine was where the walls of the the engine house carried the beam bearings, and the cylinder and other parts for the crankshaft and flywheel were fixed directly to the floor. A portable engine was one where the engine was supported on a separate wooden or iron frame, which enabled it to be dismantled and moved without demolishing the engine house structure. A variation on this was the tank-bed, or cabinet engine, where the framing of the engine was supported by an iron watertank which held the engine's condensing water.

Smaller portable engines were also developed. In 1807 Maudslay patented his table engine in which the cylinder was mounted vertically in the middle of a table-like frame, with the crank shaft running between the legs. Bell-crank and side-lever engines were originally developed for marine use but were also used for industrial applications.

The beam engine and its smaller cousins were in common use until about 1830, when the spread of different manufacturing methods, especially the application of the planing machine, made possible the introduction of the direct action engine. These were engines with cylinders, both horizontally and vertically mounted, where the piston rod was coupled directly to the crank by a connecting rod kept on a true path by means of a machined slide, instead of through an intermediate rocking beam and Watt-type parallel motion. However, these more modern engines were largely in the future when Farey was planning his book and plates, though he did refer to them briefly when he wrote the text of the second volume.

There is a difference in the topics to be covered as recorded in the synopsis for Volume 2 and the list of plates. The latter notes plates of steamboats, which are known to have been engraved.[6] The synopsis and plate list both ignore the locomotive and the steam carriage, but that is to be expected, since at the time when Farey was negotiating with Longmans (the early 1820s), neither had assumed the importance they were to have a few short years later. No doubt Farey would have modified his text to have included sections about them had his plans been fully realized.

In the section in the second volume concerning the scientific basis of steam technology, Farey noted the authorities he was proposing to use and it is plain he was aiming for the best sources available for the time. Professor Cardwell criticized Farey for his supposed lack of scientific knowledge, failing to understand that the published Volume 1 dealt with history only and that Farey was planning to deal with the science in the second.[7]

In one respect Farey was already on the verge of being out of date, since in France Sadi Carnot was formulating his *Réflexions sur la Puissance Motrice du Feu* which was published in 1824. Farey and all other writers of his time regarded the steam engine as a vapour-pressure engine, and not a heat engine as expounded by Carnot. Of all the engineering writers then in Britain Farey was undoubtedly the best placed to learn about Carnot's work, and would most probably have discussed it in the full version of his work. His brother-in-law was the engineer Colomb Gengembre, and with his brothers, Joseph and Henry, Farey also ran a branch office of his business in Paris. However, it was not until about forty years after its publication that Carnot's work became more widely known in Britain.[8]

Farey's appendix on projects and patents was meant to include everything which did not have a place in the developmental mainstream. In fact lists of patents were published by Galloway and Stuart, but Farey's goal of more extended notes was not achieved until the publication of the abridgements of patent specifications by the Patent Office. However, that was after Farey's death in 1851.[9]

An important aspect of Farey's writing is his competence as a historian. He began his story with the aeolipile of the Greeks and then proceeded to the philosophical discoveries of the seventeenth century, and the work of Solomon de Caus. Farey next dealt with the Magdeburg spheres of Otto von Guericke, and then discussed the work of the Marquis of Worcester (1663) and Sir Samuel Morland (1682) before describing the work of Papin (1681 and 1690) and Savery (1698). He described in minute detail the atmospheric engine of Thomas Newcomen (1712) and its subsequent improvement by Beighton (1718) and Smeaton (1772); it is for this section that his work is so important now. The earlier developments were written from printed sources which Farey was able to discover in the British Museum and the Bibliothèque Nationale in Paris, but the development of the Newcomen engine by Smeaton was written entirely from Smeaton's manuscripts. Farey noted that it was his intention to publish a detailed summary of Smeaton's experimental researches on the steam engine, in the style of Smeaton's published research on the windmill and the waterwheel, using Smeaton's notes and tables. He had to print a brief summary because of lack of space in this first volume of the *Treatise*.[10] Farey concluded Volume 1 with a thorough study of Watt's engine and also of the work of Jonathan Hornblower and other engineers who were then beginning to challenge Watt's supremacy.

Later writers on the steam engine have tended to adopt Farey's analysis of the early developments, and it is clear that his initial survey, especially of the atmospheric engine, has not been materially challenged. There are gaps, however, which later research has uncovered. He did not mention Dr John Allen, FRS (*c*1660–1741) of Bridgwater, who patented a steam engine boiler and a method of propelling boats by a kind of jet propulsion. Allen described them in his book *Specimina Ichnographia*, published in 1730. In this book Allen noted he was a great friend of Thomas Newcomen.[11]

By concentrating on Smeaton, Farey did not discuss at all the Newcomen engines in the South Wales and Bristol coalfields, nor did he discuss the products of firms such as the Coalbrookdale Company. This imbalance was inevitable, considering the fact that he was based in London, and that technical journalism which would have aided his research was then non-existent.

Farey intended his book to be more than just an account of the development of the steam engine; instead it was to be a manual of education for the steam engineer, containing sections on the principles of mechanics and the manufacture of iron as well as the design and proportioning of engines. He also devoted space to an account of the use of the slide rule as an aid to calculation. He held that all engineers should become proficient in its use, and advocated the pattern developed by Watt at Soho and used by the firm's staff.

Throughout Farey's book are formulae and rules for designing and constructing engines. He held that as most practical engineers lacked a sound mathematical training there was no sense in presenting his data in a form they could not understand. Instead the data was presented in the form of written rules, with the calculations being kept to basic addition, subtraction, multiplication and division. Frequently, dimensions were presented as a proportion of a basic dimension, such as the cylinder diameter.[12]

As a young man Farey had collected data about engine proportions and rationalized these into tables so as to be able to infer the principles on which the engines had been designed originally. Watt had done this, and Farey noted that all Watt's engines were produced to standard dimensions for the different sizes.[13]

Another important aspect of Farey's book is how he dealt with the various controversies which occasionally arose in the engineering world. The most serious was that relating to the Hornblower family and how they fared in the patent litigation instigated by Boulton and Watt and their sons. Farey was very judicious in his account of Hornblower's engine, and his comments betray nothing of his undoubted knowledge of the background and of the antagonism Boulton and Watt had stirred up in the engineering world. Although Farey never knew Hornblower, he was an intimate of Joseph Bramah, who had given evidence against Watt; of Arthur Woolf who had been Jabez Hornblower's assistant at the time of the Boulton and Watt *v.* Hornblower and Maberley trial of 1793; and of Matthew Murray who was the victim of a villainous campaign of harassment by the firm of Boulton and Watt early in the nineteenth century. Farey was also involved with Olinthus Gregory, author of *Mechanics*, whose publication of Jabez Hornblower's remarks about the trials brought down on him much abuse from the Watt camp.[14]

Farey did not mention the claims made by the Hornblower family that Watt's original patent of 1769 was invalidated by the prior inventions of the Rev. Henry Wood (*c*1716–1795) and by Humphrey Gainsborough (1718–1776) – the brother of the painter – though he might have intended to discuss them in his appendix of patents and projects.[15]

Farey steered a middle course between the aggrandisement and near deification of Watt by writers like Stuart and Partington, and those writers like Tredgold and Galloway who questioned Watt's abilities as an engineer and the business ethics of his company.

Farey knew Watt, and the relationship was intimate enough for Watt to call on Farey when he came to London. Watt gave Farey a draft of his comments on Robison's writings, so the relationship was of some substance.[16] Nevertheless Farey did not become so blinded by Watt's eminence as to denigrate out of hand the rival engineers who were trying to build steam engines which did not fall foul of Watt's catch-all patents. Farey was also well aware that Watt's success in prolonging his 1769 patent by means of the Act of Parliament of 1775 was due not to the purity of his case but largely through the political influence his partner, Matthew Boulton, was able to bring to bear on Members of Parliament.[17] In his evidence to the 1829 committee on patent law reform, Farey took Watt's 1769 patent as an example of a misdrafted specification and rewrote it in the light of the law to demonstrate how it might have been rendered watertight.[18]

Farey was of the view that many engineers were the victims of hostility on the part of their successors. He wrote to the mathematician Stephen Rigaud (1774–1839):

> It appears that all the inventors on steam, have been traduced in turn. Papin by Dr Robison, 3rd ed of Ency.Brit, xvii p 744; Savery by Desaguliers and all after him for ¾ of a century; Mr Watt by Bramah in a pamphlet and by Hornblower in the 1st edition of Gregory's Mechanics 180–.... Mr Trevithick has rarely and Mr Woolf never had justice done, but the latter has been shamefully treated ... There is so much narrow jealousy amongst practical men that except a few they are not inclined to do justice to any competitor.[19]

Farey's plans for the second volume came to nothing, and the book as eventually printed mainly covers Chapter X and a greatly expanded Chapter XI on the historical account of the steam engine. Whereas the Woolf engine was originally meant to be described in 60 pages, Farey expanded this to approximately 280 pages. He was able to do this because during the 1820s and 1830s the practice had grown up in Cornwall of keeping meticulous records of engine performance and publishing them. Farey thus had a ready-made statistical base from which he was able to write an elaborate study of the effectiveness of the Woolf engine, in addition to his original plan of minutely describing its construction.

Farey described first a 6 h.p. Trevithick high pressure engine and then a 12 h.p. Woolf rotative engine, for which no drawings survive. The next topic was a pumping engine erected at Wheal Vor in 1815, for which four proofs of the drawings have been discovered. There were three more drawings prepared for this series, but these have not been found. These four proofs are properly captioned but lack the lettering to key them to the text. The rest of Chapter XI was an extended discussion of engine efficiency, and took the story up to 1843. Research into other books on the steam engine published at this time has revealed that other writers illustrated engines identical to those Farey was going to describe. The evidence points to Farey supplying publishers with drawings as part of his regular business practice.[20]

PART 2

Farey began writing the text of the *Treatise* in 1820, and Henry Farey noted in his diary for that year details of the drawings he, Henry, was making for Wilson Lowry to engrave for the book.

Despite a move to Devon where John worked in Tiverton from 1821 to 1823, work on the *Treatise* continued at a reasonable rate. Early in 1824, Strahans, Longmans' printer, had composed and proofed pp. 1–40 of the introduction and pp. 21–280 of the text of Volume 1 of the *Treatise*.[21] In addition, Wilson Lowry had engraved almost all the 60-odd plates for the entire work at the time of his death on 23 June of that year.

A move to Leeds and family problems in 1824–26 occasioned further delay. By October 1826 Strahans had composed and proofed the rest of the text of the first volume of the *Treatise*, and in that month 54 of the plates by then engraved were proofed.[22]

The first volume had comprised an historical survey up to the retirement of Watt, in about 1800, while the plates and blocks for this volume only accounted for less than half of those prepared, and – most importantly from Longmans' viewpoint – paid for.

Longmans were thus justified in getting worried, and in January 1827 they wrote to Farey the first of a series of letters pressing him to get on with the work. Only the Longmans' half of the correspondence survives, but from this it is plain that Farey was a very tardy author. In fairness, however, it must be said that he had been through a very trying time, and when Longmans began writing Farey was in the middle of the preparations for his second marriage. The marriage took place on 24 February at St Pancras Church, London, and the arrangements had most likely driven his work on the *Treatise* to the bottom of his list of priorities. Longmans had written on 4 January 1827:

> We expected to hear from you on Monday agreeably to your promise. As it is now essential to fix a plan for the immediate completion of your work, we request that you favour us with a decision respecting our last communication.[23]

Farey's reply can in no way have mollified them, and on 8 January they wrote again:

> We have just received your letter which we regret is not satisfactory in as much as it does not fix a time (as you promised) for completing the whole work. We have informed you that after having spent seven years it was reasonable for us to expect that you would now enter into a positive agreement to fix a reasonable time for the completion of your work. We now have to request that you will immediately furnish us with the necessary data as to the time, that our solicitor may draw up a regular agreement to the above effect.[24]

Farey did not reply to this, and on 26 February Longmans wrote:

> On the other side you have a copy of our letter of Jany 8th to which we have received no reply and we have again to request that you favour us with a definite and agreed answer for the purpose therein stated.[25]

Communication was in some measure restored, and on 17 March Longmans wrote:

> Pray favour us with a call on Monday, Tuesday or Wednesday next between the hours of 2 and 4 – have the goodness to inform the bearer what day will suit you best.[26]

Towards the end of 1827 the first volume of the *Treatise* was ready, and it was published on 1 December (Figure 1). It was a handsome quarto of 728 pages with 25 plates and numerous woodcuts. It was printed in an edition of 750 copies and sold for 5 guineas.[27]

A compromise must have been agreed between Farey and his publisher, since the volume as printed contains a couple of short sections describing contemporary engines which Farey had originally planned to include in the second volume.

In addition to being worried about a tardy author (who had begun his work in 1820), Longmans would have been concerned about the publication of rival books on the steam engine. The decade 1822–1832 produced what Dickinson and Gomme described as a 'Galaxy of writers on the steam engine'. Farey was beaten to publication of the first volume of his *Treatise* by Partington, (1822), Stuart (1824) and Tredgold (1827). Prices varied between 8 shillings for Stuart's *Descriptive History* and 5 guineas for Farey's *Treatise*. These rivals were without exception inferior in content and in the breadth of knowledge and experience of their authors, but being cheaper they naturally attracted sales and many went into several editions.[28]

The first volume of Farey's *Treatise* was reviewed in the *Repertory of Patent Inventions*, where the reviewer complained about its bulk and discursiveness. He rounded off his diatribe as follows, exclaiming that he

> shall not be surprised to see in some newly-edited history of England the whole of the chronicles of all nations up to the Norman Conquest, not forgetting the book of Genesis, with complete illustrations of the fall of man, and the primitive nature of the Serpent, interspersed with extensive compilations on agriculture, chemistry, orthography and the use of globes.[29]

He then went on to complain about the inclusion of material about the atmospheric engine from the Smeaton MSS. The reviewer favoured Tredgold's *Treatise on the Steam Engine*, which had been published a little time before Farey's book had appeared, and which the *Repertory of Patent Inventions* had previously reviewed.

Farey wrote a rebuttal of the various charges laid against his book. He commented that the *Repertory* seemed more interested in supporting theoretical as opposed to practical investigations of engineering, and was concerned that this could lead to erroneous and ultimately dangerous practices being used in engineering as a result. Farey commented further that many of Tredgold's theoretical statements were based on simple model experiments with no guarantee they were sound when applied to real-life situations, and he listed a number of Tredgold's theorems for engine proportions which were clearly wrong. The reviewer had asserted that there was no need to include anything about the principles of mechanics or the slide rule. He also objected to the inclusion of material about Smeaton. In his reply Farey justified these sections on the grounds that there was no work available on mechanics specifically geared to the steam engine, whilst the use of the slide-rule was virtually unknown among

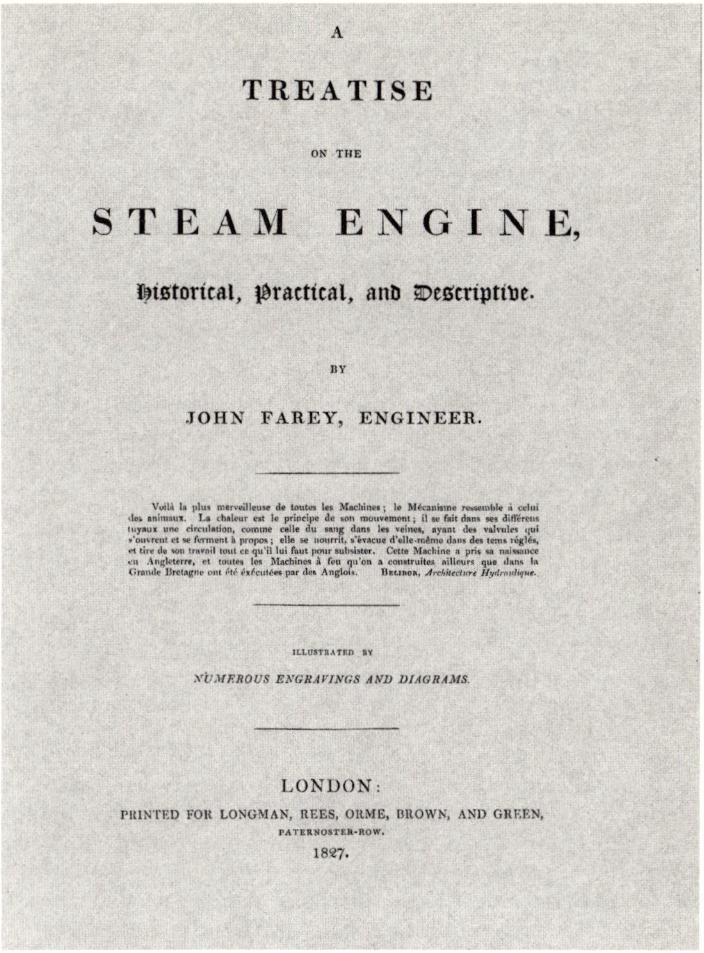

A

TREATISE

ON THE

STEAM ENGINE,

Historical, Practical, and Descriptive.

BY

JOHN FAREY, ENGINEER.

Voilà la plus merveilleuse de toutes les Machines; le Mécanisme ressemble à celui des animaux. La chaleur est le principe de son mouvement; il se fait dans ses différens tuyaux une circulation, comme celle du sang dans les veines, ayant des valvules qui s'ouvrent et se ferment à propos; elle se nourrit, s'évacue d'elle-même dans des tems réglés, et tire de son travail tout ce qu'il lui faut pour subsister. Cette Machine a pris sa naissance en Angleterre, et toutes les Machines à feu qu'on a construites ailleurs que dans la Grande Bretagne ont été éxécutées par des Anglois. BELIDOR, *Architecture Hydraulique*.

ILLUSTRATED BY

NUMEROUS ENGRAVINGS AND DIAGRAMS.

LONDON:

PRINTED FOR LONGMAN, REES, ORME, BROWN, AND GREEN,

PATERNOSTER-ROW.

1827.

Figure 1 Title page of Volume 1.

engineers. He pointed out that the material on Smeaton was relevant since a large number of atmospheric engines were still being built in the colliery districts where they were more economical to run than the Watt type, and in any case the section had been based on hitherto unpublished MSS of Smeaton, and not, as the reviewer asserted, reprinted from Smeaton's *Reports* (1812), or Rees's *Cyclopaedia* (1802–1819).[30]

Tredgold, who was a compiler of formulae and rules and not a practical mechanical engineer, meant his book to be a *vade mecum* for designers, and indeed posterity has judged his work to be the more useful, since it went into a second revised edition in 1838, and was translated into French (1828, 1838 twice) and Spanish in 1831. The first and second editions of Tredgold's book differ markedly: all Tredgold's strictures about Watt's abilities and his business ethics in the first edition were removed or re-written to express the opposite in the second. This was clearly part of the sanitizing process then in train, organized by James Watt Jr, to ensure that his father's memory should remain pristine with all criticism suppressed.[31]

By June 1828, 128 copies of the first volume of Farey's *Treatise* had been sold,[32] and Longmans again put pressure on Farey to complete the project. The next surviving letter is dated 31 July 1828 and shows their concern:

> As we wish to see you *most particularly* on the subject of your work on the Steam Engine, you will oblige us by mentioning any day next week that best suits your convenience, to favour us with a call between the hours 2 and 4 o'clock.[33]

No more correspondence survives for this year, but from the next letter, dated 13 January 1829, it can be inferred that Farey had convinced the publisher that he would complete the work within a reasonable time:

> We have been anxiously expecting to hear from you that you had all the copy of the work on the steam engine ready, agreeably to your promise when we last saw you. We beg you will, by *return of post* inform us of how you stand.[34]

The letters cease here, but an entry in the accounts takes up the tale:

> Turner and Sons for Law Expenses in urging Mr F on with the work.
> (Aug–Nov 1829) £8 12 6[35]

The events after this in regard to the *Treatise* are far from clear, but it is plain that a radical change of plan was agreed between Farey and his publisher. It seems that it was decided to publish only the remainder of the historical survey, up to and including the time of Arthur Woolf (1766–1837), with the rest of the book, with the accounts of contemporary steam engines, the material on the scientific basis of steam technology and the appendix listing all steam engine patents apparently abandoned.

Yet this does not explain why Longmans did not try to salvage something and use the plates they had already produced and paid for. The only Longmans-published book so far traced which does contain illustrations from Farey's book is John Bourne's *Treatise on the Steam Engine* (1846), but here it was a plate and some blocks from the published volume, not the unpublished one.[36] A possible explanation for Longmans' inaction may lie in the form of contract they had with Farey, which might have contained terms giving him sole rights to the plates, even though Longmans had paid for their production.

There is no doubt that Farey possessed numbers of copper plates of this kind, since the newspaper account of the salvage operations following the fire which occurred at his house in December 1844 mentions heavy iron fireproof safes containing plates which needed 20 men to move them.[37] The suggestion that it may have been publishing practice for authors to retain possession of copper plates is reinforced by the fact that the inventory of the Watt MSS at Doldowlod, made for the trial Muirhead *v.* Watt in 1870, lists boxes of copper plates engraved to illustrate Muirhead's *Mechanical Inventions of James Watt* (1859).[38] They were thus held by the author and not the publisher or his printer. On the face of it Farey should have found it easier to produce technical descriptions of engines for which the drawings had already been engraved, than to undertake the task of compiling technical statistics about the performance of the Cornish engine for the book as it eventually appeared.

The strangest feature of all is the slowness with which Farey and his publisher produced the rest of the work. No haste was evident in getting on with it, and the production was allowed to drift for almost twenty years. However, Longmans sometimes noted that the work was in progress in advertising literature bound in with newly published books.[39] The accounts show that Longmans paid Joseph Wilson Lowry (1803-1879), son of Wilson Lowry, for more plates at spasmodic intervals. By December 1838 the text of the second volume was proofed as far as page 136, and in 1839 proofs of 33 unpublished plates were made, but it was not until after 1842 that the rest of the book was completed. Farey brought his text up to date to include events occurring in that year, and soon afterwards Strahan completed the printing of 500 sets of sheets. It was not until 1849 that the costs were posted in the accounts, and these also recorded that no plates were ever produced for the book, and that the sheets were never bound.[40]

Farey died in July 1851 but Longmans do not seem to have abandoned all thought of publishing the rest of the *Treatise*, since their accounts record that in October 1856 they had seven of the plates proofed (probably relating to the volume as it was then set).[41] This plan was abandoned, and they recorded later in the year that 494 sets of the printed sheets were sent for waste paper. The remaining six sets were sent for waste in February 1862, leaving the proof copy in Farey's library (which his only son Edward had retained, with some other books and drawings) as the only one surviving.[42] The second volume of Farey's *Treatise* is thus one of the rarest books in the English language, since it was never published, and the destruction of the sheets printed is fully documented, leaving Farey's proof as the unique copy.

Longmans' accounts also show that no royalty was paid to Farey for the *Treatise* in his lifetime – which is not surprising in view of the events surrounding the publication of the book – but in June 1864 they recorded that £178 18s 8d was standing to his credit. This may well have been moneys owing him for other work he had done for them and not relating to the *Treatise*. Regrettably, the G series, or 'Gentleman's Ledgers' – Longmans' individual record of all the work done by a writer or artist – no longer survive and so this point cannot be proved. The sales of the first volume of the *Treatise* naturally had tailed off in time, but when the remainder stocks of 160 copies were sold to Bohn in 1862, 568 copies or 76.5 per cent of the initial printing had been sold. The balance remaining of 22 volumes from the original printing had been sent out as review copies or had been taken by Farey himself. This gave Longmans a profit of £718.3.1 from a total outlay of £2551.7.0.

EPILOGUE: THE SUBSEQUENT FATE OF VOLUME TWO

Edward Farey died, unmarried, on 15 October 1913 at the age of 88, and willed to the Institution of Civil Engineers 'the printed but unpublished volume of the work on the steam engine by my father John Farey'.

When the lawyer came to wind up the estate he mis-read the will, and instead of the Institution of Civil Engineers being sent 'the printed but unpublished volume' he told them they were to receive the 'printed and published book', and sent them a copy of Volume One of the *Treatise*.[43] It is assumed that the proofs of the second volume of the *Treatise* were sold in the disposal of the residue of Edward's estate, but nothing was found about this until the summer of 1915, when P.M. Barnard, a bookseller of Tunbridge Wells, Kent, wrote to the Patent Office Library offering them for sale. The only letter of this correspondence to have survived stated that Barnard had no idea of the book's previous history and he had bought it at public auction. He had no knowledge, either, of the inscription on the flyleaf – Edward Farey's name and address. This letter has a pencilled note that the Patent Office Library purchased it for £2.15.6, including postage.[44] The Patent Office Library also contacted Longmans and was told there was no record of the book ever having been printed. Their records also showed three plates having been destroyed in 1878. This cannot be substantiated, since it is not mentioned in any papers surviving today at Reading University. It is possible they consulted a document destroyed in World War II, when part of the Longmans Archive perished in the Blitz.[45]

The material the Patent Office Library bought from Barnard comprised an incomplete and extra-illustrated and annotated Volume One, page proofs of the two chapters comprising Volume Two, and, perhaps, a collection of manuscript and printed pieces including a synopsis and plate list, Farey's manuscript list of corrections to Volume One and a fire-scorched prospectus.

The acquisition was inspected by Rhys Jenkins, the patent examiner and technical historian, in December 1919, the first writer to describe the work in an article published in *The Engineer* in 1920.[46] In this, Jenkins commented on the proposals to publish the second volume, but made no mention whatever of the plate list, synopsis and corrections; nor did he mention the incomplete first volume. Jenkins referred to the annotated copy of the incomplete Volume One of the *Treatise* in this collection (which relates only to Smeaton) in the work he produced with H.W. Dickinson, *James Watt and the Steam Engine*, of 1927. He also referred to the collection in a paper on Arthur Woolf he published in *Transactions of the Newcomen Society* in 1933, wherein he remarked, 'the author's proof of the letterpress, *together with a continuation of the MS* [APW's emphasis], bound in a volume, was acquired by the Patent Office Library, some years ago'.[47]

This comment raises all kinds of questions as to whether there was more to the Farey purchase than survives today. Certainly there is nothing in the British Library now which fits Jenkins's description, and the best conclusion which can be reached is that he was mistaken, a surprising thing for someone who in his lifetime had a reputation for meticulous scholarship. On the other hand he did not refer to the synopsis and plate list, which could mean that they came to the Library from another source, and that he never saw them when he inspected the printed books in 1919. If there ever was a manuscript continuation, its rediscovery would be most welcome.

A further oddity about the acquisition is that nobody seems to have attempted to discover more about Edward Farey, whose name and address was written on the cover of the incomplete first volume. Had they done so they could have learnt of his death, discovered his lawyer and in turn been told about his American cousins, descendants of his uncle Henry and aunt Marianne. Since this period saw the formation of the Newcomen Society and the great interest in Watt, as exemplified by H.W. Dickinson's and Rhys Jenkins's monumental study, *James Watt and the Steam Engine* (1927), this indifference is all the more puzzling. It is possible that information might be found in the notes of Rhys Jenkins and Eng. Capt. Edgar Smith, both of whom researched exhaustively on technical history and biography.[48]

The only other British technical historian known certainly to have been interested in Farey at that time was Capt. R.S. Alston (1895–1939) who lived at Hambrook, near Bristol. During the 1930s he indexed Volume One of the *Treatise*, and circulated privately typewritten copies to fellow members of the Newcomen Society. On his death his collection of stationary and marine engine models went to Bristol Museum, and a number of his other books, including his annotated copy of Volume One of Farey's *Treatise*, went to Bristol Reference Library. Other books, perhaps duplicates, are known to have been sold by George's, the Bristol booksellers. Like Rhys Jenkins, Alston had the reputation for meticulous scholarship, but his notes, if they survive, have not so far been traced, and so it is not known how far or in what directions his research into the life of Farey took him.[49]

The growth of industrial archaeology in the 1960s meant that Farey's book began to be more widely appreciated, and two publishers proposed issuing a reprint of it. One of these, the Cornmarket Press, got as far as issuing a prospectus and obtaining an ISB number for it, but did not proceed. The other publisher, David and Charles, published a reprint in 1971 and 1972 comprising an unannotated copy of Volume One, and the page proofs of the unissued Volume Two, now in the British Library. The volumes are in a smaller format than the originals, with the result that detail on the plates is not as crisp as it might have been. Although the publishers made a brief reference to other manuscripts accompanying the British Library volume, they did not say what these comprised, and missed the chance of including Farey's synopsis, list of plates and corrections.

Von Tunzelmann's view of the importance of Farey's book is correct, but how much more valuable would it be for today's technical historians had Farey's original concept been fulfilled. How much more valuable would it have been if Farey had been able to publish further books describing other technologies, as he had once planned. In fact after the publication of the first volume of the *Treatise* Farey changed direction and devoted most of his time to engineering consultancy. Apart from the occasional letter to the technical and scientific press he was to publish nothing more under his own name.

APPENDIX 1: SYNOPSIS FOR REMAINDER OF THE TREATISE

Vol 2

Part II TECHNICAL DESCRIPTION OF STEAM ENGINES NOW IN ESTABLISHED USE.

High Pressure Steam Engines.

Chapter I — Mr Woolf's High Pressure condensing steam engine.

Description of Mr Woolf's single engine for pumping water for draining deep mines erected at Wheal Vor mine, Cornwall, 1815, 6 plates, 40pp. [Printed Vol. 2, pp. 103–132, pl. XXX–XXXV]

Description of Mr Woolf's double-acting rotative steam engine for turning mills, 2 plates, 12 pp. [Printed Vol. 2, pp. 75–91, pl. XXVII–XXIX]

On High Pressure steam. On the thickness of metal for high pressure boilers. The precautions necessary to avoid explosions; safety valves, feeding apparatus, etc. 8pp.

Chapter II — Mr Watt's rotative steam engine for turning mills, being on the construction termed House Built, adopted for the largest engines.

Description of a rotative engine of 20 horse power and flour mill constructed by Messrs Thwaites, Hick and Rothwell, Bolton, Lancashire, 1820. 7 plates, 40pp.

Description of a rotative engine of 30 horse power constructed by Messrs Fenton Murray and Co, 1820, 3 plates, 8pp.

Chapter III — Mr Watt's rotative steam engine for turning mills being on the construction termed portable, or independent, frames for small sizes.

Description of a rotative engine of 3 horse power constructed by Messrs Maudslay & Co., London, 3 plates, 6pp.

Description of a steam engine and mill for shelling rice constructed by Messrs Maudslay & Co. for the Island of Ceylon. 1 plate, 4pp.

Description of a rotative steam engine of 6 horse power by Fenton, Murray and Co., 1 plate, 4pp. [Printed Vol. 1, p. 691, pl. XVII]

Description of a rotative steam engine of 8 horse power applied to work a forge hammer constructed by Thwaites, Hick and Co. 1 plate, 6pp.

Description of a 10 horse power independent engine by Fenton, Murray and Co., 1 plate, 4pp.

Description of a dredging machine for removing mud from docks. 1 plate, 8pp.

Description of a 10 horse power independent engine applied to a horizontal sugar mill constructed by Jukes, Coulson & Co., 1 plate, 6pp. [Printed in Vol. 1, p. 712, pl. XXII]

Description of a 10 horse engine and sugar mill, by Rothwell, Hick and Co., 1 plate, 6pp.

Description of a steam engine of 14 horse power by Fenton, Murray & Co., 1 plate, 4pp.

Description of a steam engine of 16 horse power by Maudslay, 2 plates, 8pp.

Chapter IV On different means of trying the performance of a steam engine, and of judging what power it actually exerts.

Description of an indicator or instrument to ascertain the elasticity of the steam within the cylinder of a steam engine when it is at work. 1 plate, 8pp.

Description of a barometer to show the exhaustion with the condenser; a different kind of barometer; description of a steam gauge; description of safety valves; syphon pipes and sundry appendages to a steam engine. 12pp.

Part III THEORY OF THE PRINCIPLES OF ACTION IN STEAM ENGINES.

Chapter I On heat; on the nature and properties of heat, its production in combustion, and by friction and by the compression of bodies. Operation of heat to produce mechanical power by dilating and expanding bodies. 60pp.

Chapter II	On the nature and properties of combustible matters which are to be found in the fuels used for steam engines and the products of their combustion. The nature and properties of water and its constituent principles. The phenomenon of its conversion into steam by heat. 40pp.
Chapter III	On the velocity of motion with which steam will pass through pipes and passages of a steam engine when at work. 8pp.
Chapter IV	Detail of experimental researches by which the leading facts for the theory of steam engines have been ascertained.
	Messrs Dulong and Petit and others – the measure of temperature and of heat. 8pp.
	Dr Black, Irvine and others on latent heat, capacity for heat or latent heat. 6pp.
	Mr Southern, Bettancourt, and Clement on the properties of steam. 12pp.
	Count Rumford on the transmission of heat in fluids and on heat and combustion. 12pp.
	Mr Smeaton on mechanical power and collision, Mr Southern on friction. 6pp.
	Colomb on friction; Gay Lusac on the conducting power of a vacuum. 10pp.
	Attwood on preponderance; Captain cater [*sic*] on the pendulum; Robins on the force of fired gunpowder; also Hutton and Rumford. Banks, Tredgold and others on the strength of iron and other materials; also Barlow and Wood. 56pp.

APPENDIX

A record of projects for the improvement of the steam engine beginning with the earliest proposals and containing all the successive projects to the present time. 200 pp.

Note: All the important inventions or propositions which have led the way to the established plans now in use have been already described in this historical account, Vol 1; This appendix is intended to fill up and make the record complete and it will contain a list of all the patents with a brief account of their purport with references to books and records where more complete information may be obtained.

Farey noted that the book would thus total 610 pages and would require 37 plates. There would also have been many woodcuts needed. He had allowed 5 plates to illustrate the material forming part III and the Appendix, but remarked that possibly woodcuts could be substituted for these.

APPENDIX 2: ANNOTATED LIST OF PLATES AND WOODCUTS INTENDED FOR THE SECOND VOLUME OF FAREY'S *TREATISE ON THE STEAM ENGINE*

This list has been built up from the particulars of the plates published in Volume 1, described in Volume 2 as eventually printed, and listed in the manuscript plate list and the synopsis for the book, now with the proofs of Volume 2 in the British Library.

They are in the order of a census of Farey drawings the author is compiling, and are preceded with the initials FT (Farey's *Treatise*). Where appropriate they are cross-referenced to plates in other publications. Short-titles are sometimes used. The manuscript plate list was originally written in pencil with most of the entries being overwritten in ink, as Farey finalized his plans. The remainder appear to have been erased, but some are still legible. Farey numbered the entries, and these are recorded under the FT numbers, but the two series are not the same, since it is clear that the final order of publication was not the same as Farey's initial list.

In the Science Museum Library is a collection of large glass-plate negatives of a number of drawings Benjamin Hick provided for Farey. They are extensively annotated by Hick with technical details about engine construction. In the reproduction here of these plates, the annotations have been ignored, to permit the printing of the drawings to the largest possible size.

VOLUME 1

FT 1 (Vol. 1, Plate I)

[PL 1] Mr Savery's patent fire engine for raising water by fire.

FT 2 (Vol. 1, Plate II)

[PL 2] Newcomen's ... steam engine ... by Mr Smeaton, 1772 (longitudinal section).

FT 3 (Vol. 1, Plate III)

[PL 3] Newcomen's ... engine by Mr Smeaton, 1772 (plan and parts).

FT 4 (Vol. 1, Plate IV)

[PL 4] Pitwork and pumps of a steam engine for a colliery.

[FT 2, 3 & 4 are all of Long Benton pit engine. Smeaton's original drawings for this are to be found at fol. 72/72v–83 of Smeaton's *Designs* in the library of the Royal Society.]

FT 5	(Vol. 1, Plate V)
[PL 5]	Upright section of Chacewater fire engine ... 1775
FT 6	(Vol. 1, Plate VI)
[PL 6]	Parts of Chacewater fire engine
FT 7	(Vol. 1, Plate VII)
[PL 7]	Working gear of Chacewater fire engine
	[Smeaton's original drawings of this engine are to be found at fols 111v–125 of Smeaton's *Designs* in the Royal Society. FT 5, 6 and 7 are copies respectively of plates II, IV and III of Smeaton's *Reports*]
FT 8	(Vol. 1, Plate VIII)
[PL 8]	Newcomen's ... Steam engine ... by Mr Curr, 1790
	[This engine was erected at Attercliffe Common in 1790, and described and illustrated by Curr in his *Coal Viewer and engine-builders companion*, Sheffield, 1797]
FT 9	(Vol. 1, Plate IX)
[PL 9]	Section of Mr Smeaton's Portable engine [See FT 14 below]
FT 10	(Vol. 1, Plate X)
[PL 10]	Watt's engine for draining mines, 1788
FT 11	(Vol. 1, Plate XI)
[PL 11]	Watt's patent rotative engine (sectional elevation)
FT 12	(Vol. 1, Plate XII)
[PL 12]	Watt's patent rotative steam engine (Boiler and elevations)
FT 13	(Vol. 1, Plate XIII)
[PL 13]	Watt's patent rotative steam engine (cylinder, valves and valve-gear)
FT 14	(Vol. 1, Plate XIV)
[PL 14]	Elevation of a Smeaton Portable Fire engine, 1765 [See FT 9. This engine was described and illustrated in Smeaton's *Reports*]
FT 15	(Vol. 1, Plate XV)
[PL 15]	Projects for rotative steam engines
	[Plate reused in John Bourne's *Treatise on the Steam Engine*, 1846]
FT 16	(Vol. 1, Plate XVI)

[PL 16] [Upper] Steam engine by Fenton, Murray & Wood, 1806
[Lower] Bell-crank steam engine of 4HP by Boulton and Watt, 1802

[The bell-crank engine was patented by Murdock in 1799. Farey sketched the engine illustrated here at Jackson's anchor manufactory, London, 12 February 1805 (Sketchbook 1, 174ff). In 1809 Joseph Farey was awarded a silver palette by the Society of Arts for sending a set of plans he had prepared of it (Sketchbook 1, fol. 176, note). Henry Farey noted in his diary that he drew this plate for the *Treatise* on 22 July 1820.]

FT 17 (Vol. 1, Plate XVII)

[PL 17] Steam Engine of 4HP by Fenton, Murray and Wood, 1802

[Farey sketched this hypocycloidal engine at Brewin's tannery, Bermondsey, 3 January 1809 (Sketchbook 1, fol. 197ff). He had drawn it earlier and it was published by Olinthus Gregory in his *Treatise on Mechanics* (London 1807), Vol. 2, pp. 290–292, pl. XXXI]

FT 18 (Vol. 1, Plate XVIII)

[PL 18] Rotative steam engine of 6HP by Fenton, Murray & Co.

FT 19 (Vol 1, Plate XIX)

[PL 19] Watt's Rotative steam engine of 36HP, 1808.

FT 20 (Vol. 1, Plate XX)

[PL 20] Cylinder and valves of Mr Watt's rotative steam engine, 36HP, 1808.

FT 21 (Vol. 1, Plate XXI)

[PL 21] Steam engine and mill for rolling sheet lead.

FT 22 (Vol. 1, Plate XXII)

[PL 22] Perspective view of a steam engine and horizontal sugar mill by Jukes, Coulson & Co., London.

FT 23 (Vol. 1, Plate XXIII)

[PL 23] Watt's engine for pumping water for supply of towns, constructed by Messrs Boulton and Watt, 1803

FT 24 (Vol. 1, Plate XXIV)

[PL 24] Parts for the above engine

[This was the engine at Chelsea waterworks, erected in 1803. Farey sketched this in 1805, together with details of the Old Engine, erected by Watt in 1779. Another version of

	this drawing forms nos III and IV of the 'Steam Engine' plates in Rees's *Cyclopaedia*.]
FT 25	(Vol. 1, Plate XXV)
[PL 25]	Watt's double cylinder engine for blowing air furnaces, by Boulton and Watt, 1807

VOLUME 2

FT 26	(Vol. 2, Plate XXVI)
[PL 26]	A High Pressure steam engine of 6HP by Trevithick, 1806 [Not mentioned in the synopsis. According to the text of Vol. 2 of the *Treatise*, this was the usual Trevithick pattern with a cast-iron cylindrical boiler and an enclosed vertical cylinder. The dimensions were: Cylinder 8ins dia × 4ft stroke Boiler, $4\frac{1}{4}$ft dia × $7\frac{3}{4}$ft long Connecting rods of $11\frac{1}{2}$ft centres Flywheel 11ft dia with 6 arms Drive gear, 6ft dia, with 84 teeth Speed, 24 rpm [The only dimensioned drawings in Farey's sketchbook of this pattern Trevithick engine are of one installed at the Small Ordnance Wharf, on the Thames, which had identical boiler and cylinder sizes, but different ones for the flywheel and drive gear. The steam and smoke were not removed by a chimney fixed to the engine, but ran into an undergound flue which must have led to a separate stack. A model based on Farey's notes was made by the author for Newcastle Museum of Science and Engineering (Figure 2).]
FT 27	(Vol. 2, Plate XXVII)
[PL 35]	Woolf's rotative steam engine, 12HP [Presumably elevation]
FT 28	(Vol. 2, Plate XXVIII)
[PL 56]	Sections
FT 29	(Vol. 2, Plate XXIX) [not in plate list] [The text of the *Treatise* , Vol. 2, p. 75ff gave the following dimensions: Cylinders: 1ft 4″ x $3\frac{1}{2}$ft stroke and 9ins × $2\frac{1}{2}$ft stroke Boiler: $33\frac{1}{2}$ins dia × 10ft long Chimney: 12ins × 19ins × $28\frac{1}{2}$ft high

No other dimensions are given, but Farey noted the parts were made to the same proportions as a standard Boulton and Watt engine of 12HP, with a 20 inch dia. cylinder. Particulars for calculating these were to be found in the *Treatise* Vol. 1, Chapter VIII. A very detailed illustrated account of the Woolf engine as made by Edwardes of Paris was published in the *Bulletin de la Société d'encouragement pour l'Industrie National* (No. CLXXIV), December 1818, pp. 365–386. Another illustrated account of a Woolf-type engine adapted for mine-winding is to be found in A.M. Heron de Villefosse, *De la Richesse Minérale,* Vol. 3, 1819, pp. 87–108.

FT 30 (Vol. 2, Plate XXX)

[PL 27] Woolf's single-acting expansion steam engine, with two cylinders, 1814, Wheal Vor tin mine, Cornwall.

Horizontal plan of the buildings, showing boilers, cylinders, shaft, pump-rod, balance bob.

FT 31 (Vol. 2, Plate XXXI)

[PL 28] Woolf's Wheal Vor engine, vertical section, longitudinally through the buildings, with the end elevation of one of the boilers (Figure 3).

FT 32 (Vol. 2, Plate XXXII)

[PL 29] Woolf's Wheal Vor engine, vertical section transversely through the buildings, with longitudinal elevation of the boilers (Figure 4).

FT 33 (Vol. 2, Plate XXXIII)

[PL 30] Woolf's Wheal Vor engine. Section of cylinders with their pistons, section of valve boxes.

FT 34 (Vol. 2, Plate XXXIV)

[PL 31] Elevation of cylinders and valve gear (Figure 5).

FT 35 (Vol. 2, Plate XXXV)

[PL 32] End of the beam, with details of the parallel motion gear (Figure 6).

FT 36 (Vol. 2, Plate XXXVI)

[PL 53] Description of pumps and pit-work of steam engine for draining deep mines in Cornwall.

FT 37 (Vol. 2, in plate list only)

[PL 54] Parts.

[Plates FT 30–36 are the earliest engravings of the high-pressure beam pumping engine in Cornwall. Plates FT 31, 32, 34 and 35 survive as proofs from the engravings in the possession of the Farey family in America, and are reproduced here for the first time. Part of figure FT 32 is reproduced exactly in Christian's *Traité Méchanique Industrielle*, 1825. The only difference is the mode of the hatching of the sections. They clearly have a common original. The Wheal Vor engines were made by the Neath Abbey Ironworks Co., but no original drawings survive in the collection of company drawings now in the Glamorgan Record Office. See also Lawrence Ince, *The Neath Abbey Iron Company* (Eindhoven, 1984).

FT 38	(Vol. 2, Synopsis, Chapter 2)
[PL 30]	20HP Flour Mill engine by Thwaites, Hick and Rothwell, Bolton, Ground Plan (Figure 7).
FT 39	(Vol. 2, Synopsis, Chapter 2)
[PL 31]	Elevation of engine (Figure 8).
FT 40	(Vol. 2, Synopsis, Chapter 2)
[PL 32]	End elevation of engine (Figure 9).
FT 41	(Vol. 2, Synopsis, Chapter 2)
[PL 33]	End elevation of mill and gearing (Figure 10).
FT 42	(Vol. 2, Synopsis, Chapter 2)
[PL 34]	Section of cylinder and side pipes (Figure 11).
FT 43	(Vol. 2, Synopsis, Chapter 2)
[PL 35]	Parts, beam, connecting rod, flywheel (Figure 12).
FT 44	(Vol. 2, Synopsis, Chapter 2)
[PL 36]	Parallel motion and governor (Figure 13).
FT 45	(Vol. 2, Synopsis, Chapter 2)
[PL 37]	Section of 20HP cylinder and side pipes and slide valve (Figure 14).
	[FT 43 and FT 44 have been made by combining details from three of Hick's original drawings.]
FT 46	(Vol. 2, Synopsis, Chapter 2)
[PL 39]	30HP rotative engine by Fenton, Murray and Wood, 1820. Elevation.
FT 47	(Vol. 2, Synopsis, Chapter 2)

[PL 40] End elevation and section.

FT 48 (Vol. 2, Synopsis, Chapter 2)

[PL 41] Section of cylinders and general plan (Figure 15).

[Henry Farey made drawings of Fenton, Murray and Wood's 30HP and 50HP engines on July 3–7 1820. He noted the dimensions:

Cylinder 3ft dia × 7ft stroke
Air pump 3ft stroke
Beam ? ft
Beam gudgeon, 7″ dia × 12″ long (main) 3½″ (end)
Piston rod, 4″ dia
Flywheel 24ft diameter, rim 12″ × 6″
Necks of flywheel shaft 10″ dia × 14″ long
Connecting rod 22′6″ long
[The engine worked with a long D slide-valve and an eccentric motion.]

FT 49 (Vol. 2, Synopsis, Chapter 2)

[PL 17] 3HP rotative engine by Maudslay, 1820. Elevation.

FT 50 (Vol. 2, Synopsis, Chapter 3)

[PL 18] Section and plans.

FT 51 (Vol. 2, Synopsis, Chapter 3)

[PL 19] Parts on a larger scale.

[No particulars of this engine have been so far discovered.]

FT 52 (Vol. 2, Synopsis, Chapter 3)

[PL 20] Steam engine and rice mill for Ceylon, by Maudslay, 1814.

[The rice mill was extensively described and illustrated by Thomas Hoblyn (who received the Isis Medal) in *Transactions of the Society of Arts*, 1813, 250–259, plates XXI, XXIII. In a subsequent paper he described the hydraulic oil press made by Bramah. Neither account contains any information or illustrations about the steam engine.]

FT 53 (Vol. 2, Synopsis. Chapter 3)

[PL 20] 8HP engine and forge hammer by Thwaites, Hick and Rothwell.

[This drawing forms part of the Science Museum collection of Hick drawings (Figure 16).]

FT 54 (Vol. 2, Synopsis, Chapter 3)

[PL 57] Dredging machine by Rennie, 1817

[This is probably the machine designed by Rennie for the Liverpool Docks. The engine was made by Boulton and Watt. A.W. Skempton 'A history of the steam dredger, 1797–1830' in *Transactions of the Newcomen Society*, 47, 97ff.]

FT 55 (Vol. 2, Synopsis, Chapter 3)

[PL 24] 10HP engine and sugar mill by Rothwell, Hick and Rothwell, 1824

[This drawing forms part of the Science Museum collection of Hick drawings (Figure 17).]

FT 56 (Vol. 2, Synopsis, Chapter 3)

[PL 27] 14HP Fenton, Murray and Wood, 1820

[Henry Farey copied this drawing 24 June–1 July 1820. He noted that Murray's made this engine in 14, 16 and 20HP sizes and gave the following dimensions:

Cylinder 23″ dia × 3′ stroke
Airpump 15″ dia × 18″ stroke
Beam 10ft long, 18″ deep in the middle
Flywheel 16ft dia
Rim 10″ × 5″
Necks of the flyshaft 7″ dia × 9″ long
Cistern 3′ 3″ wide
Top of cylinder to centre of beam 9′ 9″
Outside of cylinder jacket 2′ 4″
Beam centre [gudgeon] 4″ dia

An identical pattern engine, but of 50HP size was described and illustrated in Stuart *Anecdotes of the Steam Engine* (London 1824), 443–4, unnumbered plates opposite pp. 443, [2] 444. These plates were engraved by Wilson Lowry, and it is tempting to surmise that Farey supplied them as part of his service to writers and publishers (Figure 18).]

FT 57 (Vol. 2, Synopsis, Chapter 3)

[PL 28] 16HP engine by Maudslay, 1816
Elevation, end view and plan

FT 58 (Vol. 2, Synopsis, Chapter 3)

[PL 29] 16HP engine by Maudslay, 1816
Parts, section of cylinder and condenser

[A 16HP engine by Maudslay is illustrated in John Bourne's *Treatise on the Steam Engine* (London 1846). It is the same pattern of cabinet engine illustrated in FT 56, and Bourne made the point that it was old-fashioned. The publisher's accounts for Bourne's book show that payment was made to

Farey for use of his drawings for plate 10, but in the early editions of Bourne this was a 60HP Fairbairn engine]

FT 59 (Vol. 2, Synopsis, Chapter 4) [Not in plate list]
Steam Engine Indicator
[No drawing by Farey of the indicator has been found, so the one selected here shows the pattern of instrument introduced by Farey and manufactured by McNaught (Figure 19)].

FT 60 (Vol. 2, listed in MS plate list only)

[PL 58] Old Steam Boat
Pencil note

FT 61 (Vol. 2, listed in MS plate list only)

[PL 59] Steam Boat for river navigation constructed by Messrs Maudslay & Co.
[Presumably the *Richmond*, the first steam boat on the Thames. Henry Farey noted preparing drawings for Lowry of the *Regent* and the *Falcon*, the latter reaching him on 1 October 1820.]

FT 62 (Vol. 2, listed in MS plate list only)

[PL 60] The *London Engineer* steam boat, the engines by Messrs Maudslay & Co., 1819

FT 63 (Vol. 2, listed in MS plate list only)

[PL 61] Steam engine of the *London Engineer* Steam Boat

FT 64 (Vol. 2, listed in MS plate list only)

[PL 62] Parts of Ditto, Cylinder, condenser etc., 1818

[Henry Farey worked from July to October 1820 copying drawings of this steam boat. They went to Lowry on 7 October 1820. The Science Museum Library has at least one sectioned drawing of this vessel, but it is not known who the draughtsman was.]

FT 65 (Vol. 2, listed in MS plate list only)

[PL 63] Steam Boat by Mr C. Baird, Petersburg, 1819

FT 66 (Vol. 2, listed in MS plate list only)

[PL 64] Oblique boat engine for ditto. Pencil note

FT 67 (Vol 2. listed in MS plate list only)

[PL 65] Maudslay's Boat. Pencil note

FT 68 (Vol. 2, listed in MS plate list only)

[PL 66]	Engine. Pencil note
FT 69	(Vol. 2, listed in MS plate list only)
[PL 67]	Section. Pencil note
FT 70	(Vol. 2, listed in MS plate list only)
[PL 68]	Locomotive engine. Pencil note
FT 71	(Vol. 2, listed in MS plate list only)
[PL 69]	Hick Atmospheric. Pencil note [This forms part of one of the drawings in the Science Museum Hick collection (Figure 20).]
FT 72–76	(Vol. 2, listed in MS plate list only)
[PL 42–46]	Illegible pencil notes of which only the numbers can now be read.

List of woodcuts for Mr Farey's work on the steam engine

[Volume 1]

Soloman de Caus' definition of raising water by fire, 84
Otto von Guericke's exhausted cylinder, 88
Papin's steam cylinder, 97
Desaguliers' Savery fire engine, 112
4-passage cock for ditto, 113
Joshua Rigley's or Kier's fire engine on Savery's plan, 123
Newcomen's atmospheric engine, 129
Cataract for regulating an atmospheric engine, 189
Force pumps for waterworks engine 247, 249
Boiler at York Buildings, 244
Boiler at London Bridge waterworks, 255
Boiler of Cronstadt engine, 265
Smeaton's cylinder boring engine at Carron, 294
Watt's first engine, p.322 (sectioned cylinder and condenser)
Watt's cylinder with wrought iron case, 1778, 333
Watt's cylinder, 1780, 335
Watt's cylinder of expansion engine, 1782, 347
Watt's steam gauge, 372
Watt's spindle valves, 373
Hornblower's double cylinder engine, 1781, 384
Apparatus for determining the elastic force of steam, 396
Apparatus for determining the elasticity of low-temperature steam, 397
Apparatus for determining the volume of steam from an equal weight of water, 400
First application of a crank and flywheel, 410
Parallel motion linkage, 430

Parallel motion linkage (2), 431
Circular throttle valve, 435
Circular throttle valve and governor, 436
Valve gear of Albion Mills engine, 1785, 524
Catches of working gear, 529
Logarithmic scale, 535
Soho sliding rule (2 cuts), 535, 537
Inverted rule, 541
Farey's logarithmic slide rule, 568
Parallel motion linkage, 598
Francis Thompson's double atmospheric engine, 659
Sherratt's double cylinder engine, Garrat, Manchester, 662
Cartwright's engine, 666
Murdock's valve gear, 1808, 397
Hornblower's engine at Mueux (in plate list, but not used)

[Volume 2]

Leopold's high pressure steam engine, 3
Trevithick's high pressure steam engine, 1803, 14 (2 cuts)
[This is a different pattern to the engraved engine noted above. It comprised a hemispherical-topped boiler with a horizontal cylinder cast into it. A model of this is in Vienna. Hellmut O. Janetschek, 'The First Trevithick Engine in the Austro-Hungarian Empire', *Jnl Trevithick Society* 1986, 13, 60–68.]
Woolf's substitute for a flywheel, 47
Woolf's first patent tubulated boiler, 1803, 49
Woolf's self-regulating steam valve for high pressure steam, 51
Woolf's hemp pistons, 60
Woolf's hot water pump, Wheal Vor, 112
Woolf's Shaft pump, Wheal Vor, 142
Brunton's sinking pump, *c*810, 176
[Farey referred to a paper about this pump in *Transactions Society of Arts*, 1811, Vol. XXIX, p. 98. He may have been the author.]
Double slide rule, 226

Among the papers with the unpublished Volume 2 at the British Library is a MS errata list by Farey covering pp. 1–260. There are further corrections in the annotations to the incomplete Vol. 1 in the British Library and also the complete Vol. 1 in the library of the Institution of Civil Engineers.

ERRATA sheet

[Volume 1]

P 19 Lines 5, 7 and 8 from the bottom, the word impetus is used improperly. It should read energy in four places where it occurs.

P 40 Line 7 from the top for (see p. 21) read 22

P 41 Note Acts of Parliament [illegible] 1824 Since this article was printed the standard (entitled standard yard 1790) in the custody of the clerk [illegible] for measures of length is declared to be a brass scale made [illegible] A distance of 1 yard marked upon the bar [illegible] two gold studs. The temperature being 22 degrees of Fahrenheit's thermometer. Thus is determined the Imperial standard yard of brass, and when compared with a pendulum vibrating seconds of mean time in the latitude of London in a [illegible], at a level of the sea, the length of an Imperial Standard Yard is 36 inches to 39.1393 inches. The length stated on p 22 and also on p. 40 is 39.1386 inches by Sir Geo Shuckburgh's standard, which differs a very little from the Imperial Standard.

P 87 Line 12 from the bottom. For Paris 1698 read 1661. Note the copy of Mr Pascal's work in the British Museum is dated 1698 and makes no mention of any prior edition, but the author has examined a copy at Paris which is dated 1664 and is called a second edition. The work was written in 1647 and the passage states the first edition was printed 1661 after the death of the author in [ms blank]

P 99 Para 2 from the top for dated 1709 read 1707. The same error in line 14 and other places. Mr Thomas Savery should not be termed Captain as he held no such official title and is always stated Mr Savery in the publications of his own times. The miners in Cornwall called him Captain.

P 102 Admitted Q1 read O1

P 107 P 107. Note at bottom for a gallon of water ale measure read beer measure. The gallon of either ale or beer measure was the same, but the author in the text meant beer measure because he reckons 36 gallons to a barrel.

P 108 Line 16 from the top invented by Papin 1682 read 1681

P 109 Line 7 from the top. Papin who did make that discovery in 1689 read 1690.

P 142 Line 9 from the bottom should read: It will recoil or begin to rise the moment the steam passage is opened.

P 155 Line 20 'The only effectual remedy is to add an air vessel at the side of the pump'. The words should be: the only effectual remedy is to admit air into the pump, for this purpose the miners usually etc. ... The passage as it stands refers to the engines for waterworks mentioned in p. 213, which sometimes had air vessels to the purpose, but these were not used in mines.

P 162 Line 9 should read: Multiply the square of the diameter of the base in inches by .341 lbs; then multiply by the perpendicular height of the column in feet and the product is the weight.

P 165 Bottom line should be motion per minute / 245 and 96ft.

P 179 Line 2 from the top should read: But would be rather more than 2 seconds for the working stroke and rather less for the returning stroke, because Mr Smeaton usually adjusted his engines so as to perform the working stroke with a slower motion than the returning stroke.

P 187 Line 3 from the top should read: would make their returning stroke rather quicker than the working stroke. This is correctly stated on p. 202.

P 260 Line 3 from the top should read: T is a cock to empty the boiler.

ACKNOWLEDGEMENTS

I am most grateful to the late Arthur and Edith Farey for making available copies of their family papers and for giving permission for their publication; to Messrs Longmans for permission to make copies of documents from their archives and for permission to publish extracts from them; to the Trustees of the Science Museum Library for permission to reproduce the Benjamin Hick drawings, and, in the first place, to Michael Wright and Rodney Law for bringing these drawings to my attention; to the Trustees of the British Library for permission to quote Farey's synopsis, list of plates and engravings and notes about corrections. Fellow Members of the Newcomen Society are thanked for their customary help; the late George Watkins read an early draft and Hugh Torrens and Michael Wright more recent ones. As ever, I am in the debt of my wife Jane for her efficient criticism and proof-reading.

Notes and References

1. G.N. von Tunzelmann, *Steam and British Industrialisation to 1860* (Oxford, 1978) p. 2.
2. British Library, London, Class No. E.F.20.
3. A.P. Woolrich 'John Farey jr (1791–1851): Engineer & Polymath', *History of Technology*, 1997, 19, p. 112.
4. Rees' *Cyclopaedia*, Vol. 34, part, 26 October 1816.
5. John Farey, *Treatise on the Steam Engine* (London, 1827), Vol.1, p. vi.
6. Henry Farey's diary, 1820. Various entries for drawings he produced and sent to the engraver Wilson Lowry. The diary is unpublished and is in private hands in America.
7. D.S.L. Cardwell, 'The History of Technology – Now and in the Future' *Industrial Archaeology Review*, 1978, 2, p. 108.

8. Sadi Carnot, *Reflexions on the Motive Power of Fire: A Critical Edition with the Surviving Manuscripts*, Robert Fox (ed. & trans.) (London, 1986).

9. Commissioners of Patents, *Abridgements. Class 49, Steam Engines*, 2 vols (London 1866).

10. A.P. Woolrich, 'John Farey and the Smeaton Manuscripts', *History of Technology*, 10, 1985, p. 188.

11. A.P. Woolrich, 'Mr Newcomen's Friend: Dr John Allen of Bridgwater (1660–1741)' *Proceedings. Somerset Archaeological & Natural History Society*, 142 (Taunton, 2001).

12. John Farey, *Treatise on the Steam Engine* (London, 1827), Chapters VII and VIII; Graham Hollister-Short, 'Invisible technology, invisible numbers', *Icon*, 1995, 1: 132–147.

13. John Farey, *op. cit.* (12) (London, 1827), 531–2.

14. H.S. Torrens, 'Jonathan Hornblower (1753–1815) and the steam engine: a historiographical analysis', *Perceptions of Great Engineers: Fact and Fantasy* (Liverpool, 1994).

15. H.S. Torrens, 'Some newly discovered letters from Jonathan Hornblower (1753–1816)', *Transactions of the Newcomen Society*, 1984, 54: 189–200.

16. James Watt, *The articles steam and the steam engine written for the Encyclopaedia Britannica by the Late John Robison ... with notes and additions by James Watt* (London, 1818).

17. John Farey, *Treatise on the Steam Engine* (London, 1827), p. 650.

18. House of Commons Sessional Papers, 1829, III, 415–676; Eric Robinson, 'Matthew Boulton and the Art of Parliamentary Lobbying', *Historical Journal*, 1964, VII, No. 2, 209–229.

19. Bodleian Library, Rigaud 39, f42–43, dated 26 December 1838. I am grateful to Professor Hugh Torrens for this reference.

20. This is confirmed by the fact that the French writer M. Christian, in his *Traité Méchanique Industrielle* (Paris, 1822–1825), illustrated a Woolf-type boiler which is identical, not only in the boiler proportions, but also those of the flues, with part of one of Farey's drawings for the *Treatise.* Whilst the boiler dimensions might be known from other sources, the flue details could not, and so this points to Christian's drawing having a common original with Farey's. The plate list and synopsis mentions a Fenton Murray and Wood tank-bed engine which was sketched by Henry Farey in his diary. A plate of this engine, identical in detail and dimensions, was engraved by Wilson Lowry for Stuart's *Anecdotes of the Steam Engine.* Another Fenton Murray and Wood engine noted in the synopsis and plate list was also sketched by Henry Farey. This was house-built and an engraving of this engine was used on publicity material issued by the firm.

21. Farey noted the progress of the printing of Vol. 1 of the *Treatise* in the annotated copy of the book he presented to the Institution of Civil Engineers in 1842. The same annotation appears in the incomplete copy of the book now in the British Library, London.

22. Reading University, Longmans Archives; Accounts, Divide Ledger A2, fol. 17.

23. Reading University, Longmans Archives; Letters 32B.

24. *Ibid.*, Letters 32D.

25. *Ibid.*, Letters 36B.

26. *Ibid.*, Letters 37A.

27. *Ibid.*, Accounts, Divide Ledger A2, fol. 17.

28. While Farey was writing the second volume, Stuart (1829), Ross (1829), Milne (1830), and Nicholson (1830) also appeared. H.W. Dickinson and A.A. Gomme, 'Robert Stuart Mickelheim', in *Transactions of the Newcomen Society* 1941–2, 22, p. 163; E. Wyndham Hulme, *Statistical Bibliography in Relation to the Growth of Modern Civilisation* (London, 1923).

29. *Repertory of Patent Inventions* 1828, VI, p. 192.

30. *Ibid.*, 359–64.

31. L.G. Booth, T.J.M. Sutherland, N.S. Billington, 'Thomas Tredgold (1788–1829): some aspects of his work', *Transactions of the Newcomen Society* (1979–80), 51, 57–94.

32. Reading University, Longman Archives, Divide Ledger, D2, fol. 344.

33. *Ibid.*, Letters 89D.

34. *Ibid.*, Letters 89E.

35. *Ibid.*, Accounts, Impression Book 8, fol. 191v.

36. The publisher's accounts for Bourne's book, also at Reading University, show an entry for a plate being borrowed from Farey's *Treatise* (Plate XIV) as well as the legal agreement between Farey and Longmans for their use of the material.

37. *The Observer*, 29 December 1844, p. 1, col. f.

38. A copy of this inventory is in Birmingham Reference Library. (Information from Hugh Torrens.)

39. Longmans book catalogue for 1835, p. 9, advertising Volume 1 of the *Treatise* noted that Volume 2 was in the press. I am grateful to Rodney Law for this reference.

40. Reading University. Longman Archives, Divide Ledger A2, fol. 36. The proof of the book in the British Library contains manuscript continuations to the tables of engine performance in Cornwall.

41. Reading University, Longman Archives: Accounts, Miscellaneous Expenses Ledger, A2, fol. 36.

42. *Ibid.*, Accounts, Divide Ledger 5D, fol. 139.

43. Institution of Civil Engineers Archives. Letter from Messrs Janson, Cobb & Pearson, dated 19 November 1913.

44. Barnard's letter is with the proof of Farey's book in the British Library.

45. Longman's letter is with the proof of Farey's book in the British library.

46. Rhys Jenkins, 'An Unpublished book on the Steam Engine: Trevithick', in *The Engineer*, 21 May 1920.

47. Rhys Jenkins, 'A Cornish Engineer – Arthur Woolf', in *Transactions of the Newcomen Society*, 1933, 13: 55–73, Note 1.

48. Jenkins's papers are in the Science Museum Library. Smith's papers are split between the Science Museum Library, the National Maritime Museum and the Naval Engineering School at Gosport, Hants.

49. Alston encouraged the late George Watkins (1904–1989), Britain's leading authority on the steam engine, when he was a young man. George Watkins in turn encouraged me when I began to take an interest in the field. Alston's collection of steam engine models then displayed in motion in Bristol Museum was a perennial source of interest to me during wet school half-days.

Figure 2 FT 26 Print of a model based on Farey's notes about the Trevithick engine which was made by the author for Newcastle Museum of Science and Engineering.

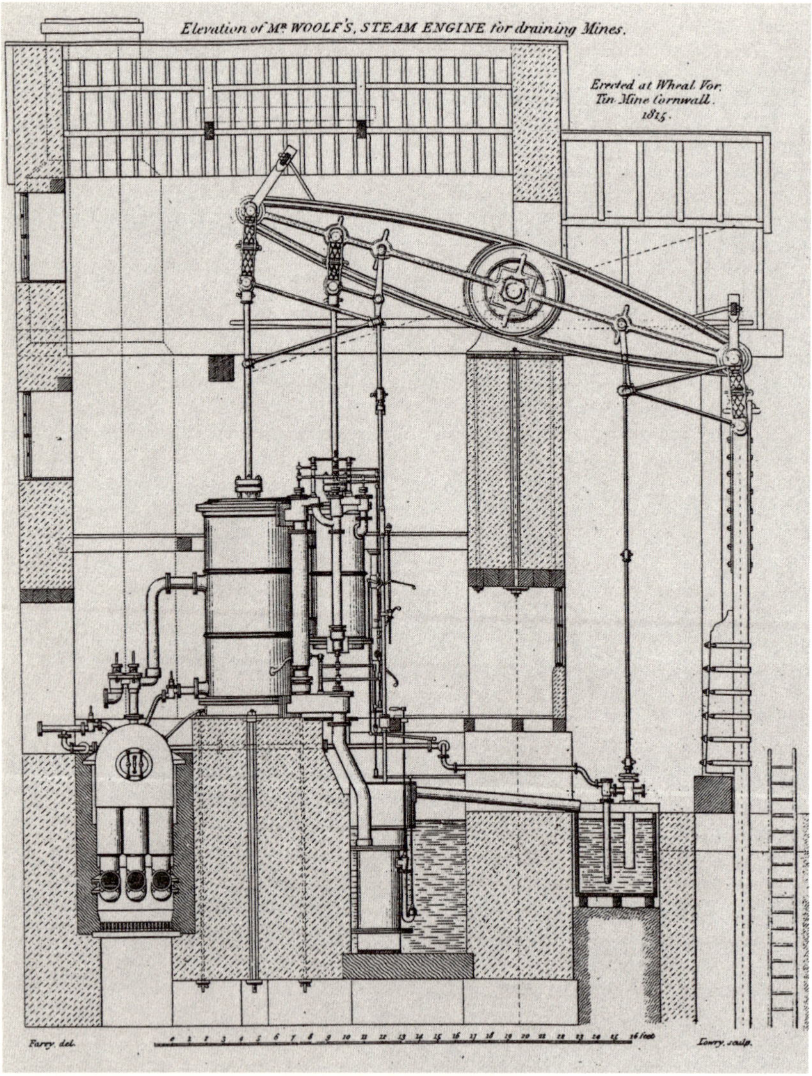

Figure 3 FT 31 Wheal Vor engine – longitudinal vertical section.

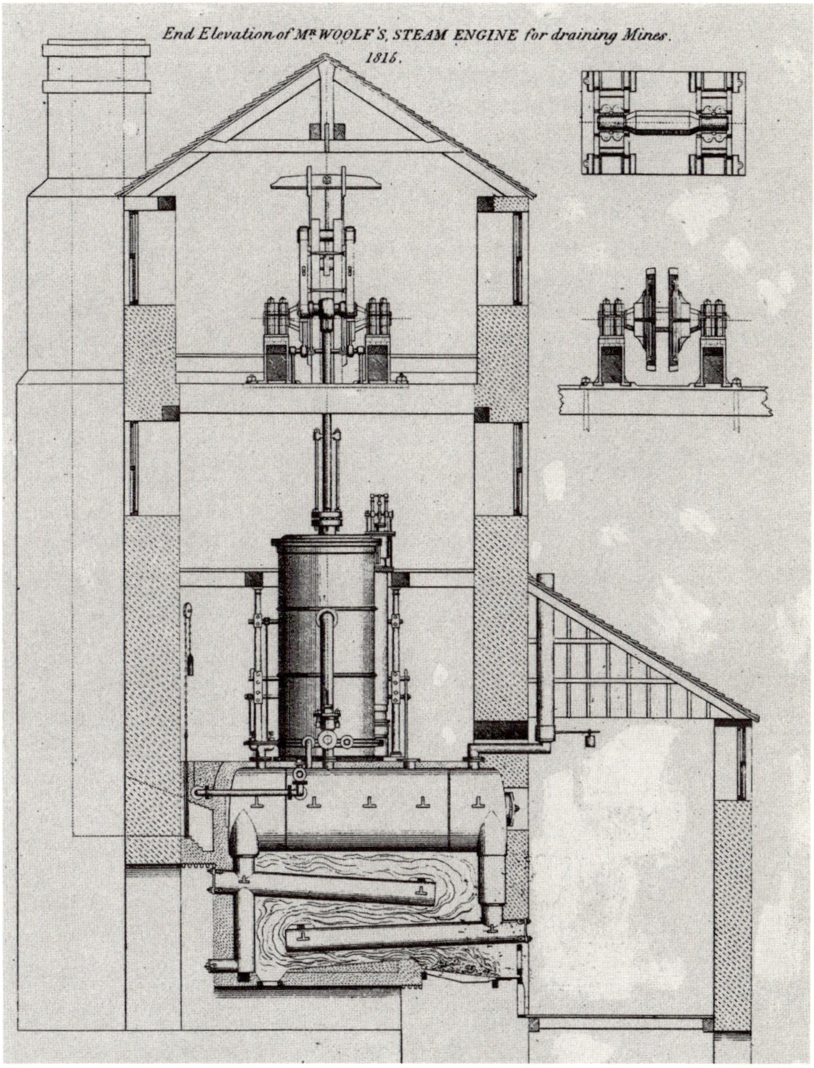

Figure 4 FT 32 Wheal Vor engine – transverse vertical section.

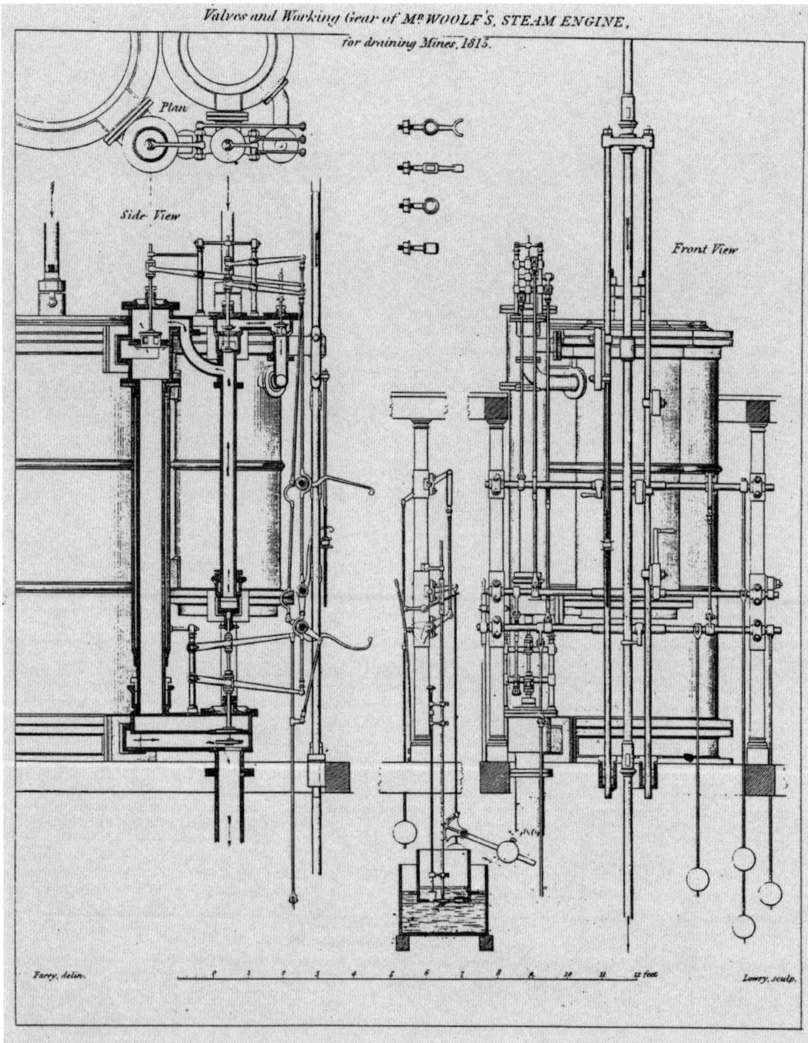

Figure 5 FT 34 Wheal Vor engine – elevation of cylinders and valve gears.

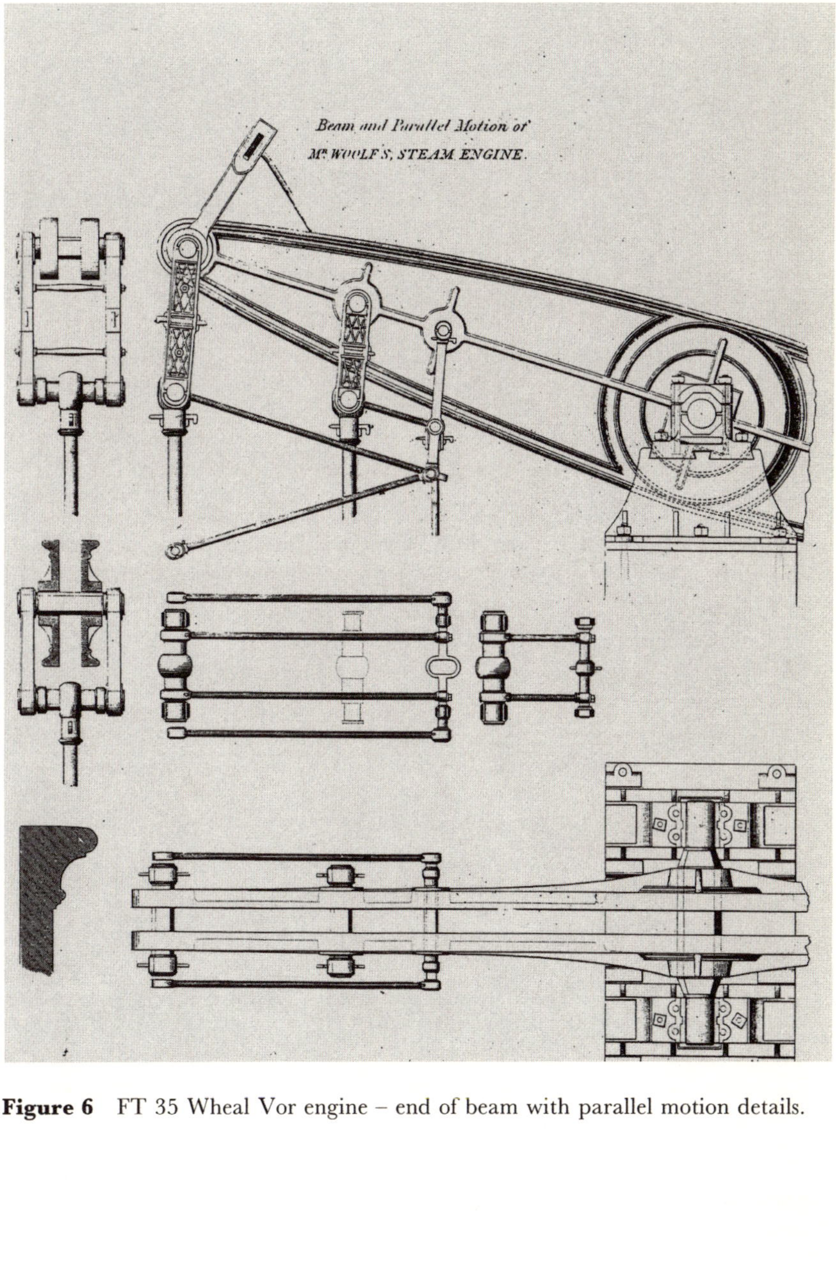

Figure 6 FT 35 Wheal Vor engine – end of beam with parallel motion details.

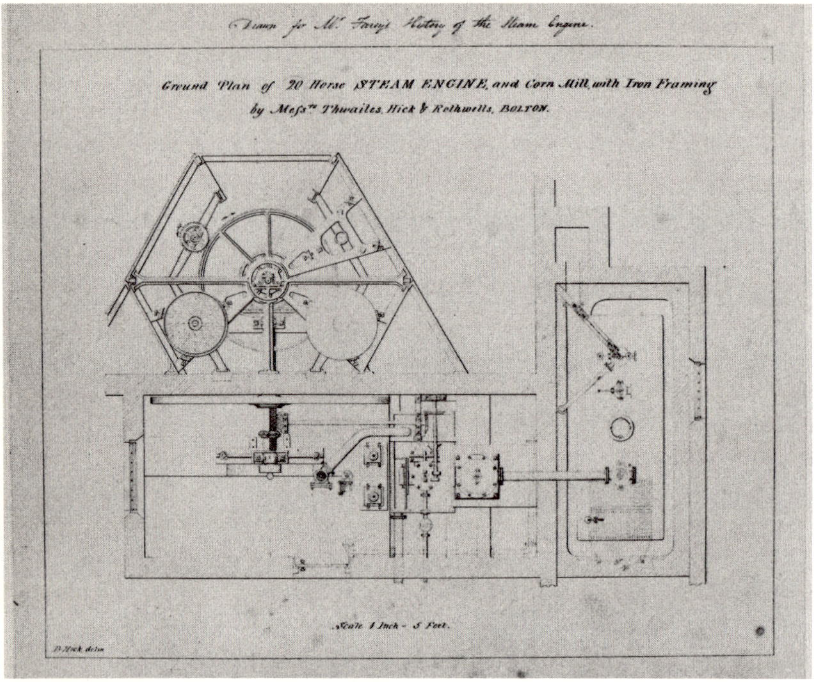

Figure 7 FT 38 20HP Flour Mill engine by Thwaites, Hick and Rothwell, Bolton, Ground Plan. Science Museum/Science & Society picture library.

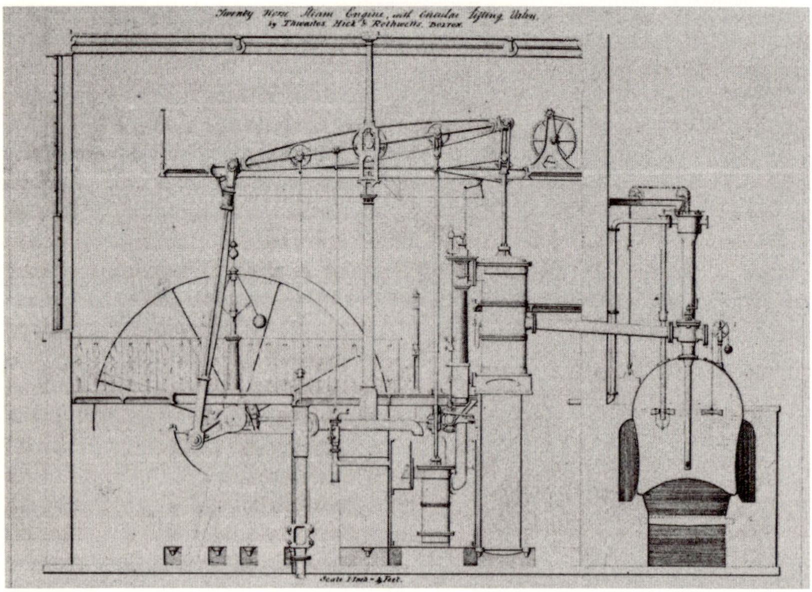

Figure 8 FT 39 Elevation of engine. Science Museum/Science & Society picture library.

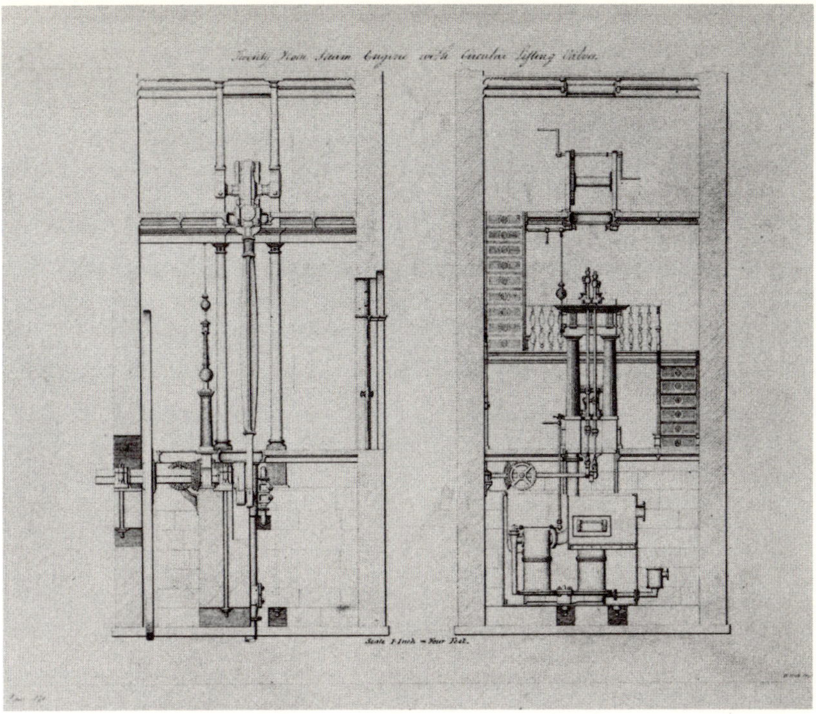

Figure 9 FT 40 End elevation of engine. Science Museum/Science & Society picture library.

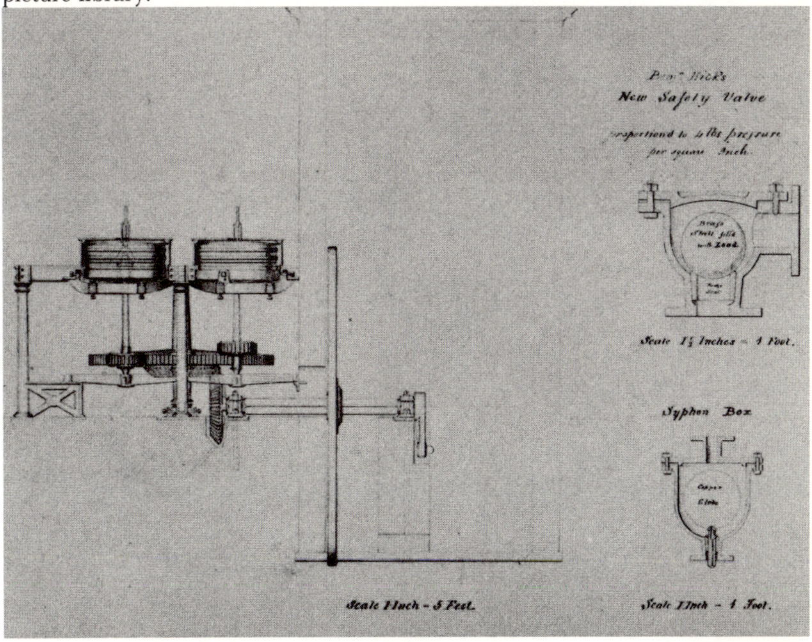

Figure 10 FT 41 End elevation of mill and gearing. Science Museum/Science & Society picture library.

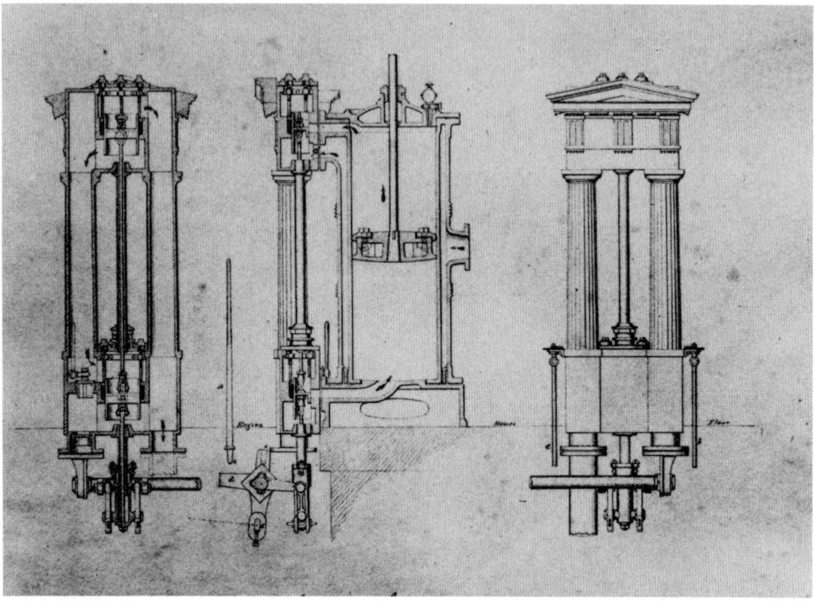

Figure 11 FT 42 Section of cylinder and side pipes. Science Museum/Science & Society picture library.

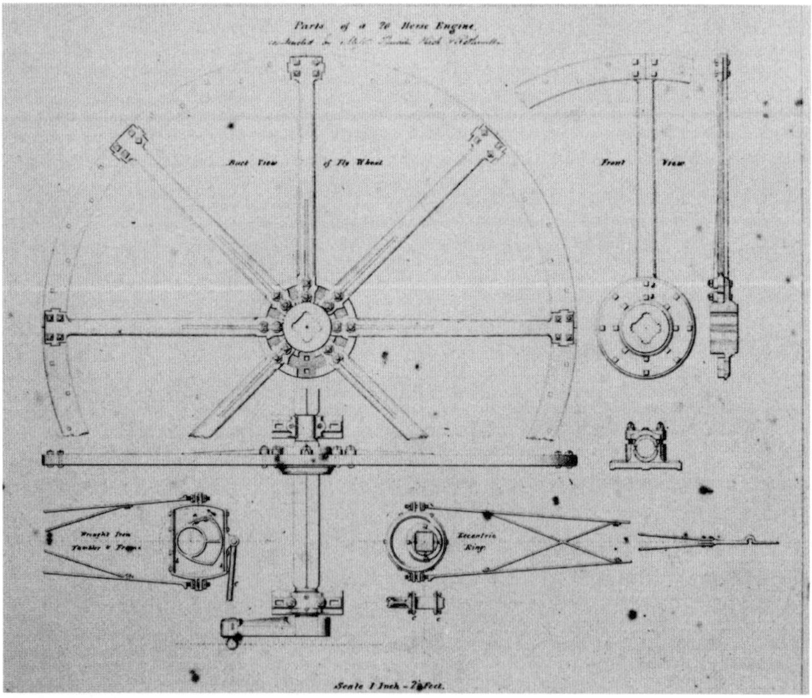

Figure 12 FT 43 Parts, beam, connecting rod, flywheel. Science Museum/ Science & Society picture library.

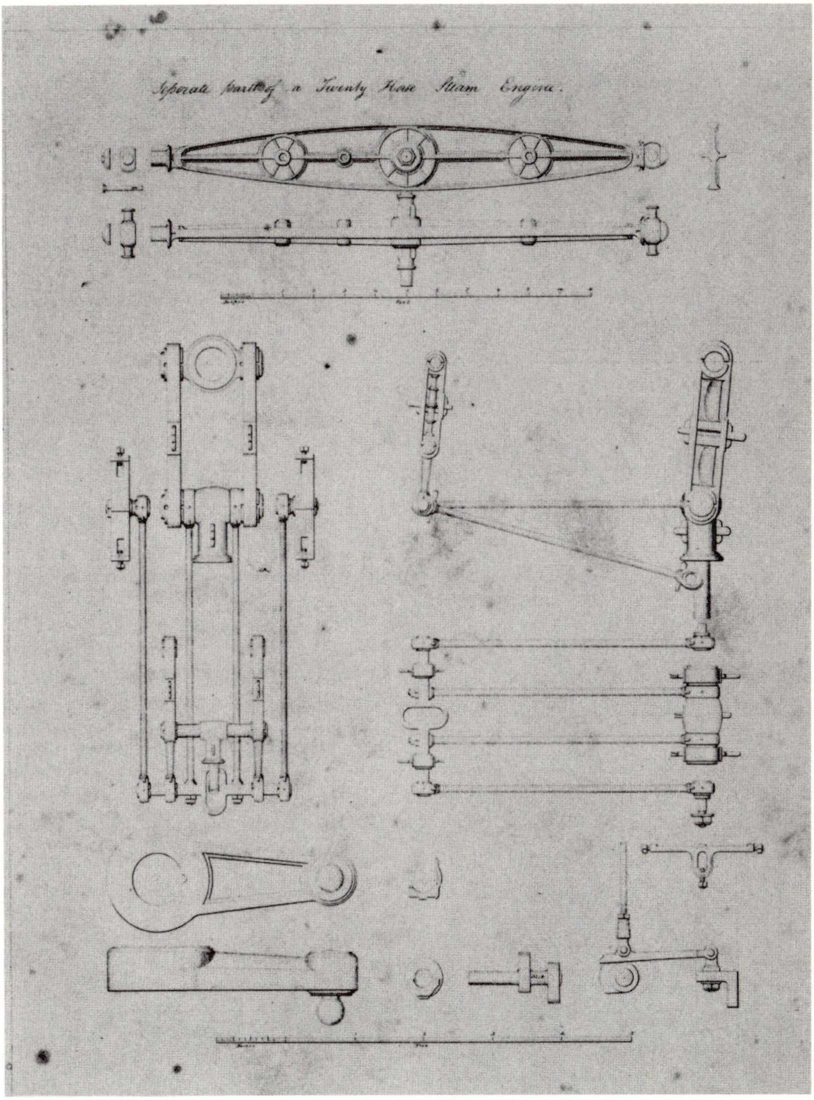

Figure 13 FT 44 Parallel motion. Science Museum/Science & Society picture library.

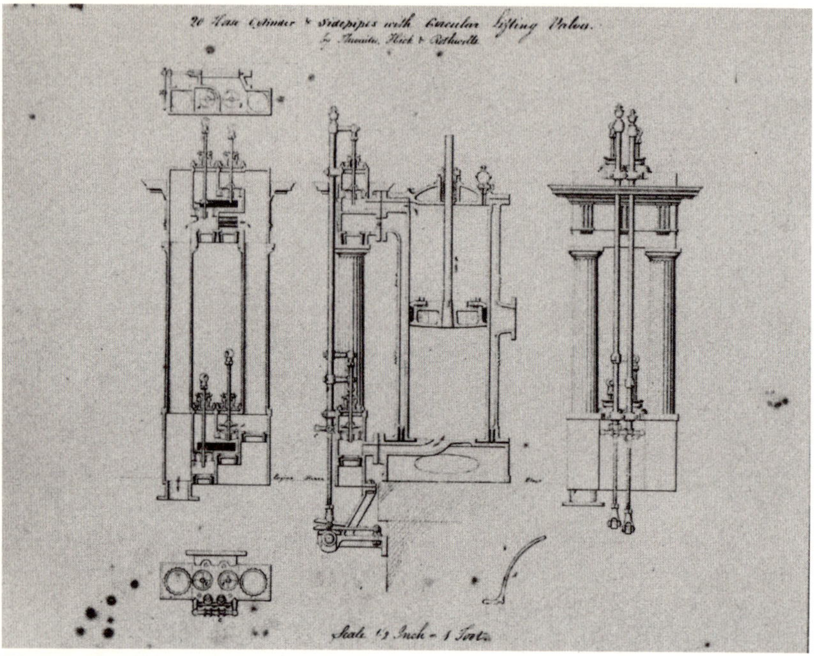

Figure 14 FT 45 Section of 20 h.p. cylinder and side pipes and slide valve. Science Museum/Science & Society picture library.

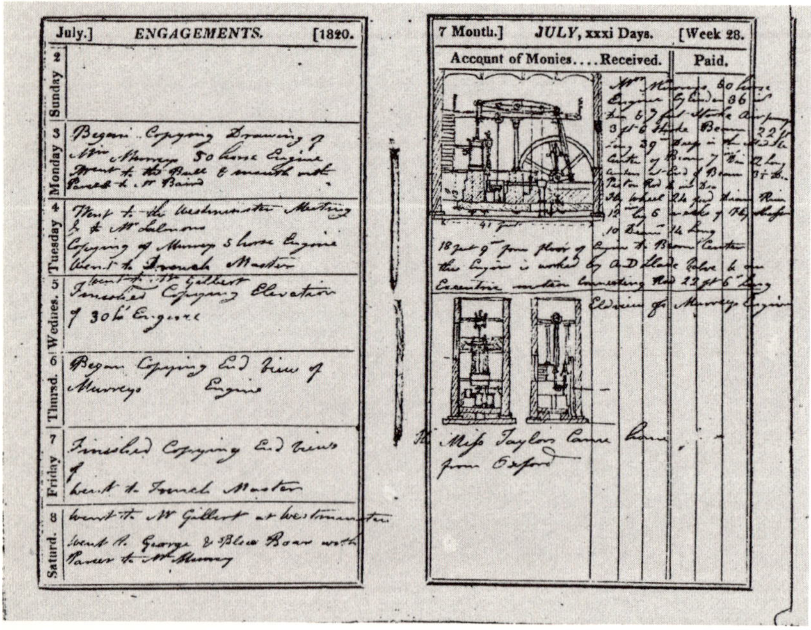

Figure 15 FT 46 30 h.p. rotative engine by Fenton, Murray and Wood, 1820. Drawing from Henry Farey's diary.

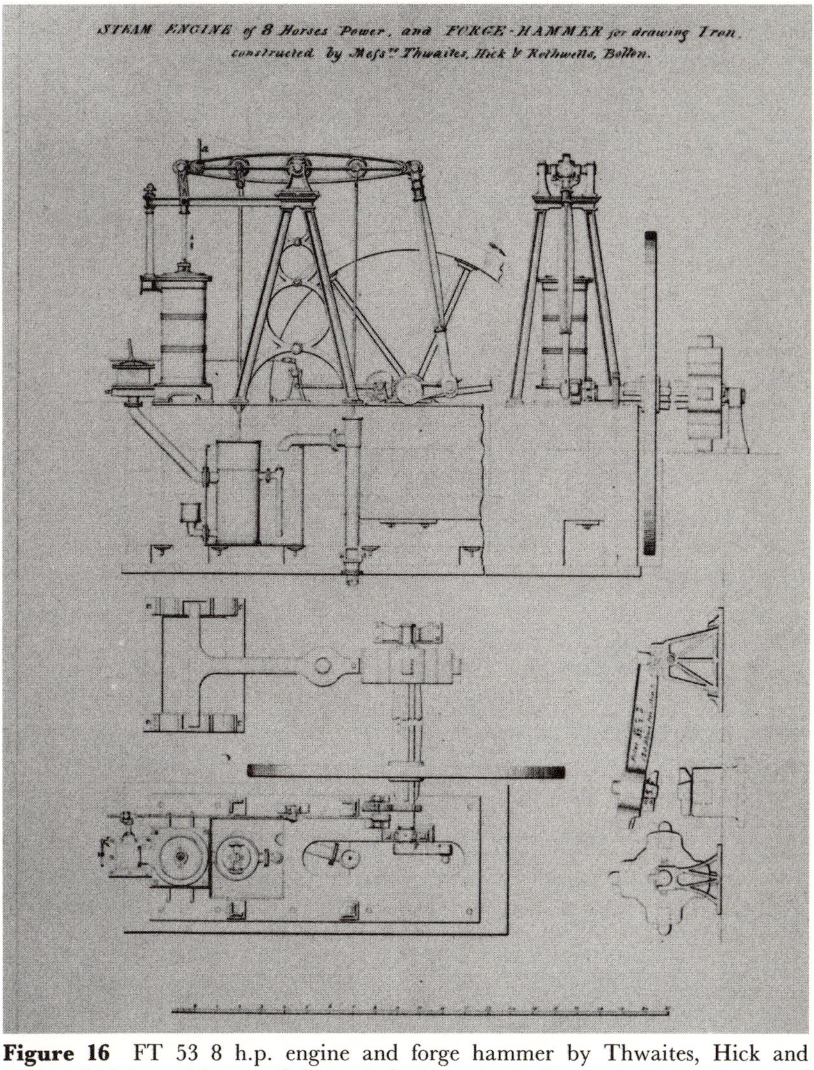

Figure 16 FT 53 8 h.p. engine and forge hammer by Thwaites, Hick and Rothwell. Science Museum/Science & Society picture library.

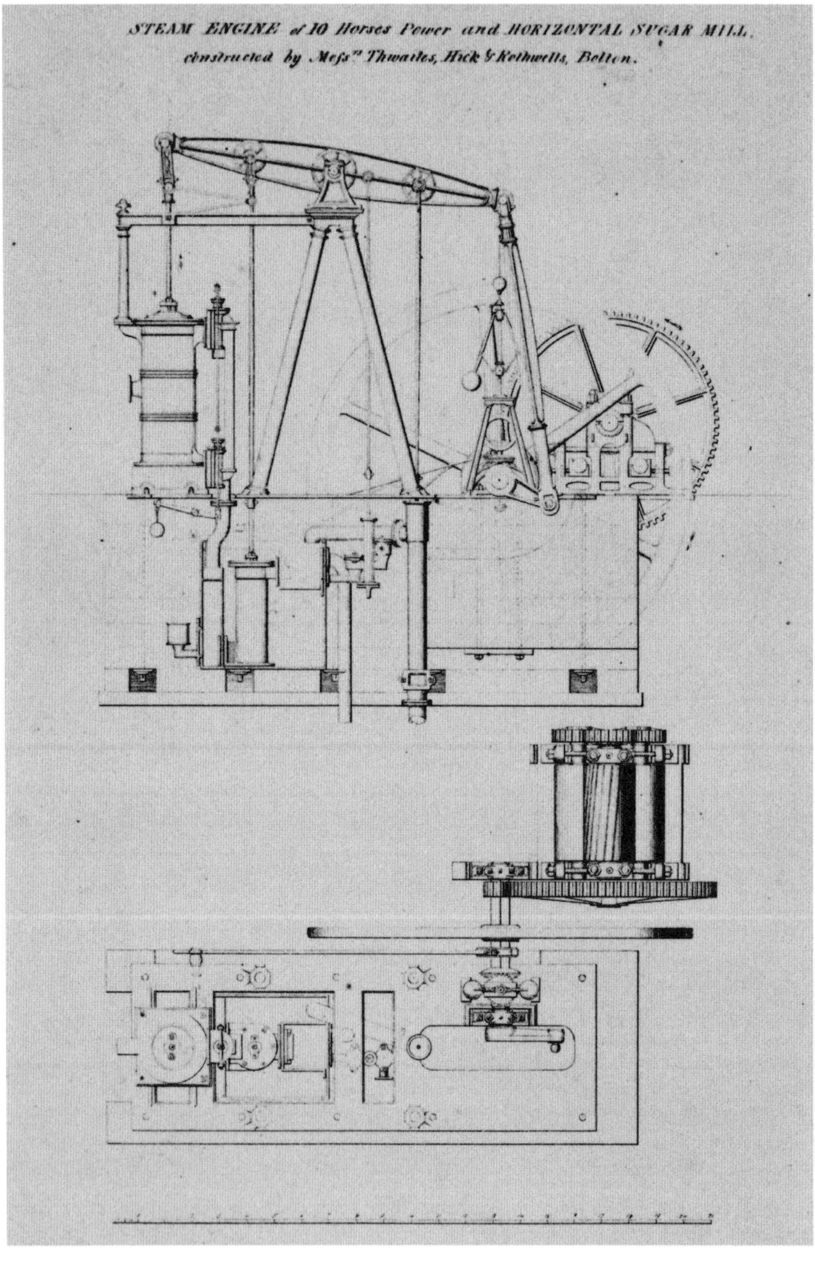

Figure 17 FT 55 10HP engine and sugar mill by Hick and Rothwell, 1824. Science Museum/Science & Society picture library.

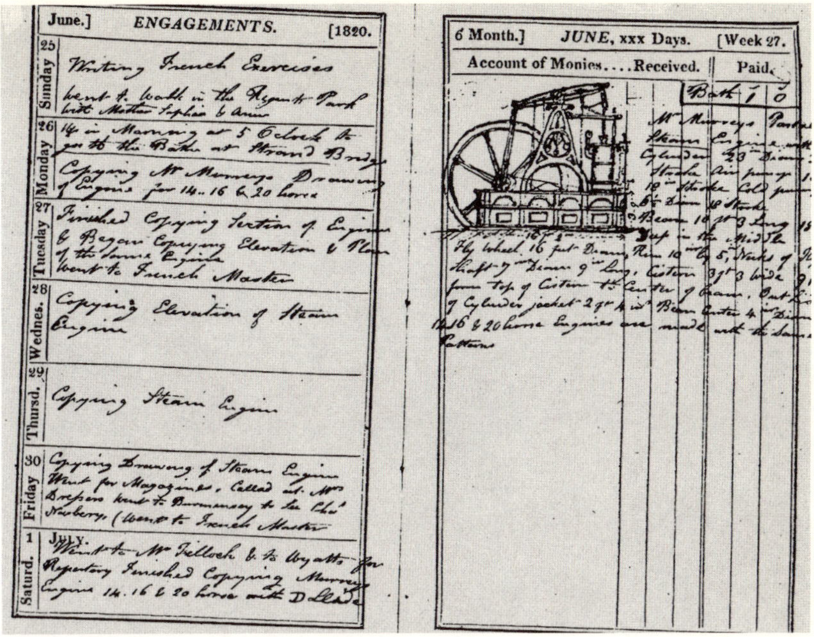

Figure 18 FT 56 14HP Fenton, Murray and Wood, 1820. Drawing from Henry Farey's diary.

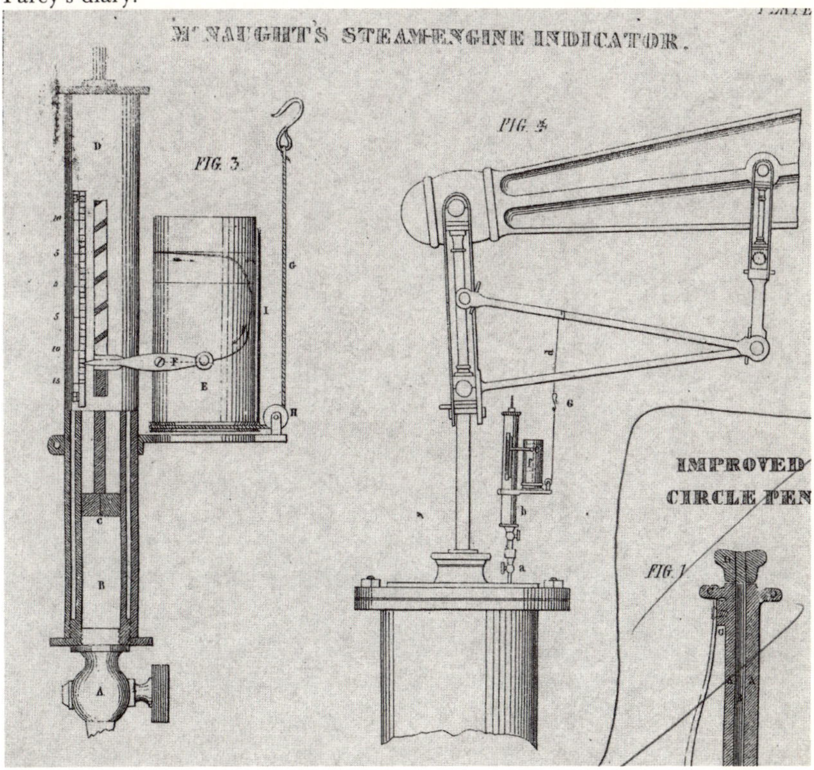

Figure 19 FT 59 [Not in plate list] Steam Engine Indicator. No drawing by Farey of the indicator has been found, so the one selected here shows the pattern of instrument introduced by Farey and manufactured by McNaught.
Source: *The Practical Mechanic and Engineer's Magazine* (Glasgow, 1842), 1, Plate 9.

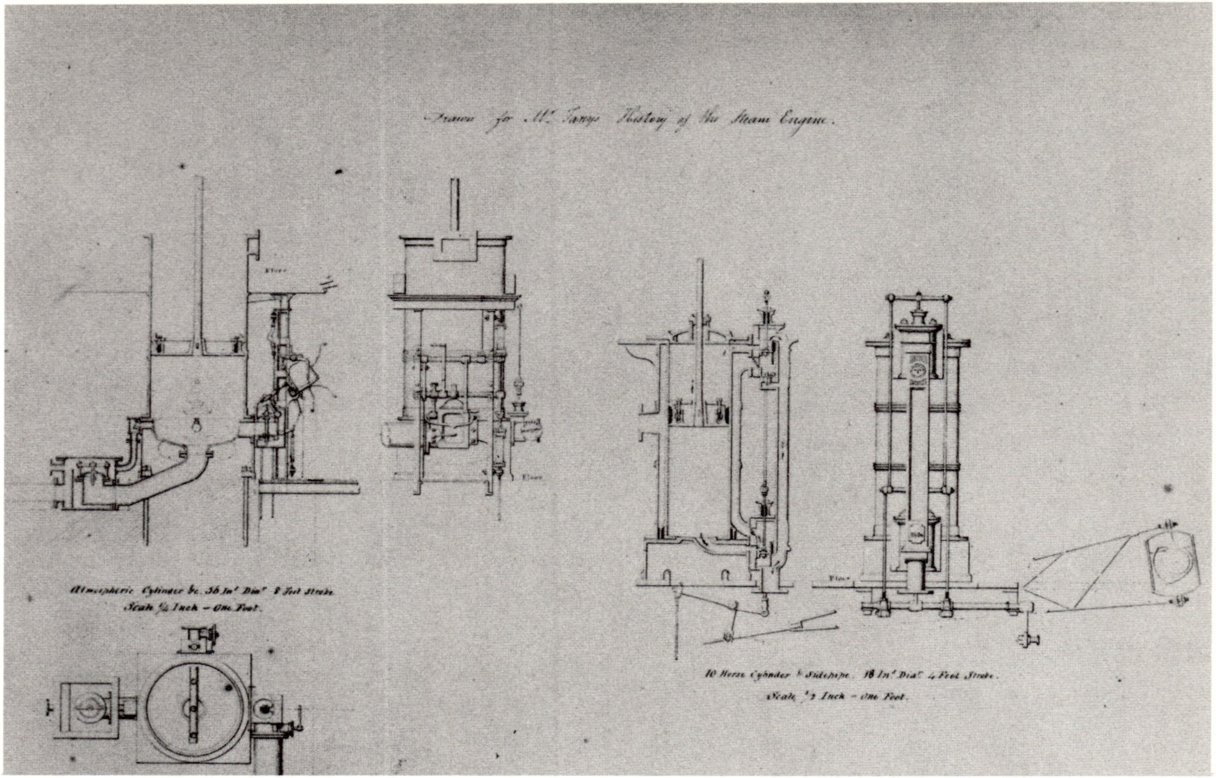

Figure 20 FT 71 Hick Atmospheric. Science Museum/Science & Society picture library.

Summer Breeze and Flushing Toilet: Deception and Reality about Our Technical Past

MICHAEL FORES

With regard to authority, it is the greatest weakness to attribute infinite credit to particular authors . . . For truth is rightly named the daughter of time, not of authority.

Bacon (1620)[1]

the historian . . . is aware that what he does to his so-called authorities is not to believe them but to criticize them.

R.G. Collingwood, *The Idea of History* (1946)[2]

1. OF ROBBERY, SKILL, MAKING, FAITH, LIFE, DECEIT AND MAGIC IN THE HUMAN REALM

In *Technics and Civilization* (1934), in a programme of work also producing *The City in History* (1962), Lewis Mumford wrote thus of the human, technical past, of which the inception of electrical engineering dealt with below forms a part: 'Robbery is perhaps the oldest of labor-saving devices, and war vies with magic in its efforts to get something for nothing – to obtain women without possessing personal charm, to achieve power without possessing intelligence, and to enjoy the rewards of consecutive and tedious labor without having lifted a finger in work or learnt a single useful skill.'[3]

Mumford had some strange ideas on such topics.[4] His 'technics' comes from a sense of awe and special respect, as shown in a chapter with an anachronistic title, 'Cultural Preparation'. How did Homo, at any time in the past, know what specific future to 'prepare' for? Under 'The Duty to Invent', we can read: 'Technics is a translation into appropriate, practical forms of the theoretic truths, implicit or formulated, anticipated or discovered, of science.'[5] In this viewing, where is personal skill, and playing hunches about the future? How wise too is a claim, say, 'X used unknown Theory Y successfully'? This is like Kuznets's absurdity, also from the USA (in the self-proclaimed New World), in his rationale for naming a 'scientific epoch' (seen as a New World itself), that Watt's steam engine was one of the first 'science-based' inventions; while this writer knew the 'base' was not then available.[6] Why should we accept 'intelligence' as the cause of gaining power over others? What of the arts of plotting, dissembling and statecraft, which may use intelligence but scarcely this trait exclusively?

Bizarre too is Mumford on magic, though this may come from 'history of science', where awe, anachronism and fudge are rife. Thus, there is Shapin, also from the New World, on *The Scientific Revolution* of Newton's day, when 'science' was rarely discussed. We read of inquiry into 'the natural world', though some of this is artificial; and 'science' is a type of action, though *Revolution* focuses mainly on changes in knowledge.[7] Meanwhile, of 'preparations' for an unknowable future: 'Between fantasy and exact knowledge, between drama and technology, there is an intermediate station ... magic... [where] the general conquest of the external environment was decisively instituted.' This is about knowing the human realm as well as moulding bulky items within it: 'the magicians not only believed in marvels but audaciously sought to work them: by their straining after the exceptional, the natural philosophers who followed them were first given a clue to the regular.'[8]

Well, perhaps so, on regularity. Newton, pious Christian, mathematician, alchemist, and hero to those boosting an Age of Reason, may have heralded, with the magicians, an orderly, but ill-named, 'natural science'. Yet magic differs. Magicians deal with items classed as 'scarce goods' by today's 'economists', being things which Homo has to move around its planet forcefully, if they are to be moved. Magicians make and move around massive items, for which a value is shown in markets, such as useful, bulky carriages from pumpkins, in which young girls can go to the ball. 'Scientists' and their forerunners, including the 'natural philosophers' of Newton's day, have fashioned a different product – articulated knowledge, an abstraction from physical reality, a valueless, non-material item which weighs little even in its written form. That said, a case can be made from *Technics* about those bulky items which 'natural scientists' study. Science (artificial like a carriage, but a 'public good') varies, as its valueless Statement S (a log and re-presentation of the world, which is not owned and is not subject to trading or robbery) does not alter its state or location on transfer from A to B.[9] For 'scarce goods' in Homo's milieu – engines, guns, jewellery, cooked food – new, valuable 'presentations' there, these conditions hold: If scheming Homo A or Collective C is not satisfied with a re-presentation of Item X, such as an artist's 'likeness' of it, or a log about X from a scholar; if A or C is not willing to pay the market price for X, when its owner will sell (and cede ownership); if A or C will not rob its owner, or hire someone with magical 'powers' to make X theirs; so, A or C will have to consider a process of making X, in the manner now garbled as 'technological' (an '-ology' whose main products are not in the form of articulated *logoi*/words/laws).

A case set out in *History of Technology* has been this: Those sailing under this flag have tended to rob Homo – an animal which 'has designs' on things, uses items 'instrumentally', strives to 'engineer' preferred 'ends', is often light-hearted and usually a believer – of technical skills. Human skill, which must seem magical to other animals, has been argued away in print. Authors have gone some way to dispatching the mark of Homo, by taking it down the valley of the shadow of death by scholarship, inventing a meta-world on paper, tending to produce a History of Marvels.[10] A first aim, then, is to expand a critique of a genteel 'history of technology', named as if engineers deal with *logoi*, as if Homo were not involved in all that noisy action – forging, forming, building, stamping, joining, fitting, fashioning – so altering a concrete reality around us, as evident from observation.[11]

Homo is a meddling constructor; and what is now 'engineering' has long been at the core of its culture: hence the Stone, Bronze and Iron Ages and an Industrial present. Making valuable engines is different from making valueless science. Thus Forbes, *The Conquest of Nature* man, who should have known better as a chemical engineer, was wrong to lecture engineers on their 'discoveries' and a need to discuss for the general good.[12] Asking -ologists of any stripe to make engines is like asking war correspondents to run military campaigns – only in dreams will this work out well. Better to tell of the Clothed Species, a forceful actor in its locality, a tool-using agent of making and mending. Here is a risk-taker, as shown in a study of engineering design,[13] an animal moaning of 'problems' faced, but lost without them, a dealer in the relatively unknown. Engines are fashioned from 'stuff' (*cf.* Homo, decked out in 'fabrics'). Engineers, such as E.W. Siemens (1816–1892), of a 'tolerably changeful existence', wanting to be a scientist – 'technical progress is only to be attained by the diffusion of scientific knowledge among technicians'[14] – have struggled fitfully to put together concrete items in saleable forms: once, returning to Berlin to study science, Siemens gave up his patents, with invention judged as too risky (38; these refs are to his autobiography). Engineers have striven in the conditions of the solids, liquids and gasses of school textbooks, those of the singular, everyday, tangible, mainly observable world of local familiarity. Call this World A1,[15] where items chosen for working may be worked-up things themselves – 'The first challenge of materials is to select the right materials for the job'; 'Few materials can be used in their raw state'.[16] This is a world whose traits are logged in such terms as hardness, density, strength, ductility, thermal and electrical conductivity, molecular composition and chemical stability, as well as the measurable 'three dimensions' (length, mass and time) of the 'physicist's' reality.

Current electricity, barely existent naturally, is interesting, given this invisible thing has, like 'force' in 'physics', notable physical effects (lighting lamps, altering the motion of inert bodies), and the first is used in power-transmission and the passage of information 'electronically'. As its name shows, current electricity is thought to be linked with 'static electricity'. It makes no difference if electric current is said in *Natural Science* to 'flow' from a 'positive terminal' of a 'battery'; or if the chic story of the day is that 'electrons', first 'discovered' at the Cavendish Laboratory,[17] go the other way. It is not important here to speculate on links between the two types of electricity, or on who invented a means for making current. Rather, note some effects of a link with 'magnetism'. Then try to recall those never-failing, risk-free 'experiments' at school with iron filings on paper, and a 'permanent magnet' making predictable patterns. Siemens worked on electricity for telecommunications;[18] and a difficulty was to make it 'flow' as textbooks now say it does. Life was like looking through a glass darkly. The idea of 'creative' Art and its Works has been in vogue in times 'modern'. As an engineer, Homo is the same in processes engaged as 'insightful Artists', dealing with bulky items in World Al.

We have had a *mélange* of logs bearing on this, some saying that a History T ('history of technology') came from a History S ('history of science'); and there is a History G (including 'world history') acting on each. None of these is discrete. All scholars, wielders of scalpels and pens,[19] examine the past of a single universe. 'Natural science' may be of the artificial; 'history' is inseparable.[20] Of our technical past it is worth excising much culture-bound clutter, defining Homo and its 'modern' state with Collingwood's disrespect as set out in the epigraph.[21] The idea of a thoughtful Homo, stemming from an Age of Faith and Imprecision, does not help. Section 2 will treat strange, piously told tales of 'discoveries' made by Homo, a moody organism of 'lively' actions.[22] In show-off Homo's world, even its salads have their dressings. Scholars may tag us all *sapiens*, stressing their own wares; but Homo has deceived skilfully, 'putting it on' (clothes, 'airs', 'acts', disguises ...); artefacts are tested, not for a means of their making – 'the seams should not show'. Rarely Humanism's individualist, Homo has gone about collectively[23] and clothed. Clothes are made-up items, used in sacred rites (priests being 'of the cloth'), not only to be 'in fashion'. The Clothed Species shows Homo facing materials, in war and peace, in good times and bad, as a creative believer.

For Bacon (epigraph), the authors erred. Of his Instances of the Road, they 'are such as indicate the gradually continued motions of nature. This species of instances escapes rather our observation than our senses; for men are wonderfully indolent ... consulting nature in a desultory manner ...'[24] Here is the legacy of 'a quest for the truth, no matter where it may lead ... [with] tentative steps [made] towards an open mind towards man and his nature...'[25] Hence Time, transcendental, personified in Section 3, 'About Time, Seams, Communications and Power in Berlin and Elsewhere'[26] – unlike 'scientific method', she never changes. 'Play something like the murder of my father/ Before my uncle. ... The play's the thing/ Wherein I'll catch the conscience of the king' (a Renaissance author).[27] With Time, daughter of Aeschylus and an unknown deity, as our guide, a play may serve. Engineering concerns bulky, scarce goods, doing for one dollar what any fool can do for two. Engineers, unlike scholars, have been engaged commercially, worrying about costs and clients, often in areas named now as Finance, Marketing, Sales. Poetic mimesis aids a grasp of the past. Time has seen it all, and will tell ... For the Cambridge *History*, the 'new industries – such as the electrical industry' of the later nineteenth century, as in Berlin, 'originated wholly in scientific discovery'[28] – 'Closer relations of science and technology'.[29] For an English-speaking Mumford, it was Michael Faraday in London, not Siemens in Berlin, who invented the dynamo.[30] Electric current is a made-up item.[31] Study of Siemens's life shows engine-wrights welcoming, as 'problematical', workable items, not as folk 'conquering nature'.[32] We may catch him at his first invention (in electrolysis), laying cables, at war: 'Although my inclinations drew me always to purely scientific investigation, yet I perceived that I must first of all turn my whole energies to technical work, as its results alone could procure me the means and opportunity for scientific work – and did in fact procure them.' (80). So electrical engineering started, to one of its main participants, contrary to the terms of the thesis spelt out in the Cambridge *History*, and to those of 'technics based on science' of *Technology, Science and History*.[33] With futuristic ideas of this type in print, Homo has lost touch with itself.

2. OF UNIVERSALS, 'DISCOVERIES', HOMOCRAFT[34] AND THE CLOTHED SPECIES IN VOGUE

'If I were you, I wouldn't start from here.' This 'Irish joke' has a useful message. The *History* starts from the European Renaissance, giving birth to an ostensibly secular Art.[35] Take its stance, of the 1890s from Acton, in dividing the past into Ancient, Medieval and Modern, rather as the newer *History* named a possible 'turning-point in the history of western society' at the time of that Renaissance.[36] The *History,* introduced by a Bishop of London, devised as part of 'a new science of history' (1957) when this subject was first offered for teaching and study, was to be a 'universal Modern History': 'the two main features of modern history are the development of nationalities and the growth of individual freedom' (re *The Renaissance*, a 'historical fact of signal importance').[37] Compare Plumb on 'the astonishing vigour and originality' of fifteenth-century Italy: 'Art reflects society ... and mostly it [Renaissance art] possessed a universality of appeal that shows little sign of diminishing ... the originality of its greatest artists lies ... in their concentration on the uniqueness of the human temperament and the significance of individual experience.'[38] A study of History G, made by 'the cooperative principle', was put out as superior to national stories: 'the long conspiracy against the revelation of truth has gradually given way'. 'A higher process', readers were told, was engaged about the past and its 'influence [on] the common fortunes of mankind'.[39] Hence, the *History*'s first chapter on the Age of Discovery, and the Cambridge Museum of Archaeology and Anthropology on Voyages of Discovery: 'From the 15th century Europe expanded beyond its medieval frontiers.'[40] With nations downgraded, contributors must have been asked, by Acton and others, to tell of more human freedom.[41] In Cambridge-speak of heroic abbreviation, say, 'Europeans no longer hemmed in: discoveries of distant lands and old bounds loosened'.[42]

Yet no one need have forged, in or outside Baconry, an 'empiricism', or need have done more than ponder critically 'authorities' in Collingwood's mode (epigraph),[43] to wonder: if the *History* is of the full human past, why start from the Renaissance, specific to European experience?[44] Why be so parochial to class, as heroic Voyages of Discovery, only those starting from there? Why be so keen on progress to 'individual freedom' for Homo, when major parts of successful life for all animals have involved coping with nameable unfreedoms? Limitations of bodily and mental abilities; the operation of the Laws of Nature, whereby massive things are most tricky to deal with; how animals must deal with the wilful aims of others on a crowded planet.[45] For Homo, there are the restrictions too of faiths and cultures, providing guides to action and thought; how people often yearn, expressing love for them, for the constrictions of 'being at home' in a single place. There is the trait of the Clothed Species, unfree to go naked in public, or to sport the dress of rival groups, happy normally to abide by the limits of using syntax and the meanings of words (subjects agree with objects; '-ologists' make and tend *logoi*; a cat is a cat, except in fairy-tales) – parts of the Iron Cage[46] of human habit and culture. Then why, as well, try to dress up a 'universal' *History*'s second chapter in those histrionic terms of the New World,[47] as this effect was so for folk 'discovered' by marauding, crusading, 'pioneering' Europeans of *c*1500? As a yet-to-be-named America was not a 'discovery' to (say) those of 'the Sioux nation' (mistakenly called Indians by Europeans), inhabitants of the same world as such mariners as C. Colombo and F. da Magalhaẽs (Magellan), neither of them cartographers pursuing 'geographical discovery'; they were, to the Sioux, European ways and outlooks which were newly found. Thus Europe's Old World was a New World, within our single planet, to those suffering from being 'discovered' by Europeans.[48]

Of the Cambridge *History*'s whiggery – 'indeed all history must tend to become more whig in proportion as it becomes more abridged' (Butterfield) – there is the irony that, while one trait of this school was, for this writer (himself, a prominent Protestant), that 'progressive' whigs have backed Protestants against 'regressive' Roman Catholics, Acton, criticized as a whiggish historian, was a prominent Catholic.[49] But the reason for citing it here is simpler: with the 'modern' clerks in unholy orders drifting from Homo, in 'histories' the grandest statements are often items forged to make the unbelievable seem true.[50] Take an idea of change from Historians T, most active in the USA, the self-styled Land of the Free whose ideologues use the stagy New World tag to claim separation – given the Cambridge *History* puffing Bacon, with 'the most powerful mind of modern times', who had studied at this Cambridge. Bacon, 'this prince among thinkers', was interested in America with its newness. The Old Europe of *c*1600 was a place of cultural malaise, making it hard to see things correctly: 'America was certainly "the greatest birth of time"; Bacon applied these words to the philosophic system of which he was the founder.'[51]

We have been told by Edwin Layton, from the USA, as part of a tradition of writing for what might be called the Theatre of the

Bourgeoisie, that in the nineteenth century there was generated a 'scientific technology', puffed in the spirit of the Cambridge *History*'s universalism, with a new mode of action forming a technical New World.[52] Yet there has been no obligation, explicit or implied, on those making and mending engines (bulky, scarce goods) to log on to paper (or on to other two-dimensional surfaces) what is used of such abstract, valueless science-as-knowledge, which no one can own. If such knowledge of 'science' is used, this may be so 'without thinking'. 'Scientific technology' cannot then be identified for science-as-knowledge. If – Hey Presto! a magician has waved his wand! TRANSFORMATION! – 'science' is sought in science-as-an-activity, as implied from the title of one of Layton's articles in *Technology and Culture*[53] from the Land of the Free, the cold morning light shines awkwardly. Since there are no norms telling Homo how to 'be scientific', a 'scientific technology' is not discernible.[54] Besides that, asking for this is like expecting warriors to run their campaigns like war correspondents. To Layton's dismay, the present writer 'rejects the idea that the work of engineers such as Watt produced a scientific transformation of technology [here, an activity]'.[55] The idea of 'scientific technology' is thus well classed, in Collingwood's terms, as a thing of uncritical 'belief'. In Bacon's terms, it is the progeny of 'authority', not of 'time' (epigraph).

'Art, like science [each a Western affair], is international' (*The Golden Century*); and both are likely to find favour with those schooled in 'universal Modern History'.[56] The Renaissance, often said to have been the forge of 'modernity', has been linked with 'the discovery of the world and of man' – women, countryfolk and workers need not apply – and with a fiction, 'Columbus's discovery of America'.[57] It has also been linked with the onset of an overtly secular Art and/or 'fine art' – plays at the National Theatre are not started by a Bishop of London – whereby 'the Artist' is assumed, by some of our species, to have special powers ('Art reflects society'). Martian anthropologists might wonder at logs set down on paper of 'discoveries' made by Homo – the *History* on 'scientific discovery'; Forbes's technical 'discoveries'; the puffing, in 'Art history', of a Renaissance-era 'discovery' of perspective – for Gombrich, a 'momentous' act, part of 'the conquest of reality', with the start of 'the architecture of a new era', one of 'lightness and grace' – a technique of representing 'reality' (a subject) as an object (a Work).[58] Compare comment on a painter 'Masaccio's art' and 'the perspective frame in which he placed his figures. We feel we can almost touch them . . .'[59] In the Earthlings' Natural History Museum in London, SW7, there is (there was, at the time of writing) a Discovery Centre for its young. Concrete topics of potential finding lie waiting, many to have a 'cover' removed and to be touched physically after an act of discovery. Revelation! Dis-closure! 'Dis-covery on Planet Earth', the Martians might judge, 'is the opposite of obscuring, shoving things under the carpet'. As is the case across Exhibition Road in the Discovery Wall in the silver Gallery of the Victoria and Albert Museum (the V & A), the National Museum of Art and Design. Slide a panel, blink an eyelid, and extant Artefact A is revealed. Disclosure! Appearance! Slide the panel back, blink again, and A is not to be seen. Magic! Disappearance![60]

Here are acts of an instant, akin to the forcing apart of a lid from a box, to show the box's concrete, physical, non-abstract contents. 'There it is!' Eureka! Why then try to tell tales of generic 'scientific discovery', as its 'philosophers' have done so,[61] when observation (scientists do this) shows that statements of science do not exist before a first articulation; and such 'discovering' typically takes more than an instant? Why should Daniel Boorstin, writing from the USA of *The Discoverers*, have chosen to tag anthropology as something 'discovered', and the author of *The Greeks* have suggested: 'Mathematics are perhaps the most characteristic of all the [''classical''-era] Greek discoveries'?[62] At what exact instant, of day or night, did these special events occur? 'Have those clothed Earthlings', the Martians might well ask, 'invented a Cult of Discovery, to avoid pondering the mischief they have wrought themselves on their once-green planet?' 'Newton discovered the Laws of Motion!' Then again, from a future biographer of Newton, and from the Cambridge press, in a chapter with an anachronistic title for the times, 'Organization of the Scientific Enterprise': 'While he was the Lucasian professor [of mathematics and in the five years before in Cambridge, Cambs] . . . Newton discovered the calculus, the composition of white light, and the law of universal gravitation.'[63] Where exactly did all these inventive actions occur? By the leafy banks of the sacred Cam, where the named items had been guarded from the indignity of viewing by the diligent Cam-maidens?[64] An answer to these queries lies with a sacred status awarded to items of Western culture.

In those times and places which the self-conscious 'moderns' have tagged 'pre-modern' and 'non-modern', the loudest noises (thunder, church bells), the most vivid colours (the sunset, religious paintings), the most beautiful sights (a rainbow), the tallest things in the locality (mountains, the spires of consecrated buildings), the most attractive smells (a rose) have been assumed sacred. A link has held that things made by Homo have been classed as profane; only those classed locally as 'natural' would normally be thought sacred, thus:

nature : culture /sacred : profane

The USA, dusted off as if it were, within our single planet, a New World, is an invention of the campaigning Clothed Species, as much so as Newton's Laws of Motion – each puffed as 'discovered' in an 'early modern' past. Their 'discovery', as noted in histories, makes them seem to have been natural in their first instance. Assume Modernity has not arrived. There was no Renaissance, with a shift in ideas of time's passing, when after-folk 'had to believe that not providence, or some blessed *fortuna*, but rather man's virtue and will and dedication control the outcome of events' (*The Renaissance Discovery of Time*).[65] It is an act of ascription, that each is natural and sacred in the culture, which has granted America, Newton's Laws, Time, Laws of Perspective, (parts of) Mathematics and Anthropology a protected status, as awesome things not to be dealt with in Collingwood's disrespectful, 'critical' mode. 'Olivier *is Hamlet* at the Vic! Newton's Laws *have* been discovered', readers of 'serious' scripts affirm faithfully, '*since so many important authors have said so*'. Normal disbelief has been suspended. We face, from apparently sober writers, an effect like that of 'Hamlet Without the Prince', on the Death of Skill.[66] It would be difficult, about human competence, for guileful Homo to 'make up' intentionally a greater charade than this idea of the Discovery of the Artificial. If Artefact A had been discoverable, why bother to spend valuable time and effort trying to make it? Hence a need to redefine the imaginative meddler, who has insisted on serving up such fare.

Of Homo's actions after an ill-tagged 'scientific revolution', I used the name Homo Faber to help strip away some mysteries.[67] In the recent West, 'human rationality has been totemized along with science'. Machines, predictable force-transmitters, act dynamically, so alternative paths are chosen from rules and information available *ex ante*. Adaptable, artful Homo, some of them said to be 'gifted', is prone to passion.[68] A good-humoured joker and a dark-hearted plotter of plans, believer and dissembler, an organism breathing, 'falling in love', waging war, praying, consuming (food, 'consumer goods'), defecating – so Homo has to take up the rest of actions shared with Machina, as the specialist in facing the unknown. Is Homo 'as good as gold' or 'as tricky as a wagon-load of monkeys', more warlike or peaceful, believer or sceptic, fun-loving or serious? We have sought self-knowledge with ideas of nature ('human nature'), reason ('rational science') and action ('thought *v.* action'). So what of advice given by A to B to 'act naturally' in unfamiliar places? This often signifies the opposite of what is said, 'do as the local culture indicates'. If A is told to 'be rational', this means 'do as I say, and don't bother to reason why'. A 'contrived argument' is inapt; though all arguments are, in fact, contrived by Homo. A 'fabrication' in words is judged untrue; yet we tend to wear clothes made from 'fabrics'.[69] Humans go in for 'active' engagement with their milieu; yet 'acting up' implies trickery – 'Actors may be brilliant virtusos, capable of astonishing us and moving us profoundly; but in their very virtuosity they are weavers of illusion and deceit.'[70] 'Practical' folk are looked down on, but not those working in a 'doctor's practice'. 'Engineering work' means the trains are not running; but Art Works are puffed as worthy.

What's in a name? For years, those seeking a glimpse of their kind have found 'serious' tales told of *Homo sapiens*, 'developed' from 'primitive' forms.[71] While it may flatter to see this; while it is agreeable for scholars to see qualities they offer: reckless portraiture is about. Homo is clever, with its large brain, and artful (good at joining things); but two World Wars in a century in its 'civilized' areas hardly hints of wisdom. Nor does Global Warming – the 'developed world' has been so forcefully capricious as to pump unsuitable gases into the single atmosphere of a planet, to the detriment of all its organisms. Have we been truth-seekers or tellers of tales? Homo, for whom to Shakespeare 'I love you' may be a prelude to rejection, or even to murder, has made many artefacts, concrete and abstract, more than in Forbes's *Man the Maker*; with its chapter on 'The Conquest of Distance' – is the shortest way from A to B now nil? Why not, too, The Conquest of Time?[72]

Homo has, in forging its milieu – as a confectioner operating outside the cakeshop, as a composer making other things than musical pieces – made Works of Art, roads, fields (agriculture), arms, craft (vessels), houses, other containers, 'laws of motion', other laws, money, institutions, medical remedies, mathematics, other languages, other signs, rites, machines, ceremonies, clothes, fabrics, toilets, anthropology, fairy-tales, histories, baked beans ... mainly in World Al. Treating 'decision-making' as a declension from 'rational' action (as under Management Science) involves a non-grasp that machine-dynamics are predictable, as opposed to Homo's. Yet crusading Homo has produced a household tale, 'How [rationally] Greek we were in the lab. today, my children', a fudging of reality – with adult work painted up as planned out.[73] If there is talk of 'scientific rationality', Homo is like a machine. If it is 'scientific discovery', there is little tricky to do. If it is 'scientific genius', it is too mystical for words. 'The scientist as a non-human. Discuss.'

Homo has made things judged locally as 'sublime' and has amused itself skilfully. There are Aeschylus's plays and the rumba, Newton's insights and tales of powerful witches. Focus here the fallacies of 'the rational scientist', 'Artistic genius' and *The Naked Ape*.[74] A campaigning maker, Homo has – whatever Age its historians say prevailed, whatever the conditions – rarely gone uncloaked: 'Man is born naked but is everywhere in clothes (or their symbolic equivalents) ... The adornment and public presentation of the body, however inconsequential ... it may appear to individuals, is for cultures a serious matter: *de la vie sérieuse*, as Durkheim said of religion.'[75] Clothes, a boxing-in of the body, are for show. They have been used, as in 'the Middle Ages',[76] to identify folk of classes, tribes, working groups. Students in jeans, soldiers 'in uniform', 'stuffed shirts' ... This has been the norm, not a fad of the day. For gregarious Homo, which can have fun and be 'deadly serious' (Art-love) – some of which claim, with the zeal of cultery, Individualism – some are 'uninterested in fashion'. But this is the organism distinctive for its sense of the collective, with its made-up cultures, its group rites, its making – as in a sporting of pennants, banners, emblems – of the symbols of belonging. Note the many words for collectives – team, mob, society, school, order ('ordained priests', 'of the cloth').[77] Then imagine those human pastimes, organized warfare and sport – the first a long-standing one, with the captains and the kings riding out in a fine expression of show, the second newer, though each passionately exercised – without the colours, bunting, chants ('WE WON!') and bodily clothing to signify 'which side you are on'. Homo naked has felt 'ill at ease'.

'What is this thing called love?' sang out a poetic Porter. 'This funny thing called love?' What too is this funny thing called the Clothed Species, assailed by Porter's 'mystery' of love? This is forceful Homo, prone to aggressive acts on others, often excused in terms of 'peace'. This is believing Homo, whose cult-adherents are thrilled by tales of selfless action, of saints and martyrs. This is self-deceiving Homo, claiming to love freedom, but scurrying off to the unfreedoms of culture, home and other loci of belonging. It is useful to amplify on means to make and tend the valueless science, given that artefacts (Engine E, Work A, Statement S) are the items to be assessed, not a means of their making.[78]

Science is part of a wider Show Business: 'We observe P ... analyse Q ... show R' – thus:

subject: object / the past: history / engines: design / the human milieu: science[79]

Science and its predecessors have been valueless re-presentations of specified subjects. 'Observation' has been of past events or subjects. With 'the present' a fleeting moment, gone before you can note its arrival in a log, there are only a past and a future. If scholars of any stripe claim, of parts of World Al, 'the reading will be 0.345 tomorrow at 0956', others will class this as 'evidence' only of faith. Hero may have been female; but Homo has celebrated male heroes, such as those of 'progress', Bacon, Newton, Watt, with their English-sounding names; but, in the sobriety of active work, 'historians' and 'scientists' have faced the same past of one universe, with the natural and artificial often inseparable. Newton's Laws of Motion and the Periodic Table of the Elements (some non-existent naturally) do not belong to Physics, Chemistry or History. Members of all such groups have sported them at will. The scholarly tags, 'physics', 'biology', do not have the discriminatory power, in the human milieu, of names such as 'dog', 'cat' and 'banana'.

Our species is tagged here as a group of believers. The idea of 'scientific method' comes with a sense of transcendental awe. Time, personalized or not, passes as well as tells. Scarce goods alter with use and time's passing. Individual Homo matures, learns new skills, ages and dies. Scientists and their forerunners have used all the wiles, moods, emotions, senses and traits available to humankind. It is inconceivable that there has been a 'way to be scientific', for a 'scientific technology' to be identifiable; but, with Homo opting to use the analogy of machines (base, force, structure ...) for its own dynamics, this animal has been painted up, as in the operation of the historians' apathetic fallacy, as if it were a machine, in an effect which may be called mechanomorphism.[80] So, a range of things has been shaded out of view – sinning, the will to have fun, to find fame, to seek redemption, to humiliate; a sense of pity, awe, fear .... Meanwhile, back with Art – a creature of the Western bourgeoisie who believe 'Olivier *is Hamlet*!', while others show less will to suspend their disbelief – to do with opinion, on 'the importance of a Work', or whether W is a Work at all: 'Beauty is in the eye of the beholder.' Despite some denials from 'the Art world', notions of Art and Beauty are close, subject to the fads of fashion. Is the Clothed One more a strip-show merchant of truth-revelation, or of fudge and cover-up? If sport prepares for life, with its role of deception of 'the other side', the arts of dissembling are key ones. If the effects of awe are neglected, we might grasp Homocraft.

3. ABOUT TIME, SEAMS, COMMUNICATIONS AND POWER IN BERLIN AND ELSEWHERE: A PLAY[81]

The first part of Section 2 has dealt with views from the mountain top, by Homo of Homo. Here is Butterfield on the curse of whig 'abridgement': 'the more we are discussing and not merely enquiring, the more we are making inferences instead of researches, then the more whig our history becomes if we have not severely repressed our original error'[82] – as with the anachronism of telling of a Scientific Revolution; and, in what follows, Time will urge readers to examine some of the nooks and crannies of human life and lot, aspects of these neglected by the merchants of generalization. Engineering may be compared with Art. A case here is that the profane status of Art in the historians' Age After the Age of Faith is illusory. For one anthropologist, 'the study of aesthetically valued objects [subjects of awe] constitutes a path towards transcendence. In so far as such modern souls possess a religion, that religion is the religion of art, ... whose shrines consist of theatres, libraries and art galleries, ... and whose dogma is the dogma of universal aestheticism.'[83] Work W is made so the seams should not show. Marks of contrivance should fade from view. 'Art reflects nature.' 'This is the face of a real human being. There is no trace of a pose', for Gombrich, in that confident, 'serious' mode of Western 'art criticism', on 'The Mirror of Nature', about a Work of Rembrandt van Rijn of Leiden – for Ashley, a 'genius' who 'lifted Dutch provincial art on to a higher plane' in *The Golden Century* – a man schooled into posing as a 'serious artist' for self-portraits.[84] Hence the problem of the Artist. The seams should not show. The seamster is an unwelcome spectre at the feast, a Banquo's-ghost reminder of what Art-lovers do not wish to ponder, human agency in a making. In scholarship, again re-presenting the world, marks of wielding the scalpel of dis-crimination must go.[85] True, there has been Gombrich's 'riddle of style' – different Artists act differently, while tackling one subject.[86] For a Work (here, an object), an Art-lover may see the mark of brush-strokes; but a 'genius at work' must not be thinking – as spirited Homo tends to do – of football, how repetitive tasks bore, sexual intercourse with his girl-friend, a boys' night out, going to the toilet, taxes, lunch.[87] Horror to think of it!

Regarding science, the other Great Invention of a 'modern era', the previous section suggests that our species is a believer, grown up with fairy-tales told about its past. Thus, Pope's Newtonry, 'God said, Let Newton be! and all was light', is a horror story for real-life, clever Homo, attracted, as a moth seeks bright lights, to dark places and to risky enterprises. By meddling there, its ability is tested. 'If the Renaissance ushers in the modern world, we should find in it the roots of that [its] most characteristic feature ... a preoccupation with physical science. Yet ...

humanism neglected science' (the Cambridge *History*, seeming to deny Acton's Grand Rationale).[88] Historians may now believe less than before in the importance of the Renaissance; but in the V & A they still talk of Renaissance Art.[89] Of Science Museum affairs, over the road, Newton and Watt loom large as 'discoverers' changing human lives: 'Watt's discovery [steam engine] was no accident.'[90] (Where did he find it?) Gone is the England of oak, ash and thorn, when all could see into his forge and watch the village smith! Now is an 'industrial' present – taken out of code, 'skilled' Homo is a 'manual worker' (pianists and surgeons need not apply); and scholars do not get out of their studies regularly to observe Homo working.

From Werner Siemens[91] we can see he was haunted by two images, ganders and gutta-percha of an exotic name. Take the second, he resigned a military commission early (in 1849) to go into business reluctantly, as an eldest son with a large family (eight siblings younger than himself) to help support, after the early death of his parents in 1839–40. (30, 34). As an ex-student of the Berlin Artillery and Engineering School – time spent there, from 1835–38, 'I reckon to be the happiest of my life.' (23) He was already an inventor, as with a mode of electrolysis, devised in a prison cell in Magdeburg, the town where he had joined the Prussian army, and where he was later confined for illegal duelling: 'I consoled myself with the thought that I should have a good deal of leisure time for my studies' – aided, as he was, by a 'friendly young fellow' in a local chemist's shop, who agreed to smuggle useful equipment in. (32) This proved to be so.

Lord Bacon, who has been raised to the aristocracy as Werner von Siemens will be, author of the Instances of the Road, is Time's godfather; and he is on London Bridge to wish us well on his road. He seems happier with his own books under his arm, than with affairs of state as Lord Chancellor. We must log down not only the dramatic and the interesting, but all we can grasp – as in 'putting Nature to the question'.[92] There is a powerful, 'ordering' device at work, he says, 'the human understanding', distorting our grasp of the world, about whose detail 'man always believes more readily that which he prefers'.[93] If there has been a 'cultural legacy of the Scientific Revolution', which some say kindly he foretold, Bacon goes on, this has been what Whitehead said: After the changes, even the grandest of abstractions ought to be subject to the test of detailed evidence. Elaboration confuses.[94]

Bacon's god-daughter tells us, as we leave, we are to meet in Berlin, E.W. Siemens, a native of the countryside near Hannover, a product of *Mittel-Deutschland*, 'a true child of this age',[95] educated at home with his siblings until the age of 11, who walked part of the way to Berlin when he was 17 and found autobiography difficult: 'I have repeatedly tried in Charlottenburg [in Berlin] to continue my task, but there, where everything is pressing forward, I have not succeeded in persistently looking backward.' (226) We will meet him in 1866, after he has resigned from the army and after the death of his first wife, Mathilde. Many born in other lands have served the Prussian state well, whose capital has been

Berlin since 1701; and this particular Siemens, the eldest of various brothers finally working in the family firm, was the key player in it. As speakers of English, we may know more of brother Wilhelm (1823–1883), another inventor, who went to live in England, married over there, anglicized his first name, was knighted by Queen Victoria, and had his biography written before Werner; but Werner was the instigator, the mover and the shaker of the whole commercial and technical affair. He even taught the younger brother mathematics. From five to seven in the morning in Magdeburg, when Wilhelm was 16, they were found striving away in Werner's lodgings: 'I . . . was rewarded by the particularly good examination he afterwards passed in that subject.' (29–30)

Meanwhile back in urban Berlin – and in rural Mecklenburg, where the family had moved – Siemens had started manufacturing, as well as physical inquiry, before 1849, linked with his work in electrical telegraphy: 'in . . . 1847 I induced the mechanician J.G. Halske, with whom our common labours had bound me closely, to hand over his business to his partner and to start [with Georg Siemens, a cousin and lawyer, as a partner] a telegraph factory, into which I reserved . . . the right of entry . . . (52). Werner Siemens had published scientific papers before, such as his first in 1845, on the theory of air-engines (40), in the year when he was also a founder of the Berlin Physical Society, with whose other members he always kept close contact. Notably, too, after his 'withdrawal from all official duties . . . It lay with me now by good work to raise to the utmost the business which already bore my name, and to obtain recognition in the world as a man of science and as a technician. Although my inclinations drew me' to science (Section 1), Siemens engaged first in technical work, as essential in allowing such inquiry to occur. (80) As with current electricity. 'In all the history of the Firm, the Engineer had been in command. The founder had looked upon himself as an engineer', a *Techniker* not, as suggested in the Cambridge *History*, a *Wissenschaftler*[96] – the latter of which the present writer was described to be in Berlin in 1979, in Tempelhof then a colony of the USA, with US Army tanks seen on the streets, a mile or so from where Siemens and Halske first set up, living virtually 'above the shop' in the Schönebergerstrasse near the Anhalter Bahnhof.

Siemens had been interested in telecommunications from his army days; and electrical current had seemed as capricious as humans often are. To err is human, to forgive divine, Time reminds us, lest we humans on the road had forgotten! Those unfamiliar with the subject might have expected, from an armchair, that Siemens and Halske's main worry, as pioneers of telegraphy (writing at a distance), as with the early case of the Berlin-Potsdam railway, had lain with the affairs of those predictable, dramatic players, Current, Tension and Resistance, as measured in Amps, Volts and Ohms. These to be matched, say, with the three components, Length, Mass, Time, of the physicists' CGS system, named above. Yet, a main difficulty had lain, as they sought to establish their reputations, with finding an insulator for underground cables, and with keeping a system of telegraphy in good order, as with the physical protection which cables proved to require. Errant, real-life Homo stole overground lines. Rodents ate insulating materials underground, 'even the protective lead-covering', if not laid deeply (more depth, more cost); and workmen damaged cables trying to repair faults. Even in 1851, 'there was lacking a properly trained staff to keep the extended network of wire in good order' (89). There were rivers to cross, too, with protection needed for cables from the hazards of navigation – say, from dragging anchors.

And there were always the clients! Clients and customers haunt the scene of engineering, as much as considerations of cost. Here, some were governmental clients, of the trickiest of them all, often dictatorial but uninformed. Although Siemens had retired from the army high in its telecommunications service, his company received no contracts for many years from the Prussian government, due to a dispute about competence (89). All this is why looking through a glass darkly was proposed above, for a general mode for Homo in technical work. Hoping for one outcome and finding another. Peering into the future, using a faulty telescope. Thus, the Siemens concern stumbled on gutta-percha by chance, from Wilhelm in England, where he had been sent to market Werner's inventions. But that 'finding' was not all. The Berlin concern had first to devise, and make, an engine to Werner's design, a mechanical press, for forging this material into a suitable insulator for electrical transmission (51), in a time of low voltages and little heating of conductors. Of the gander, Siemens had, while young, to conduct sister Mathilde, some three years older, through a farmyard, to have her knitting lesson with the parson's wife. Such a ferocious beast lurked therein: 'Then Werner shall go with you, and I hope he has more courage than you have.' (C.F. Siemens, father) Armed with a large stick, fearful and with the resolve of 'juvenile heroism', he had achieved his aim 'with eyes shut' and had recalled the lesson about persistence, 'its striking effect' on forging a human character. 'It is curious what a deep and lasting impression this first victory made on my childish mind' (11), that of a species which tends to assess events in such stark terms as those of defeat and winning the day, for which personal knowledge is often more important than that which is published.

This has been what Time has told us, in introduction to our trip to nineteenth-century Berlin, along with a denial of her discovery in Europe *c.*1500. She has been with us always on Earth, not only as 'the grim reaper', but telling us things, cracking jokes, healing the scars of our emotional disasters, wiping the world's bottom ... but our preparations are interrupted! Drama! Life is like that: 'for a man who proposes to act, the thing that is important to think about is the situation [his master, his oracle, his god] in which he stands ... he is not free at all' to choose actions, since 'the empty space in front of him, which he proposes to fill up with activities for which he accordingly now begins making plans, will be very far from empty by the time he steps into it'.[97] In the Danish war of 1848, at issue is the security of Kiel harbour, where Siemens's sister now lives with her husband, a professor of chemistry, in the role of local politician. 'Sister Mathilde wrote me in great anxiety and almost envisaged her house [by the harbour] in ruins, since it was particularly exposed to any assaults by the Danish men-of-war.' (56) Mathilde had had to be protected from the Gander. Now the not-so-great Dane loomed. On such chance links, the world works. Then, too, a history of engines, put together without dealing with 'engines of war' would, Time stresses, be only a partial account of the past of Homo and such devices. For forceful Homo, the spanner of the 'fitter' used to fix an engine can be used as well to kill a foe. We are off to Kiel to take a closer, and a more-involved, look: 'the historian must re-enact the past in his own mind.' (Collingwood) Further: 'The historian is something more than the merely passive external spectator. . . . By imaginative sympathy he makes the past intelligible to the present.' (Butterfield)[98] We must do something of the same.

Schleswig-Holstein has uprisen, a German-speaking area ruled by Danes; so, when Lt Siemens came to Kiel, in his Prussian uniform, after a delay in gaining leave from his superiors in Berlin, he was welcomed. Here, an initial problem is that, while threatening Kiel harbour with their warships, the Danes have left a garrison in a fort at its entrance. A solution to these difficulties has been Siemens's idea of using underwater mines to protect the harbour.(56) His first plan had been to load explosive powder in special canvas bags, made water-tight by a material called caoutchouc (57), then to fire them from the shore by wires carrying electrical current produced from batteries. But the bags are not here. We must use caulked barrels instead. Siemens says this will be the first instance of the use of underwater mines for this purpose; and Time agrees. But some of our number, unused to such heavy work as filling barrels and moving them about, do not appear to appreciate his case in point. Though, surely, all our ears prick up, when the deviser of these engines of war tells us that gutta-percha has been the chosen insulator for the electrical conductors designed to fire the explosives. The mines will not, in the event, be fired in anger, Time says; though Lt Siemens will lead his troops into action, with two of his brothers there to help, to storm the fort we can see across the harbour. 'Go home, Danes!' But the measure of the success involved will be found as much in the non-use of Lt Siemens's inventions. When they are taken up in two years' time, they will be found still operable. Why then Time and Berlin, the latter being where Karl Marx had studied and Max Weber would grow up, a city now, unusually, with a suburb named after an inventor, Siemensstadt in Spandau borough, named after the author of *Inventor and Entrepreneur*?

In a script known in English as *Aristotle's Poetics*, but mainly about what is now called 'tragic theatre', itself 'one of the distinctive functions of Athenian democracy'[99] – as linked with the school of Peripatetic philosophers and probably not written by someone who would have been surprised to be known as 'a Greek called Aristotle' – there is something germane to our case; and this has latterly been as garbled as the name and nationality of an author.[100] We have been assured of the opinions of a man known for his strict scholarship, a physician's son from the north of Hellas, who was not a member of the 'demos' of Athens when teaching and working there, something like this: 'Aristotle says the truth may better be gained from Art than History.'[101] (Note a use of the present tense of awe about the past, as used too about Art: 'Leonardo is telling us ...') Aristoteles may well have thought what *Peri Poetikes* (literally, *About Things Made*) suggested; but today's notion of an exclusive Art is anachronistic and inappropriate for the popular theatrics of about Aeschylus's time, on which *Peri Poetikes* dwelt. (Cf. 'Art ... is international', as cited above, about Europe in the seventeenth century and on.) In the script named, a contrast was proposed between two modes of inquiry, identified as those of *mimesis* and *istoria* with the latter word used (as in today's 'natural history') to denote scholarly inquiry generally.[102] Further, what we choose to know now as 'Greek Theatre' (from Hellenic, *theorein*, to view) – a setting of poetic mimesis – was, in its day, a noisy affair, part of a local, religious festival of song and dance with robed priests in attendance, such as the City Dionysia, paid for by public funds. Thus Aeschylus, an Athenian from Eleusis, would, in the fifth century BC, have acted in his own shows; the original word for poetry, *aoide*, implies music and song; the name of the tragic Muse, Melpomene, indicates singing rather than speaking; the original Hellenic name for a dramatist was 'dancer'; and the plays performed were probably more like today's rock concerts and football matches, than our 'seriously staged', theatrical shows, from which some of the original emotional content has been extracted – here was an affair, perhaps with no stage and little scenery, produced in the open air for thousands of viewers, to encourage Hellenes into virtuously Hellenic conduct.[103] It was not, say, 'work of universal genius', a notable anachronism of latter-day piety – 'Olivier *is Agamemnon*! Isn't he *wonderful*!' Back in the passing of time, in the shadow of their own special acropolis, treasury and citadel of Athenian rule, the mimesis involved was about trying to re-produce didactically, in a story and a show, every aspect of Hellenic life and lot. Medes vanquished, Hellenes heroic, right done by 'our' folk .... 'Athenian tragedy was from the beginning inseparably bound up with the material and social advancement of the Athenian people.' (*Aeschylus and Athens*)[104]

Now, what had Time been warning us, as we came off the boat at Calais? Yes, it was to tell us, when in doubt, to believe nothing claimed in the West about Art or science, topics of special Awe and Celebration. As Collingwood, a philosopher, had put it, what 'the historian ... does to his ... authorities is ... to criticize them'.(epigraph) Quite so! When Kenneth Clark, as logged in his book on *Civilisation* was starting his televisual epic about the Western Arts, on the aptly named Pont des Arts in Paris – praising the 'reasonable' buildings near by and the 'civilized' book-buying of those found crossing the bridge for centuries – he implied he could see the book-stalls by the Pont St Michel.[105] But you can no more see these book-stalls from there than you can talk with the Seine-side buildings as Our Hero seemed to be doing. 'Bonjour, M. le Louvre!' The Pont Neuf (a very old bridge), for one, is in the way. Only 'Artistic licence' could have allowed this moving tableau to seem to millions to be fitting. For Clark himself, to be talking to inert, mute stones and to be found praising, as 'civilized', only restricted traits of human living. As for a powerful, anthropomorphic 'science' said to have achieved anything, as with inert knowledge dolled up as a vibrant activity in the operation of the historians' pathetic fallacy, this has often amounted to specific humans going in search of a meal-ticket. Of myth generally, this 'is not concerned with human actions at all.... And the divine actions that are recorded are not dated events in the past: they are conceived as having occurred in the past, indeed, but in a dateless past which is so remote that nobody knows when it was.'[106] A 'union of science and technology', then, has been mythic: no one can say where and when it occurred. Who were the bridegroom, the witnesses, the celebrant? The Very Rev. C.M. History, as vicar? 'For truth is ... the daughter of time.' (epigraph) 'The human understanding is active and cannot halt or rest, but even though without effect, still presses forward.'[107] Which is what we must do now, in physical terms.

Lt Siemens, fresh from protecting his sister again, wishes to take us further back in time from the occasion when, in Berlin, he will invent a dynamo, whose main role Mumford sought to rob from him. He wishes to go back to his invention of a method of electrolysis sold to a local tradesman, one of the happiest moments of his life. (32) At 21, Wilhelm 'went to England to exploit his brother's gold-plating process';[108] and the firm gained. But Time overrules him. She is always doing this, we have noted on the road! Homo has been a notable con-structor and a trans-porter of things.[109] Electrical trans-mission is the source of the fortunes of the Siemens companies; and she has not finished preparing us on this. If the nineteenth century is well tagged, as some say, as an Age of Progress, some say 'Railways and Telegraphs, habitually mentioned in one breath' are symbols of the New Age on the way, an era after that in which 'Germany, as Heine has said, could be heard snoring'.[110] Which is not even to get on to the dynamo, and to the chances of sending current electricity about the planet, in order to power other, powerful engines (electric motors, which came to be used in countless locations). As suggested from London, SW7, in 1982: 'Electric motors make the world go round!'[111] Might not Siemens and Emil Rathenau, the forge of AEG, in Berlin, therefore, be as important in some Grand Scheme of Things as Watt and Stephenson, earlier steam-power pioneers from over a strip of sea?

Borders are irrevocable things, and names are signs of their presence. Germany is not England. Science, a 'body of knowledge', is not an active process too. Such distinctions have been part of Homo's attempts to grasp and understand its surroundings by naming. 'It did not take clever Homo long to realize', Time went on as we crossed the German border near Aachen, 'once current electricity had been made in large enough amounts to be studied, it had a link with magnetism. We are dealing with materials within a concrete, everyday reality of massive items (cf. mass, length and time, Section 1). This is where engine-wrights have always waded about, usually collectively, such as with Siemens and Halske in electrical engineering ... facing bulky and valuable items, seeking to mould them at will at a moment, meddling, contriving, forging, milling, grinding, soldering, planing, assembling, fabricating, fitting ... trying-out solutions in their workplaces, unsure of final outcomes.' This is engineering, as conceived in one of its parts in *Designing Engineers*, from a study of a chosen quarry: 'the process of designing is a process of achieving consensus among participants with different "interests" in the design ... There is no overriding perspective, method, science, or technique ... The process is necessarily social [collective] and requires the participants to negotiate their differences' directly.[112]

Time's idea of the forceful Clothed Species – always involved in projects, trying to 'engineer' solutions and aiming for specific 'ends' – is from her father's cultural heritage, she tells us, fetchingly clad today in a power-suit, with a teasing grin sported on her face. This is a heritage, as transmitted via ideas of what is often called Renaissance Europe, about a whole animal, not just its parts. Not minds and bodies, the physical and the spiritual. Though, she has gone on, she disagrees with another feature of this heritage: 'the universe, both the physical and the moral universe, must be not only rational, and therefore knowable, but also simple; the apparent multiplicity of physical things is only superficial.' (*The Greeks*) That is an arrogant nonsense. It is a typically human conceit, she says firmly enough (as only half-human herself) for none of us to want to disagree, for human beings to assume that those items about which we are knowledgeable are necessarily the key items in a whole![113]

'The Clothed Species, for ever loaded up with its tools and instruments (spanners, pens, words, ideas ...)', Time has stressed teasingly, 'and with its well-developed sense of the dramatic, has been a propagandist – spicing up neatly wrought, scholarly tales such as those of the Renaissance and Scientific Technology, as well as popular puffings of the Land of the Free and a New World within a single planet of habitation. There is puffery too about "the fashion world", rather as "the Art world" is celebrated locally; but being "in style" seems to have been endemic to this species, "lively" like the fictional Elizabeth Bennet, of *Pride and Prejudice*, often capricious, not normally a drone, with a sense of poetry in its soul. Who is insulted by the designation (cf. "designer clothes" and tags) "being stylish"?' The tag, the Clothed Species, she has chosen partly metaphorically.[114] Thus, Homo has cloaked up stories about self and milieu, as judged agreeable locally, puffing the 'rational scientist' and 'professionals' acting as rationally as machines.[115] 'Puff, puff! Has the Age of Steam not gone?', Time asks us with a smile. As for bodily clothing, 'putting it together', 'putting it on' and 'making it up' have loomed large for showy Homo. Yet, if Garment G – not necessarily 'a suit', or by a well-known Armani – 'suits you'; it may be 'becoming' to you, and look 'natural' on you: so the metaphoric seams should not show, between you and a garb. 'Make-up' on a face may be a part of 'cosmetics'; but some clothes are worn cosmetically too, 'having no other function than to beautify'.[116] Here is Artistry again, Beauty's bed-fellow and a topic of personal opinion.

Near Potsdam, at Glienicke by the Havel lake, where Siemens, as an artillery officer based at the pyrotechnic factory in Spandau, had once put on a successful, colourful fireworks display for royalty (33) – was this a Work of Art? – we wonder about Time's mother. Bacon, forgetful, as befits a Great Philosopher, was wrong about Truth. Time tells us Truth and Beauty are, in fact, her sisters. Time is ever-young like Lord Hamlet, and too busy to have given birth to Truth or to America. The thought of being Truth's mother makes her laugh. America too? 'No darn comment!' Time is no daughter of the Age of the Serious. But she is wearing a stern look, as we meet Werner Siemens in Berlin. The day is billed as that of the invention of the 'dynamo electric machine' (226–231); but there are, in the Markgrafenstrasse workshop in 1866, some unforeseen troubles, which, Time tells us with a wry smile, even she, the daughter of Aeschylus, did not expect.

The city is in turmoil between wars; Halske is retiring; Siemens's first wife has died, leaving him, a good family man, in the familiar condition with young children (four, now) to support;[117] and as vexing as all this is the competition. For engine-wrights, there is always the competition! Who else is up to similar tricks? The pressures of the fray, which engineers must live with, or go under! 'It must be remembered that at that period he [Werner] was still his own chief planner, designer and draughtsman and that no one can exercise these functions adequately as a side line and by spasmodic fits of attention.'[118] So Siemens and Halske are very sorry, but they did not wait for us. The 'first problem' was what others might have classed as mechanical, not electrical, 'the construction of an effective electric exploding apparatus without steel magnets' still in use at the time of Siemens's own writing (229). They invented the dynamo yesterday, a device which would be crucial for electrical engineering; and Georg Siemens has written thus of when Werner first read a paper about this: '17th January, 1867 ... the date of the birth of power current engineering'.[119] Now, they are involved in the trickier affair of claiming priority. 'There goes devious Homo, up to its usual tricks!' Herr Siemens, who has given over some time to talk to us, tells us he has had his priority stolen before in this wicked world – this has been not only by scribblers from the USA, such as Lewis Mumford. We will go back some years, instead, to observe more of his company's pioneering exploits in telegraphy.

We are back to Berlin, therefore, in 1848 to consider some tricky, 'interesting and exciting' problems of the Berlin to Frankfurt-on-Main telegraph, for sending ungarbled messages through the wilds of Mark Brandenburg. What is happening? We peer about at some 'altogether unexpected phenomena'. It is difficult to work things out at first. 'Halske found first of all that with shorter lines our self-interrupting dial telegraphs [their own invention] acted with much greater speed than corresponded to the resistance of the line.... The better the lines were insulated, the earlier this, at the time inexplicable, phenomenon occurred. This induced Halske deliberately to impair the insulation of the line by the addition of artificial watery bypasses.' (73–74) Funny stuff electricity! Prevent it from flowing easily down the line towards Frankfurt, and you may win the day! Time says humans are capricious too – though she is not so herself, being transcendental on her mother's side. We must seek then to understand this wilful, unpredictable, annoying . . .

Now, we are off to Russia, where Werner has sent brother Karl to support the business: 'The year 1852 formed a decided turning-point in my personal as well as in my business life', given that he married in Königsberg another Mathilde, a distant cousin; and he was 'called to St Petersburg to undertake the construction of telegraph lines throughout the Russian Empire'. (96) What a spot! Drunken drivers, a suspicion that Siemens was a spy, and the 'strangeness' of the city. 'The higher government posts were at that time for the most part held by Germans from the Baltic provinces.' (101–107) We catch up with Siemens as he is with Count Kleinmichel, chief of the Ministry of Public Works and Communications, 'the most powerful man in the empire'. (103) Why is Siemens being treated as a spy? We will him on! The Berlin police chief has sent a testimonial 'in regard to the police telegraphs laid down by us'.(105) In the 'state of things', Kleinmichel's backing would be crucial; and luckily Siemens makes a good personal impression. It seems later that the Danes had tagged him 'as a dangerous character, on terms of intimacy with the democratic professors of Kiel'. (105) Family wheels whirring again! And the new Mathilde Siemens was a relative even before her marriage to Werner . . . but there is much to be done with the erection of the State Telegraphs (107) and family matters are never out of sight. Karl, for the record, 'became a Finnish subject in order to become a merchant of the first guild', in support of the family concerns. (112) What human beings will do for a cause!

As for current electricity, Time has gone on, while we picnic by the Spree, as Siemens and Halske grappled with it in their Berlin manufactory, a piece of metal was 'magnetized' by placing a coil of wire around it. Through this, current electricity flowed and was seen to produce its 'magnetic field'. Like 'magnetic poles' repel. Unlike poles attract. Here is the potential for movement, of the type Newton had mapped out years before, with the equation now sometimes written, P = mf. ('How boringly predictable, those inert bodies in motion! Please change the subject!') With a grasp of these links, and some meddling ingenuity of construction, some 'bright spark' made an engine to produce electricity by 'converting the mechanical energy' of a rotating shaft into what is now 'electrical energy'. And vice versa, as Homo has striven to obtain greater power in trans-mission, in sending-across from A to B, over the years. That is, energy for sending from Point A to a different Point B on or near the earth's surface. And it matters little, for constructing the present case, who was (were) the pioneer(s), from within our competitive species, to have made the achievement(s) identified. It does not matter now, either, who was the ingenious originator of what Siemens claims to have been his own invention, partly because the world has taken up a version of his name, 'the dynamo-electric machine'. (230)

'We are still dealing with the advent of electrical engineering,' Time says as we cross the Glienike Bridge, leaving Berlin, whence, as a son of the manor, Siemens had gone at 17. Are not our fields fertile enough, the locals had asked, as they bade 'so good a lad' farewell, recalling the sands of 'that famine-stricken land Prussia' to the south? (20) For years, we can have read of electricity in *Natural Philosophy*, *Natural Science* and *The Principles of Physics*. Once, this was a salon marvel. 'The artefact in a Natural World. Discuss.' Recently, its role – the opposite of 'scientific technology'[120] – is notable by citing Lenin on 'electrification'. From about 1877, for Bowers, writing from London SW7: 'The first company to exploit the commercial potential of electrical motors on a large scale seems to have been the German branch of Siemens.' 1881 saw Werner Siemens building, at Lichterfelde near Berlin, the world's first permanent, public electric railway. 'In 1882 at the Technische Hochschule Darmstadt the first chair of electrical engineering [in Germany] was established, and by the turn of the century' all nine THs in Germany, including TH Berlin in Charlottenburg, had broadly followed suit. (König, writing from Berlin) Overall: 'The electrical engineer has wrought great changes in the last hundred years [1882–1982], and his ingenuity has not yet reached its limit.' (Bowers)[121]

4. BACK HOME TO LOCALITY

Our trip to Berlin has encouraged us to be good historians, mindful that 'it has been easy to believe that Clio herself is on the side of the whigs'. We must then avoid anachronism, such as with Historians S harping on about their beloved Revolution: 'It is here that we reach the second fault in the whig method of approach [after an incipient anachronism, from treating the past in terms of the "the present"]; for by its over-dramatisation of the story it tends to divert our attention from what is the real historical process.' (Butterfield) There have been, say, tales told of a progressive, freedom-loving Martin Luther, cloaked up after his time as a 'modern' tilting at the dark forces of a regressive pope, himself flying the standard of 'medieval' times. Whereas, as claimed by a Protestant writer: 'It is not by a line but by a labyrinthine piece of network that one would have to make the diagram of the course by which religious liberty has come down to us ... [via] strange conjunctures ... [and] purposes marred perhaps more than purposes achieved ...'[122] So, the whig historian infers about the past, rather than studies its detail; and, for a virtuous, transcendental Luther of classic whiggery, read here that great, fudged-up Agent of Major Change, the thing(s) called Science, as discussed above. Of the simplifying whig: 'It is characteristic of his method that he should be interested in the agency [as with "science-based technology"] rather than in the process [such as that of engineering practice].'[123]

To some 'economic historians', current electricity was vital, from *c*1870, in the generation of a Second Industrial Revolution of the 'modern era', linked with utility of the thing(s) called 'science', whereby one effect of a slow-moving First Revolution of the series was overturned.[124] If one engine (a bulky, valuable, electric motor) was dreamt up and constructed, another (the dynamo) would not be far behind, as a device producing the reverse conversion. (For Siemens, the dynamo was a motor operating in the reverse mode.[125]) It is worth noting, then, how the advent of one such engine (the cheap, commercial motor) has helped – albeit slowly and over many decades – to alter factory layouts, ergonomics and active human work, irrevocably. Railways, Telegraphs – and Electricity as well – have indeed shaped an age! So lovers of dramatic phrases might wish to express it. But Time flies; so we must do so too, to keep up with what she is saying ...

Matthew Boulton, Watt's partner, had had this to say in 1776 (the year of what is puffed, from the Land of the Free, as the American Revolution), as displayed in the first gallery of the Science Museum, two of whose main exhibits (Watt engines) show what a poor choice of name was made for this collection of engines: 'I sell here, Sir, what all the world desires to have ... POWER.' Homo, Time stresses, as one who is only human on her father's side, seems 'powerful', when it is active with machines.[126] An electric motor powered by a dynamo has an output in the form as the dynamo's input. Hence 'power station', for German *Kraftwerk* (cf. 'witchcraft' in English). Hence, the sense in classing both devices as machines – predictably working devices, transmitting force or energy from one point to another on the earth's surface, to be delivered to something else – a cooker to cooked food, a power-saw cutting a tree.... Hence, too, widespread workplace change, as the use of cheap electric motors brought the fall of Wattery, the collapse of the House That Marx Built. No more, there would be the machine-dictator of *Capital*, Vol. 1(1867), the rumbling, central power-source of 'modern industry' – such as a stationary-steam engine in 'the works' – as this had been the case from the time of the water-wheels of Lancashire cotton ('dark satanic mills'), from the end of 'putting out' in textile-making.[127]

With the advent of electrical energy in workplaces (factories, offices, workshops, houses ...), human work could be arranged more flexibly, especially after the making of heavy-duty cables – with new insulating materials, new forms of protection and a capability to use higher voltages than those of 1850s telegraphy – in the likes of the Siemens factories.

When we stay in Magdeburg coming back, where Lt Siemens had once been garrisoned, we are asked by Time to read three of Mumford's books, then to assume the opposite of what this New World author wrote, as of Siemens and the invention of the dynamo by a physicist in London. 'Mumford's writings as a Work of Art. Discuss.' *The City in History* is not about cities, in the way most people know them and have lived in them. It is a song of progress to the suburb, as idealized in the USA, in the wake of a long-lasting, local virtue of living near Nature on the 'frontier'. *The Transformations of Man* , with its ideas such as those of Axial Man and New World Man – 'Science was the great tool refashioned by the New World mind ...' – maps out changes which have not occurred to the Clothed One's character.[128] *Technics* is a hymn of praise for progress to an ideal future of human imagining. Real human beings do not yearn to live in a

Land-of-the-Free state: they welcome being boxed in by limits of clothing, faith, custom, location, working specialism and a will to treat the 'problematical'. The Clothed Species has been a forceful meddler-'improver', including its clerks in unholy orders, wielders of scalpels and pens. These people have given us various theories about Homo, in which aspects of faith and disbelief, nature and culture, subject and object, have been confused and compounded.

No loitering. Please wash your hands before leaving. By order. We are off again. 'What have we learnt on the road,' Time asks? And the intellectuals of our number are loath to accept what we saw. '"Technology" is what folk discuss, who do not have the gall and personal experience to tell of engineering, with its real-life obsessions with costs and customers, as exercised by members of an animal species which has learnt to take risks successfully. You must consider first what you can see – substantial, tradable items, made from materials – how they operate, and the conditions of their making', she had stressed as we passed Brandenburg, going outwards, a reminder of when Berlin was on the German marches. Engines – from ploughs to bridges, lathes to mousetraps, houses to electronic computers – have, unlike the re-presentations of the Artists and scientists, been entirely new items in the human milieu. They have been fashioned commercially, with considerations of cost vital, even before there has been any worry about customers and the competition. Thus, Vincenti, the *What Engineers Know* man, a 'research engineer' not an engine-designer, was wrong to treat engineering 'design' (making two-dimensional schemes about future engines) as an assault on uncertainty,[129] given that uncertainty is of the essence of the need to use and employ engineers, as problem-solvers. In contrast, Bucciarelli, the *Designing Engineers* man, has been a more reliable author, stressing inherent uncertainty in the complex, collective processes of engineering: this is, in places where lively human beings contrive away at work, where neither knowledge available, nor the demands of customers, determine the character of an artefact.[130]

'Surely', Time added, as we had passed Brandenburg coming back – and we had seen it in Berlin – 'technicians use their minds and creative imaginations. But buyers of engines do not wish to lay out their hard-earned cash on designs on paper, on other ideas about the future, on technology-as-knowledge (Layton), on an "intellectual method" (*Story-tellers*), or other abstractions.[131] More of the Clothed Species have made and mended material things, than have studied, fought or prayed. Engineers work up parts of a bulky, material world, which is World Al of the introduction, a vast set of possible locations in which such work may take place; and the Grand Ab-stractions of the historians have often been a dis-traction to outside understanding. These are part of a confusing fourth dimension.' Thus the main trouble to be found in the historical arts has lain with attempts made, by human agency, to order stories told on to paper. The architecture is faulty. Writers, say, have garbled the advent of electrical engineering, by tagging it as a 'science-based' affair. Homo, a species which has so misgrasped its technical past, as has been the case over electrical engineering, has misgrasped itself. Layfolk would do better to get out on the road with Time, than read books. Or they may ponder her idea of the artful Clothed Species, often a deceiver, which even goes merrily into self-intoxication, for whose activity 'the devil is in the detail'. This has gone on, in countless localities, as Homo has made anew features of Earth, often recklessly – gardening, clothing itself, trading, scouring the planet, making-over a self-image, praising awesome things (Art and science, included), stitching things up like a seamster – then wrapping it all up in self-congratulatory tales.

From what Gellner has called the 'vision' of Enlightenment Rationalist Fundamentalism, this 'desacralizes, disestablishes, disenchants everything substantive: *no* privileged facts, occasions, individuals, institutions. In other words no miracles ...' (*Postmodernism*)[132] With this stance adopted by Homo towards trying to understand the world about us, there have been the critical outlooks of Bacon and Collingwood of the epigraph – no special respect given to any authors, nothing pronounced awesome – as well as a human tendency to use names as defining tools in support of the profanation of the human milieu. A cat is a cat is a cat, outside fairy-tales told to the young of our teeming species. Here is a way chosen for dealing with World Al, as identified in the introduction, for which most people will have assumed that any static, concrete Item X will remain in Location L unless a new force is applied to it; Item X, typically and over short time-spans, will not alter its substance and form. There will be bulky, concrete things in view, including tricky, artful Homo, Homo's colleagues, products, production processes, food, drink, components, buildings, 'raw materials' – these being valuable, scarce goods, as defined above, none of them being abstractions. Tales, in contrast, told of an ill-defined 'technology', with its close links with a fudged-up 'science' and its 'discoveries', have added a rogue, sacred element to the whole affair – from a story spun out of magical transformations and occult provision for Homo. Cinderella's pumpkin changing its substance; a shoemaker getting rich as he idly slept; knowledge seeming (as in the adoption of a name, Science Museum) to have produced valuable, scarce goods without the agency of skilful human effort.

What, if anything, has been 'regular' (Mumford, introduction, above) about active Homocraft, as this has been treated in the second section here, as concerning the skilful contriving of a forceful and meddling organism which typically becomes heavily involved in the detail of its various doings? Consider Time's Instances of the Workshop, often rather a noisy place, which may also be called the Instances of Operations.[133] The Clothed Species – rather a poseur, from R. van Rijn, upwards, downwards and sideways – is a cross between an ant, a tiger and a fox, the last of these said by humans (not by herself) to be sly and cunning. (For her, humans are the sly ones.) What is called 'invention' is untypical of engineering by definition and in the law; and most people do not have their biographies written. Generally, humans confront, collectively or singly, in ways not much talked of by outsiders to the workplace, what are identified as detailed 'problems' in the immediate surroundings at the particular time. Then, like a worker-ant, humans crawl over a subject of possible 'improvement', such as a mouse-trap not working well. They are getting familiar with, 'getting to know' by diligent, personal grasp of local detail, such items as its functions, the materials from which it is made, the way components work together, their durability – detailed issues of static and dynamic concern.

This exacting examination may, or may not, suggest an apt solution for the future – perhaps to be dignified as a 'design' – to what may later have become a new set of problems. An analogy here is 'doing' a crossword – with cryptic hints, coded messages, half-recalled patterns of other solutions. Something 'obvious' in hindsight did not appear so *ex ante*; rather it was an intriguing affair to those involved, where problems are welcomed for potential solution – a 'consuming' affair, like a good book 'which you can't put down'. 'Crosswords tend to have unique solutions', Time has gone on; 'but, in most workplaces, trying to identify a "best solution" *ex post,* let alone *ex ante*, is like whistling on the wind for a miracle.' Agents of change may decide to play the part of a pushy tiger. With issues of cost in mind, the mouse-trap may be rejigged in part or overall, with new materials chosen to work old functions – a plastic spring, not corroding on floors. Or the trap is torn apart, if only metaphorically, with the forging of new functions.... The ant is the tiger of the piece, with tasks often done together. In all of this – which we can readily grasp, if we disperse a fog of false human pride about human action – Homo works cunningly, not well advised by outsiders, who tend not to have a good grasp of all the detail involved. Note Butterfield, as cited above, on 'strange conjunctures ... [and] purposes marred perhaps more than purposes achieved' in the human past.

As a rationalization for what the ant and the tiger should best do foxily in the workshop, for the mouse-trap to operate adequately, they are dealing with the same material things everyone faces in locality and life – wood, plastics, metals, cheese, vermin ... A solution to the mouse-trap problem – or resolutions of a set of problems arising during work – must be 'sold' to a boss, or to those paying the future costs, knowing little of what is going on in detail – this, while ant and tiger rarely have a grasp of what will be the cheapest outcome, of any envisaged. It is rather like a 'customer' of commerce being confronted by a specialist trader explaining what might be done, and for how much, in fixing up a car. The payer knows part of that payment is for the intuitive judgement of the payee. Nothing is certain in Homo's future, even for a research council funding 'scientific research'. 'Remember', Time assures us, human only on her father's side, 'Homo deceives skilfully. This is part of the species's special box of tricks.' In life, where people do not normally recognize a 'best solution' if they trip over it, part of the art of playing the game is to appear confident, even when you are really unsure about things. This is why the bookish, human author of *Plough, Sword and Book* trying to characterize his species' condition of Industria as the Age of the Book, was wrong to have isolated, named and praised the victory of some type of a 'universal' culture over 'local cultures'.[134] 'To the relativists, one can only say ... As an account of the realities of our [Western] world and a guide to conduct, your position is laughable.' (*Postmodernism*)[135] On the ground, however, Homo, perforce, has only succeeded actively through operating locally. Scholar/dissectors of the campaigning Clothed Species have chopped up the human past in aid of cognition; but they have too often confused marks made by their own scalpels with items and events belonging to that past. The problem with trying to understand the world and its past, using the array of the favourite abstractions of those professing a 'universalism' of outlook – Scientific Method, the Renaissance, Art, Industrial Revolution, Technology – is that use of these notions implies the advent of conditions, for the Clothed Species, which have never been detected, outside the written page, on the single planet of human habitation.

Finally, if any readers remain unconvinced of the case attempted here, against a reification of historical ideas to which the concrete processes of human making have been subjected in print, let them consider this: while the volume of the Cambridge *History* cited above, expressly about 'material progress', has logged, in its chapter on 'economic conditions' that '1867 saw Werner's invention of the dynamo'; in its chapter on a foggy 'science and technology' at the end of that century, there is no mention of Siemens, one of its major pioneers – this, despite the stress laid on how electrical engineering was one of the 'new industries'.[136] Note the introduction of a piece in this series, 'Uneven Mirrors', the phrase from Bacon on the Idols of the [Human] Tribe.[137] Any tagging of such items as 'the great geographical discoveries' of *c*1500, and of the 'demographic disasters' of a European 'Middle Ages' (no graphing in view in each case), also any tagging of 'economic growth' generally (operations of *nomoi*/stewards not in view), smacks of an incipient gentrification by scholarship. Perhaps the lack of mention of Werner Siemens in the specific chapter has stemmed from an authorial, and an editorial, distaste for any real-life human experience, which might deny a favourite notion of the advent of a type of 'science-based' practice in 'technology'.[138] Nice story, shame about the lack of evidence to support it.

ACKNOWLEDGEMENTS

The author would like to thank Michael Wright, Roy Porter, Neil Brown and Graham Hollister-Short for their comments on versions of this paper.

Notes and References

1. Translated from F. Bacon, *Novum Organum* (London, 1620), Bk I, Aphorism 84.
2. R.G. Collingwood, *The Idea of History* (Oxford, 1946), 282, in sect. on 'History as Re-enactment of Past Experience'. This book was published posthumously from manuscripts, much edited by T.M. Knox.
3. L. Mumford, *Technics and Civilization* (New York, 1934), 83. The link with work on cities is noted in the 1963 edition. Quotation from ch. 2, 'Agents of Mechanization', in sect. 'From Game-Hunt to Man-Hunt'.

4. Cf. my critique of '*Technik*: Or Mumford Revisited', *History of Technology*, 1981, 6: 121–137.

5. Mumford, *op. cit.* (3), 52, in ch. 1 on 'Cultural Preparation', from a passage also about 'mainly empirical inventions, like the steam engine' (p. 52).

6. S. Kuznets, *Modern Economic Growth* (New Haven, 1966), 9, 10. The hiatus over a 'base' appears on a single page (p.10), in a passage contrasting 'empirical' with 'science-based' working, as broadly in Mumford's scheme (note 5, above). The author gained a Nobel Prize in economics for work on this topic.

7. S. Shapin, *The Scientific Revolution* (Chicago, 1996), 9, has 'science' as 'a historically situated and social activity'; but, when 'revolutionary' items are listed (p. 13), the main change is in knowledge. Cf. ch. 1 of W.C. Dampier, *A History of Science* (Cambridge, 1929), 1–64, on 'Science in the Ancient World'. In this book, 'science may be defined [skittishly, as two different things] as ordered knowledge of natural phenomena and the rational study of the relations between the concepts in which those phenomena are expressed' (p. xi). Of awe for an indistinct, apparently powerful 'science', see M. Fores, 'Hamlet Without the Prince: The Strange Death of Technical Skill in Histories', *Hist. Tech.*, 1994, 16: 161–184, pp. 162–3, 167–9: scholars have 'breathed life into inert science-as-knowledge .. *Homo*, whom we all meet daily ... is scarcely to be seen' (p. 177).

8. Mumford, *op. cit.* (3), 36, 37, from ch. 1 on 'Cultural Preparation', in a sect. on 'The Road Through Magic'.

9. Cf. Fores, 'Hamlet', *op. cit.* (7), 166–171. On public goods, see *idem.*, 'Transformations and the Myth of "Engineering Science": Magic in a White Coat', *Technology and Culture*, 1988, 29: 62–81, pp. 70–71.

10. Cf. my papers, 'Hamlet', *op. cit.* (7), 165, 172–177, and 'Uneven Mirrors: Towards a History of Engines', *Hist. Tech.*, 1997, 19: 85–109. Cf., too, the last sect. of M. Fores, 'Homo et Machina: Of Divided Labour, Revolutions and a History of Marvels', *History of Science*, 1994, 32: 63–87, pp. 83–85. Technical skill is part of 'knowledge' in *Tech. and Cult.*, if J.M. Staudenmaier, *Technology's Storytellers: Reweaving the Human Fabric* (Cambridge, Mass., paperback edn., 1989), 114–120, is correct.

11. E.g. R.J. Forbes, *The Conquest of Nature: Technology and its Consequences* (London, 1968), 4, in a subsect. on Homo Faber, as published in the Britannica Perspectives series: 'By the beginning of the Neolithic Age, man had at his disposal an arsenal of specialized tools.' Also, Mumford, *op. cit.* (3), 11: 'Tanning, brewing, distilling, dyeing have been as important in man's technical development as smithing or weaving.'

12. Cf. 'Epilogue: Technology and Progress', in R.J. Forbes, *Man the Maker: A History of Technology and Engineering* (London, 1950), 326–329. E.g. 'The new science has established an international brotherhood of scientific thought....we have found no rigid barriers of race, color, creed, or language between engineers and scientists of good will.... If we take a long view of the story of science and engineering, it is clear that no discovery can be kept a secret ...' *Ibid.*, 327. The other book named is Forbes, *op. cit.* (11).

13. 'My goal in this book is to probe the function of the discrepancies between the social process of design as I have observed it, full of uncertainty and ambiguity, and the ideal image of an instrumental process according to which participants claim to work.' L.L. Bucciarelli, *Designing Engineers* (Cambridge, MA, paperback edn., 1996), 13. The idea of forceful Homo is from Fores, 'Hamlet', *op. cit.* (7), 172–177, about 'the most forceful, wilful and meddling organisms of them all' (p.161).

14. He also claimed he had directed his 'labour for the full development of the Age of Science'. Quotations are from W. von Siemens, *Inventor and Entrepreneur* (London, 1966), 9, 39, 277, the first in the text from a personal note, dated Harzberg, June 1889, and that in this note from the last sent. of the book. This is the second English edn of *Werner von Siemens Lebenserinnerungen* (*c*1892), about 'how a young man ... solely through his own industry, [can] rise and do something useful in the world' (p. 9).

15. This is a Type A world of the scheme of M. Fores, 'Constructed Science and the Seventeenth-Century "Revolution"', *History of Science*, 1984, 22: 217–244, p. 229, as opposed to Type B worlds which scholars have dreamed up and fabricated, as re-presentations, to make sense of the human surroundings.

16. Cf. 'the properties of materials' of textbooks. Quotations are from the Materials Gallery of the Science Museum, London, SW7 (above the Watt engines), where key items in

the development of human culture (copper, steel, glass, concrete, rubber, polyethene, silicon ...) are listed. See also the 'materials' listed in the Things Gallery, stressing some are 'natural' (wood, wool), some are 'synthetic' (plastics, concrete).

17. 'One hundred years ago this year Cambridge physicist J.J. Thomson discovered the electron. It was a discovery that rewrote the map of science and heralded the electronic revolution.' Heading for 'Atom Smasher', *CAM* magazine, Lent Term 1997, 23–26, p. 23, about a heroic head of the Cavendish Laboratory and a future Master of Trinity (Newton's old college), 'the Electronic Century' and a plaque unveiled on 21 March 1997 in Free School Lane by the President of the Royal Society of London.

18. 'This contact with electric telegraphy was to be of decisive influence in his [Werner Siemens's] life.' Georg Siemens, *History of the House of Siemens* (Freiburg/Munich, 1957), Vol.1, 12. This is an English translation of a German edition of *c*1947.

19. On scholars as 'clerks in unholy orders', of the clerisy of the scalpel and the pen, see Fores, 'Uneven Mirrors', *op. cit.* (10), 95–96, 102, citing E. Gellner, *Plough, Sword and Book* (London, 1988), 17, 21.

20. Cf. Collingwood, *op. cit.* (2), 9, on how the historical process is 'the attempt to answer questions about human actions done in the past'. Current electricity has been a product of such actions.

21. The secure naming of things, in the manner of profanation, has gone with attempts to prevent them from seeming awesome and sacred. Cf. my piece, 'Another Dark Labyrinth', typescript in revision.

22. Cf. my sketch of 'pushy' Homo, said to 'have designs' on things, as it has scoured and refashioned our planet, in 'Uneven Mirrors', *op. cit.* (10), 89, 91; also Fores, 'Hamlet', *op. cit.* (7), 161, 172–177. Of 'lively' Elizabeth Bennet of *Pride And Prejudice*, her father was quoted in 'Homo', *op. cit.* (11), 65, to point to a mode of human action, which can be contrasted with that of idealized laboratories.

23. Note the scores of English words used for human groups, as discussed further in sect. 3 below.

24. Scholars should become familiar with all aspects of such as a seed planted, germinating, and so on. Cf. Bacon, *op. cit.* (1), Bk. 2, Aphorism 41, cited in Fores, 'Uneven Mirrors', *op. cit.* (10), 105.

25. J.H. Plumb, 'Introduction' to *The Penguin Book of the Renaissance* (London, 1964), 12: 'As the old certainties slowly dissolved or grew more limited in application, curiosity took their place ...' (p.12). This book was published first as *The Horizon Book of the Renaissance* (New York, 1961).

26. The last sect. of my piece, 'Uneven Mirrors', *op. cit.* (10), 103–105, appears as 'It's About Time'.

27. W. Shakespeare, *Hamlet*, *c*1600, Act II, Scene 2. For R.J. Quinones, *The Renaissance Discovery of Time* (Cambridge, Mass., 1972), 178, to this Renaissance-era author 'man is placed in the reign of time, where the values and social alignments that served in the past are no longer valid in the present, and man must proceed darkly, moved by his energies and ambitions, and trusting in his own resources'.

28. T.I. Williams, 'Science and Technology', in *The New Cambridge Modern History* (Cambridge, 1962), Vol.11, 76–100, p. 76. Cf. E. Layton, 'Conditions of Technological Development', in I. Spiegel-Rösing and D. de Solla Price (eds), *Science, Technology and Society: A Cross-Disciplinary Perspective* (London, 1977), 197–222, p. 208: 'by the latter half of the Nineteenth Century science stimulated many inventions leading to the growth of science-based industries, as in electricity and chemistry'.

29. Quotation from Contents, *History*, *op. cit.* (28), Vol.11, v (relating to pp. 76–7). The first sentence of Williams's ch. is: 'Science and technology were far more closely related between 1870 and 1900 than in any earlier period.' (*ibid.*, 76), though the last sect. of 'Hamlet', *op. cit.* (7) argued they were, in fact, drifting apart then in some ways. The *History* has used dramatic phrases freely in its Contents pp.: 'Earlier history of geographical [*sic*] discovery' of the finding (not by -graphers) of new lands; 'Earlier history of the New Continent; lack of records', *The Cambridge Modern History* (Cambridge, 1902), Vol.1, ix, x (relating to pp. 7, 37); and 'The Renaissance as a Turning Point in History [the Past]: Criticism of the View', *The New Cambridge Modern History* (Cambridge, 1957), Vol.1, vii (relating to pp.1–4).

30. Cf. Mumford, *op. cit.* (3), 217–218, in ch. 5 on 'The Neotechnic Phase', when 'the main initiative comes, not from the ingenious inventor, but from the scientist who establishes the general law ... It was Henry who in essentials invented the telegraph, not Morse; it was Faraday who invented the dynamo, not Siemens ... The translation of the scientific knowledge into practical instruments was a mere incident in the process of invention. ...the new inventive genius worked on the materials provided by science [scientists]'.

31. So it is a proper subject for historians: 'If we raise the question, Of what can there be historical knowledge? the answer is, Of that which can be re-enacted in the historian's mind.' Collingwood, *op. cit.* (2), 302.

32. On 'conquest', see Mumford cited above on magic, also the theme of Forbes, *The Conquest of Nature*, *op. cit.* (11) The word 'wright' (cf. 'playwright') is from the root used for the idea of 'working materials'.

33. D.S.L. Cardwell, *Technology, Science and History* (London, 1972), 2 as criticized in 'Hamlet', *op. cit.* (7), 180–182. Cf. T.I. Williams, *A History of Technology* (Oxford, 1978), Vol. 7, 1471: 'Science ... concerns itself with elucidating the mechanisms of the universe, and technology with turning this knowledge to useful ends.' Also, M. Fores, 'Newton on a Horse', *Hist. Sci.*, 1985, 23: 351–378, pp. 373–374.

34. This is, Homocraft, as discussed in my piece, 'Hamlet', *op. cit.*, (7) 165, 172–177.

35. Cf. J. Burckhardt, *The Civilisation of the Renaissance in Italy* (Basle, 1860), 1, on 'The State as a Work of Art' (a carefully constructed affair): the Renaissance was 'the mother of our own' civilization. Page ref. is to the English Phaidon ed. (Oxford, 1945).

36. Cf. divisions of the past, as, for instance, adopted in the various vols of CUP histories, such as *The Cambridge Medieval History* (Cambridge, 1911–1936): 'The principles on which the work is constructed were laid down by the late Lord Acton for *The Cambridge Modern History*.' General Preface, *ibid.*, (1911), Vol.1, v. The editors of the first of these volumes claimed, in 1902 about Lord Acton, in the Preface to *History*, *op. cit.* (29), Vol. 1, v, that 'the spirit of his design' had been retained. Quotation in the text is from *History* (1957), *op. cit.* (29), Vol.1, 1.

37. *History* (1902), *op. cit.* (29), Vol.1, v, vi, 3. The other quoted phrase is from Sir G. Clark, 'General Introduction: History and the Modern Historian', *History* (1957), *op. cit.* (29), Vol.1, xvii–xxxvi, p. xvii.

38. Plumb, *op. cit.* (25), 9, 10.

39. *History* (1902), *op. cit.* (29), Vol.1, v, vi.

40. Cf. the title of ch. 1, 'The Age of Discovery', by E. J. Payne, *ibid.*, 7–36. Quoted phrase, seen in April 1997, second floor of the museum in Cambridge, Cambs. Cf., too, Forbes, *Conquest*, *op. cit.* (11), 17–18, in a sect. on 'Man and Machine', linking 'The expansion of Europe, beginning with the discovery of the New World', with 'the growing mechanization of the crafts' there (p.18) and an increased use of energy sources, such as with 'water wheels' (p.17). The implication that a landmass called Europe expanded, with (say) The Conquest of America, is as unbelievable as the story from the same author (note 64, below) about the possibility of the 'discovery' of the water-wheel in 'classical' Hellas.

41. 'Modern history ... resembles the chief branches of Natural Science; before the results of the last experiments can be tabulated and arranged in their relation to the whole knowledge of the subject, new experiments have been commenced which promise to carry the process still further.' *History* (1902), *op. cit.* (29), Vol. 1, 4, in the Introductory Note by M. Creighton, Bishop of London.

42. Cf. my criticism in 'Uneven Mirrors', *op. cit.* (10), 85–86, of the ill-designed, genteel, tagging, in the newer Cambridge *Modern History*, of the great 'geographical [landmass] discoveries' from Europe.

43. Cf. Collingwood, *op. cit.* (2), 269, in a sect. on 'Historical Evidence', on how Bacon's invocation, 'put Nature to the question' had 'laid down once for all the true theory of experimental science', given inquirers must pose questions and 'torture' nature until 'she can no longer hold her tongue'; second: 'It is also, though Bacon did not know this, the true theory of historical method.' In fact, Bacon warned of the pitfalls of believing the 'empiric', as much as 'the sophistic ... and superstitious', philosophers. *Op. cit.* (1), Bk. 1, Aphorism 62.

44. H. Butterfield, *The Whig Interpretation of History* (London, 1931), 13–33, had a ruling whig history as a story told about how 'the past' had turned into 'the present' of any time of writing, with the historian often choosing to play the part of 'the avenger' (p.1).

45. Cf. Collingwood, *op. cit.* (2), 316, in a sect. on 'History and Freedom': 'The rational

[human] activity which historians have to study is never free from compulsion: the compulsion to face the facts of its own situation. . . . For a man about to act, the situation is his master, his oracle, his god', rather than any prior intentions for action. Cf. my piece, 'Uneven Mirrors', *op. cit.* (10), 89, 91, on how one person's 'design' may be another person's horror-story.

46. From Max Weber's *Protestant Ethic* of 1904–05, as discussed in M. Fores, I. Glover and P. Lawrence, 'Professionalism and Rationality: A Study in Misapprehension', *Sociology,* 1991, 25: 79–100, p. 86, in criticism of the Rational Ethic of the West. For Weber: 'The famous Iron Cage of "the modern economic order" . . . should have been just a "light cloak" ' worn by our species (p. 86).

47. Title of ch. 2, 'The New World', by E.J. Payne, *History* (1902), *op. cit.* (29), Vol.1, 37–66. On its first p. we can read that: 'Its [the New World's] modern historical period, in fact, coincides very nearly with that of the Old World's "modern" history . . .', partly about 'keeping historical records' (p. 37).

48. This uplifting notion of 'the new world', apparently a separate planet, was used as well in *Plough, op. cit.* (19), 65, in discussing '*our* [Western] world [of 1988] . . . a unified orderly world' (pp. 63, 65), as criticized in Fores, 'Uneven Mirrors', *op. cit.* (10), 100–103. Emphasis in original text. Gellner wrote as a professor of anthropology in Cambridge, from the old Cavendish building in Free School Lane.

49. Butterfield, *op. cit.* (44), 7, 12–13, 109, 130. Quotation is from p. 7. Cf. A. Watson, 'Introduction', to H. Butterfield, *The Origins of History* (London, 1981), 1: 'The trouble was that in all of them [grand schemes of history, including whiggery] the theory or interpretation or diagram came first.' That is, for Butterfield, who became one of Acton's successors as Regius Professor of Modern History at Cambridge.

50. Cf. the third section, on 'Distance, Industry and "the Modern World" ', of my piece, 'Uneven Mirrors', *op. cit.* (10), 97–103.

51. 'These instances by no means represent the full influence exercised by the New World on the most powerful mind of modern times, and through him on ages which have realised his ideas without adding anything to their transcendent scope and penetration.' *History* (1902), *op. cit.* (29), Vol. I, 63–65, quotations from pp. 64, 65, in the ch. on 'The New World'. Hence Bacon, cited in the epigraph on Time.

52. E.g. E. Layton, 'Scientific Technology, 1845–1900', *Tech. and Cult.*, 1979, 20: 64–89, the idea criticized in my piece 'Hamlet', *op. cit.* (7), 166–177. In Fores, 'Newton', *op. cit.* (33), 355, 365–373, the Theatre of the Bourgeoisie was devised in criticism of the *History* – modish playwrights unfold plots, generating a covert eschatology, to please their special 'audience'. Cf. Collingwood, *op. cit.* (2), 54: 'Eschatology is always an intrusive element in history. The historian's business is to know the past . . . and whenever historians claim to be able to determine the future in advance of its happening . . . something has gone wrong . . .'

53. E. Layton, 'Science as a Form of Action: The Role of the Engineering Sciences', *Tech. and Cult.*, 1988, 29: 82–97, composed in response to my piece, 'Transformations', *op. cit.* (9). For a fuller discussion of some possible links between an indistinct 'science' and 'technology', see W. König, 'Science-Based Industry or Industry-Based Science? Electrical Engineering in Germany before World War I', *Tech. and Cult.*, 1996, 37: 70–101, pp. 70–74.

54. Cf. the case about transcendental awe for 'science' among those proposing a 'scientific method', a will-o'-the-wisp of scholarship, in my piece, 'Homo', *op. cit.* (10), 65, 75–76.

55. Layton, *op. cit.* (53), 85. Note, in a piece of fudge, worthy of History S, Layton's 'science' of his title ('a Form of Action') is different from that of his chosen subtitle (about 'the Engineering Sciences'), this being a form of articulated knowledge.

56. M. Ashley, *The Golden Century: Europe 1598–1715* (London, 1969), 43, at the start of ch. 4, 'Art and Society: the Age of the Baroque', in a book in a series on History of Civilisation.

57. Man was an invention of the Humanists of the day; and Colombo's non-discovery was this, because no continent was so named before his death. Cf. Fores, 'Uneven Mirrors', *op. cit.* (10), 86, 99–100.

58. E.H. Gombrich, *The Story of Art* (London, 1950), 13th edn, 169–171, in ch. 12 on 'The Conquest of Reality: the Early Fifteenth Century': 'Brunelleschi [1377–1446] was not only the initiator of Renaissance architecture. To him, it seems, is due another momentous discovery in the field of art [not even achieved by "the Greeks"], which also dominated the art of subsequent centuries – that of perspective.' (p. 171) Cf. the scheme of Collingwood, *op. cit.* (2), 236, where

actions of the 'imaginative' artist are analogous to using 'the historical imagination': 'a landscape-painter ... may fancy that he is reproducing in his own medium the actual shapes and colours of natural things; but ... he is always selecting, simplifying, schematizing, leaving out what he thinks unimportant and putting in what he regards as essential.'

59. Gombrich, *op. cit.* (58), 173, again in ch. 12 on 'The Conquest of Reality'.

60. Cf. a travel advertisement enticing people to 'Discover the Hot Beaches of Crete', and a 'fashion' ad. for Givenchy Organza – 'Now Discover the New Perfumed Bath Line'. If the named items were not there, to be found by 'consumers', companies would run foul of trade descriptions laws.

61. E.g., for S. Toulmin, by some precepts 'one must distinguish between discovery and justification. The term discovery refers to all the stages in a scientific inquiry preceding the formulation of the new explanatory arguments that are its final outcome.' 'Philosophy of Science', in *Encyclopaedia Britannica* (Chicago, 1992), Vol. 25, 652–669, p. 663, in a subsection on 'Discovery and Rationality'.

62. D. Boorstin, *The Discoverers* (New York, 1983), 650: 'Tylor believed himself not the prophet of a dogma but the discoverer of a science.' Also, H.D.F. Kitto, *The Greeks* (Harmondsworth, 1957), 190–191, in ch. 10 on 'The Greek Mind'. Kitto's book was first published in 1951.

63. R.S. Westfall, *The Construction of Modern Science: Mechanisms and Mechanics* (Cambridge, 1977), 107. Cf. Westfall's biography of Newton, *Never at Rest* (Cambridge, 1980), 1, 39, 58, as criticized in Fores, 'Constructed Science', *op. cit.* (16), 217–221, 225, to have confused aspects of subject and object, of 'nature' and logs made of same, of Worlds Type A and B. The earlier book was published before in the USA.

64. Other unbelievable stories of 'discoveries' made by Homo include that of Forbes, in *The Conquest of Nature*, *op. cit.* (11), 12, where the water-wheel was so identified. What an amazing feat of non-ingenuity! The reinvention of the wheel, now as a non-invention! On other such mystic sightings, see Fores, 'Constructed Science', *op. cit.* (15), 221–225, and 'Transformations', *op. cit.* (9), 78–79, the latter about a transcendental tea-party in Dampier's house (cf. note 7, above) in Cambridge, Cambs.

65. Quinones, *op. cit.* (27), 176, contrasting a condition of After'dom with that in which people generally 'looked at history [perhaps, the past] theologically, as the manifestation of God's providential plan ...' The idea of this author's title is, in effect, 'the Renaissance discovery of secular time'.

66. The title of Fores, *op. cit.* (7) represents its main theme.

67. Fores, 'Constructed Science', *op. cit.* (15), 233–241, esp. pp. 234 and 239.

68. Cf. Fores, 'Homo', *op. cit.* (10), 65, 71–77. Quotation from 'Constructed Science', *op. cit.* (15), 239.

69. E.g. French, *fabrique*, and German, *Fabrik*, are equivalent to English 'factory'. Also, Homo has forged the material conditions substantially in which it survives, but a 'forged document' is an illegitimate item.

70. Similarly 'what a performance', for Homo playing a part outside a playhouse (or suchlike theatre), implies faking-up, 'putting it on' as a 'show' to others. Quotation from E.L. Scieffelin, 'Problematizing Performance', in F. Hughes-Freeland (ed.), *Ritual, Performance, Media* (London, 1998), 200.

71. Cf. names used in the Human Evolution display in the Natural History Museum, London, SW7.

72. Forbes, *Maker, op. cit.* (12), ch. 11, on changes in transport, 1750–1930, has this title, with a sect. on 'The Conquest of the Air', implying that greater speeds changed all. Speed is, in the 'dimensions' of physics, length/distance travelled, divided by time elapsed – hence the remark about time.

73. Cf. Fores, 'Homo', *op. cit.* (11), 65, 75–76, on a 'transcendental' (p. 65) notion of 'scientific method', as something taken by its celebrants out of the sea of change, the typical location of profane things.

74. D. Morris, *The Naked Ape* (London, 1967).

75. T.S. Turner, 'The Social Skin', in J. Cherfas and R. Lewin (eds.), *Not Work Alone* (London, 1980), 112–140, pp. 112–114. The first quoted sent. starts a ch. in a sect. on Action Play (pp. 69–140), where it is claimed 'being in harmony with the fashion' may, more than religion, effect personal well-being (p.114).

76. E.g. J. Huizinga, *The Waning of the Middle Ages* (1924).

77. Also, there are these words: union ('workers'), staff (in schools), group, cohort, regiment, committee, conspiracy, convention ('a convention of the like minded'), tribe, nation, polity state, hoard ('rampaging Vandals'), republic, club, team, side (sport), troupe (theatrical), congregation, concert (the Concert of Vienna), army, body ('a fine body of men'), confederation, faculty (scholars), collective, organization, faculty, corporation, assembly, corps (the Corps of Engineers), platoon, audience (theatres), spectators (shows), parliament, crowd (football), gallery (golf), family, profession, trade, industry, fellowship ...

78. Note how Code Homo is used to demean and praise. If a metaphor is from artisanal work, if X has been 'stitched up', here is un-grandeur – X has been 'simply thrown together', without care. If it is 'science', Homo does not want to use its taxes to support those 'cobbling' away in labs, 'throwing together' science.

79. Amended from Fores, 'Uneven Mirrors', *op. cit.* (10), 96.

80. As opposed to anthropomorphism, as discussed in Fores, 'Homo', *op. cit.* (10), 64, 71–78, 84–85. Cf. the case put earlier at *idem.*, 'Constructed Science', *op. cit.* (15), 239.

81. Cf. the last sect. of my piece, 'Uneven Mirrors', *op. cit.* (10), 103–105, on how those who want to compose a reliable History of Engines should get out on the road with Bacon and Time.

82. Butterfield, *op. cit.* (44), Introduction, esp. p. 7.

83. A. Gell, 'The Technology of Enchantment and the Enchantment of Technology', in J. Coote & A. Shelton (eds.) *Anthropology, Art and Aesthetics* (Oxford, 1992), 40–63, pp. 41–42.

84. Gombrich, *op. cit.* (58), 331, in ch. 20 on 'The Mirror of Nature: Holland in the Seventeenth Century'. The Artist seems to live on: 'Rembrandt needs [needed] hardly any gestures or movements to express the inner meaning of a scene. He is [was] never theatrical' (p. 333). Also, Ashley, *op. cit.* (56), 51.

85. Cf. Fores, 'Uneven Mirrors', *op. cit.* (10), 95–97, on how the scholarly arts have been to do with breaking things up. Exponents may be constructive, devising and making such items as 'scientific papers'; but their main skills on offer have been those of dis-section, the dis-tancing of things from each other, dis-assembly into component parts, dif-ferentiation, dis-crimination, dis-cernment ...

86. The introduction to E.H. Gombrich, *Art and Illusion* (London, 1960) is on 'Psychology and the Riddle of Style', before the author has gone on to discuss 'likeness'.

87. Note how examples of this human type, like those of the 'tribal' magician, have normally been male.

88. D. Hay, 'Introduction', in *History* (1957), *op. cit.* (29), Vol. 1, 1–19, p. 2. N.b. a summary, in the Contents pp. of this vol., on *The Renaissance 1493–1520*: 'The Renaissance as a turning-point in history [the past]: criticism of the view' (p. vii, relating to pp. 1–4).

89. E.g. *The Renaissance at the V and A: A Handbook for Teachers* (London, 1997).

90. This museum, said officially to be part of the National Museum of Science and Industry, has the name Science Museum up three times on its main face towards Exhibition Road. Quotation is from J. Birdsall and C.M. Cipolla, *The Technology of Man* (London, 1980), 158, as criticized in my piece 'Uneven Mirrors', *op. cit.* (10), 98.

91. His autobiography was, in part, on 'technical developments, important in my judgment, when the priority of discovery, invention, or first application demonstrably appertains to me', *op. cit.* (14), 86.

92. This is also the general mode of historical scholarship, for Collingwood, *op. cit.* (2), 269. Cf. note 43.

93. Bacon, *op. cit.* (1) on the Idols of the Tribe, Bk 1, Aphorisms 41, 45–49. Quotation is from Aphorism 49: 'The human understanding resembles not a dry light ...', but belongs to a wilful, passionate being.

94. Quotation is the last phrase of Shapin on the topic, *op. cit.* (7), 165. Cf. A.N. Whitehead, *Science and the Modern World* (Cambridge, 1926), ch. 1.

95. Siemens, *History*, *op. cit.* (18), Vol.1, 11, where it is suggested that Werner Siemens had an 'early aptitude for the natural sciences, linked, as appeared later, with a technical ability bordering on genius'.

96. Quotation from *ibid.*, 159, where it is suggested too that 'Wilhelm von Siemens [Werner's successor and son] had been brought up in this tradition' at least until *c*1867. In contrast, the Cambridge *History* (1962), *op. cit.* (28), Vol. 11, 735, has had, in its index, Werner Siemens, like his brother Wilhelm, skittishly to have been a 'scientist'.

97. Collingwood, *op. cit.* (2), 316, in sect. on 'History and Freedom'. Cf. John Lennon, in his song for his son Sean, *Beautiful Boy*: 'Life is what happens to you/when you're busy making other plans'.

98. Collingwood, *op. cit.* (2), 282, in sect. from which the epigraph quotation was taken; also Butterfield, *op. cit.* (44), 92, in ch. 5 on 'The Art of the Historian'.

99. G. Thomson, *Aeschylus and Athens* (London, 1946), 1. This book was first published in 1941.

100. Use of Greece and Greek, for Hellenes who spoke Hellenic and lived in Hellas, shows an alien stance. On authorship, see W.K.C. Guthrie, *A History of Greek Philosophy* (Cambridge, 1981), Vol. 6, 50.

101. Cf. Butterfield, *op. cit.* (44), 93, on a need to use poetic insight in composing histories. Aristoteles, like Bacon, grew up at court, where his father was physician to the king of Macedon. Bacon's father, Sir Nicholas, was effectively Lord Chancellor, from the time of the son's birth to that of the father's death.

102. *Peri Poetikes*, ix, contrasts the modes of poetic mimesis and scholarship, after the claim, at *ibid.*, iv, that Homo has had an instinct for *mimesis* from childhood. It is about *mimesis* centrally; though English translations giving the word various meanings, such as 'art' and 'representation', barb the point.

103. Cf. W.B. Stanford, *Greek Tragedy and the Emotions* (London, 1983), 3, 7, 49, 88. For Kitto, *op. cit.* (62), 96, 'the popular entertainment, that which corresponds to our [1955] cinema, was the loftiest and most uncompromising drama which has ever existed'. For Thomson, *op. cit.* (99), 245, Aeschylus 'made his *début* at the City Dionysia in the year 500, but did not win the prize until fifteen years later'.

104. Thomson, *op. cit.* (99), 194, from ch. XI on 'Tragedy'.

105. K. Clark, *Civilisation* (London, 1969), 1.

106. Collingwood, *op. cit.* (2), 15. Cf. some earlier comment: 'It is a measure of the conceit of the self-styled "moderns" that they imagine that their own pronouncement of modernity would have led to the abolition of myth.' Fores, 'Constructed Science', *op. cit.* (15), 240.

107. Bacon, *op. cit.* (1), Bk. 1, Aphorism 48, on the Idols of the Tribe.

108. Siemens, *History, op. cit.* (18), Vol. 1, 11.

109. Cf. the model set out in Fores, 'Uneven Mirrors', *op. cit.* (10), 92.

110. Siemens, *History, op. cit.* (18), Vol. 1, 9, 13.

111. B. Bowers, *A History of Electric Light and Power* (Stevenage, 1982), 264, at the end of ch. 16, 'Motors and Traction': 'Electric motors are now made in a vast range and types and sizes' (p. 264). The author was in charge of electrical engineering at the Science Museum.

112. Bucciarelli, *op. cit.* (13), 159, as proposed to be the 'thesis of this book'.

113. Kitto, *op. cit.* (62), 169–179, in ch. 10 on 'The Greek Mind'. Quotation from p. 179, cited as 'a permanent feature of Greek thought'. Cf. the Articulation Trap, in 'Homo', *op. cit.* (11), 71–77: a celebrated ideal of 'Greekness', linked with 'rationalism' and 'humanism', has not been about typical human conduct at all, as argued using Collingwood, *op. cit.* (2), 41–42 (p. 74).

114. E.g. with the song-title, *The Lady in Red,* no one will think of the subject in a red car, a red house, a red room or painted in red: the lady is wearing red clothes.

115. Cf. Fores *et. al., op. cit.* (46).

116. *Collins English Dictionary* (London, 1983), on the example of 'cosmetic illustrations in a book'.

117. 'Werner had promised his mother on her deathbed to be a father to his younger brothers and sisters ... Mathilde Siemens, Werner's wife, had died of an incurable disease in 1865. It affected him deeply, as he was truly devoted to her, and was imbued with a deep sense of the blessings which happy family life bestows.' Siemens, *History, op. cit.* (18), Vol. 1, 32, 59.

118. *Ibid.*, 44.

119. *Ibid.*, 79.

120. Layton denies 'the use of the terms "empirical" and "scientific" to characterize two historical epochs in technology [here, an activity] is such a grave error as Fores would have us believe'. 'Science', *op. cit.* (53), 89. Cf. the scheme of classification cited from Mumford and Kuznets at notes 5 and 6, above.

121. Bowers, *op. cit.* (111), 248, 249, 268. The first two quotations are from ch. 16, 'Motors and Traction', cited above; the last comprises the last para. of the book. Also König, *op. cit.* (53), 74.

122. Butterfield, *op. cit.* (44), 8, 39, 45. The last two quoted passages are from ch. 3 on 'The Historical Process', the first is from the last sent. of the Introduction. The author was a Methodist lay preacher during most of his adult life.

123. *Ibid.*, 50, from ch. 3 as well.

124. This is sometimes given as the second, general such revolution of 'modernity'. E.g. C. M. Cipolla, in the Introduction, *The Fontana Economic History Of Europe* (London, 1973), Vol. 3, 7–21, p. 18. Cf. the critique of M. Fores, 'The Myth of a British Industrial Revolution', *History*, 1981, 66: 181–198.

125. Siemens, *History, op. cit.* (18), 78.

126. This, as the pair has altered our surroundings. Cf. Fores, 'Uneven Mirrors', *op. cit.* (11), 93, on the use of machines. force-transmitting devices, with Homo classed as 'especially spirited, wilful and forceful'.

127. Re 'cotton mills', William Blake's 'dark, Satanic mills' and their effects on working patterns, see, K. Marx, *Das Kapital* (Hamburg, 1867), Vol. 1, in its long ch. 15 on 'Machinery and Large-Scale Industry'.

128. L. Mumford, *The Transformations of Man* (New York, 1956). Quotation from p. 99, 1972 paperback ed. The names cited, as given to types of 'man', are the titles of chs 4 and 6 of this book.

129. W.G. Vincenti, *What Engineers Know and How They Know It* (Baltimore, 1990), 7, 11, 16–17, 44–50. Ch. 2, 'Design and the Growth of Knowledge', is on shifts in 'airfoil technology' from *c*1900. Early on: 'No realistically useful theory existed, and empirical knowledge was meager and uncodified. Design was almost exclusively by simple cut-and-try ...[subsequently] Using relatively complete (though not yet finished) theories, supported by sophisticated experimental techniques and accurate semitheoretical correlations of data, engineers design airfoils to specific requirements with a minimum of uncertainty. Little cut-and-try is needed by a skilled professional' (p. 50). This ch. appeared first in *Tech. and Cult*, 1986, 27: 717–58, under the subtitle 'Uncertainty and Growth in Engineering Knowledge'.

130. *Op. cit.* (13), 13–20.

131. E. Layton, 'Technology as Knowledge', *Tech. and Cult.*, 1974, 15: 31–41; Staudenmaier, *op. cit.* (10), 120. Technology-as-knowledge and the idea of Layton's title, science-as-action, *op. cit.* (53), invert most people's grasp of the world, where 'science' is a type of knowledge and 'technology' is a type of action.

132. E. Gellner, *Postmodernism, Reason and Religion* (London, 1992), 80–81.

133. Cf. the two names given to Bacon's Instances of the Road, also known as Itinerant and Jointed Instances, *op. cit.* (1), Bk. 2, Aphorism 41.

134. Cf. Fores, 'Uneven Mirrors', *op. cit.* (10), 102–103, in sect. 3 on the Distance thesis, in reference to Gellner, *Plough, op. cit.* (19), esp. chs 4, 'The Tension', and 5, 'Codification'.

135. Gellner, *op. cit.* (132), 96.

136. Quotation is from C. Wilson, 'Economic Conditions', ch. 2 of *History, op. cit.* (28), Vol. 11, 49–75, p. 60. Ch. 3, by T.I. Williams, of this vol. has been cited at the end of Sect. I above, from a book on *Material Progress and World-Wide Problems 1870–1898*.

137. Bacon, *op. cit.* (1), Bk. 1, Aphorism 41.

138. This is the main theme of the last section of Fores, 'Hamlet', *op. cit.* (7) in this series.

The Establishment of the Society of Chemical Engineers in Pre-war Japan

YUJI JIDO

ABSTRACT

The Society of Chemical Engineering of Japan was conceived, established and developed in the 1930s. During this time the number of members constituting the academic engineering community gradually increased, and so, subsequently, did publications in industrial chemistry and other areas relating to engineering. In the chemical industry interdisciplinary co-operation between various professionals became important for the design, construction and development of chemical apparatus and plant. Accordingly it became inevitable that co-operation between industrial chemists and mechanical engineers in the training and education of engineers would become a reality. The organizational elements for such co-operation in research, publication, education and production constituted a quasi-institutional framework for the empirical advancement of chemical engineering as a whole. The Society of Chemical Engineering was situated at the centre of this institutional framework.

In other respects academic notions as well as the methodology of chemical engineering were introduced from the USA, and studies on unit operations were also begun at Japanese universities.[1] Nevertheless, the conceptual framework for the academic advancement of chemical engineering was conceived in Japan in a distinctively independent manner from the USA. Given all this, it was necessary, prior to the introduction of unit operations, for industrial chemists and mechanical engineers to grasp the notion that the apparatus used in chemical engineering was different in its nature from that used in mechanical engineering. The arguments that prevailed even in the process of naming the Society reflected the confusion surrounding the conceptual framework of this science. The problem of how chemical engineering developed in pre-war Japan will be examined against this background.

INTRODUCTION

The first reference to chemical engineering in Japan occurs in 1909 in the work of Kotaro Shimomura.[2] He coined the word *kagaku-kogaku* for 'chemical engineering'. *Kagaku-kogaku* would become a formally accepted expression in chemical engineering terminology after World War II.[3] In 1919 a lecture course in chemical engineering began to be offered in the Department of Applied Chemistry of Tokyo Imperial University. In 1922 a course in chemical engineering was also introduced at Kyoto Imperial University. This course, however, was named *Kagaku-kikai*, meaning, literally, 'chemical machinery'.[4] Between 1929 and 1930 prominent leaders involved in the organization of the Society (of Chemical Engineering) such as Saburo Kamei, Shiroji Hatta and Shun'ichi Uchida visited the Massachusetts Institute of Technology (MIT) to study chemical engineering.[5] The organization of the Society of Chemical Engineering was put in hand in 1935, and the Society officially established in 1936.[6] Its name was finally decided upon as *Kagaku-kikai Kyokai*, meaning 'Society of Chemical Machinery'. Why was the term 'chemical machinery' adopted in place of 'chemical engineering'? In examining the reason for this, we shall discuss some of the factors present in pre-war Japan that played a part in the establishment of the Society.

We shall also focus our attention on the emergence of two notions relating to the establishment of the Society. One is the key scientific notion of unit operations, essential to the academic advancement of chemical engineering as a new discipline. The importance of this notion in such advancement has been discussed in a number of articles.[7] We shall here stress the importance of the second notion, essential to interdisciplinary co-operation between industrial chemists and mechanical engineers (as well as for other professional organizations), namely a notion relating to 'organizing principles for interdisciplinary co-operation' as will be discussed below.

In an article on the evolution of unit operations Warren K. Lewis stated that 'a serious handicap in the recognition of elements common to the various industries' was a fundamental cause in delaying the development of chemical engineering as a discipline.[8] He further stressed the importance of physical chemistry as 'a tool for handling data' and drew attention to the work of Hausbrand of 1893 on distillation. He judged this work as '... process engineering in a highly developed state' because this work included analyses of both the material and energy balances in the system. The scientific concept of the common elements, as well as being a tool for handling data, plays an important role in interdisciplinary co-operation. Furthermore, Jean-Claude Guédon pointed up the importance of the idea of unit operations as 'an organizing principle' in the development of chemical engineering.[9]

CONCEPTS OF CHEMICAL ENGINEERING

We shall discuss here two different notions of chemical engineering which arose during the period of the establishment of the Society. Some of the professors who had studied chemical engineering in the USA contended that it was different from both industrial chemistry and mechanical engineering, and was indeed a new science whose field was that of research in processing in the chemical industry.[10] Most members of this group preferred as a name for the Society *kagaku-kogaku kyokai* meaning 'society of chemical engineering' and this indeed is the name presently used in Japan. Other professors and engineers thought of chemical engineering as essentially interdisciplinary, comprehending both industrial chemistry and mechanical engneering.[11] This group preferred the name *kagaku-kikai kyokai*, meaning 'society of chemical machinery'.

The former group came closer to the essence of the matter, judged by what is understood by the term 'chemical engineering' today. It has to be said, however, that when we look at the history of chemical engineering even in the developed countries, there were then many differences of emphasis. Guédon has provided examples from Europe. According to him, Germany was the only country in the world which succeeded in developing a chemical industry without, however, embracing the concept of chemical engineering.[12] In place of the idea of a unified discipline of 'chemical engineering' Germany adopted a method amounting to a 'division of labour' for ensuring necessary co-operation among the various professionals and for the training of engineers and workers in the chemical industry. Bartholomé has also contributed to the discussion:[13]

> Whereas in Anglo-Saxon countries chemical engineers are part of the engineering sciences and are trained completely from other chemists, German educators believe, at least in the early phase of study, in a curriculum in which, in chemistry and physical chemistry, the same courses are offered for chemical engineers as for chemists.

Guédon remarks on 'organizing principles' that:

> The notion of unit operations ... can be approached then as an organizing principle incorporating the compromises in the process of the negotiations just alluded to.[14]

He goes on:

> As David Noble had stated, A.D. Little had done for industrial chemistry what Taylor had done for mechanical engineering. In fact, the tenets of scientific management, particularly those concerned with a redistribution of the division of labor along modular lines, seems to have influenced deeply A.D. Little's conception when, in 1915, he first defined unit operations.[15]

In Germany DECHEMA, 'Deutsche Gesellschaft für Chemisches Apparatwesen', was an institution organized for just such co-operation, and played an important part in the transition from the stage of 'Apparatwesen' to that of 'Chemie-Verfahrenstechnik'.[16] In the USA the relatively new scientific notion of 'unit operations' played a role in establishing co-operation in research into chemico-physical processes in the chemical industry. Whether the factors essential to interdisciplinary co-operation take institutional form or are more loosely conceptual in nature depends on the social conditions and historical experience of each country. If we therefore confine our discussion of the history of chemical engineering in Japan to a history of 'unit operations', we might fail to discover the factors inherent in the transitional period from industrial chemistry to chemical engineering in the country. During that period the institutional framework played as important a role as scientific ideas in the organization of interdisciplinary co-operation. Attention therefore needs to be paid to the ideas that were raised at the time when the establishment of the Society was being discussed.

These ideas did not necessarily approach the academic essence of chemical engineering but it is likely that the arguments reflected the practical needs of industry in those days.[17]

SOME FACTORS INVOLVED IN AN INSTITUTIONAL FRAMEWORK FOR CHEMICAL ENGINEERING

A first factor was a quantitative growth in academic society, that is, growth in the number of members of the Society of Chemical Engineers who were professors and associate professors at universities and technical high schools (the academic members) and also in the quantity of published material on industrial chemistry.

Figure 1 shows numbers of academic members. The first rise in their number occurred in the early 1920s, resulting from two plans for the reform of higher education, of 1918 and 1923. The second rise occurred in the 1930s, most of the increases taking place mainly in the technical high schools. The plan of 1923 gave priority to the promotion of technical high schools as centres for the training and education of engineers and workers for employment in industry.

There was also a quantitative growth in publications. Figure 2 shows the number of articles published in the *Journal of Industrial Chemistry* (*Kagaku-kogyo Zasshi*), the main journal of the Society of Industrial Chemistry of Japan. From the late 1920s to the early 1930s the growth in the number of academic members and of engineers in industry resulted in an increase in the number of articles published in the journal, as also publications in the field of the engineering sciences. The steep rise in papers relating to industrial chemistry and the engineering sciences in the 1930s is of particular interest. The figures show that the period of the 1930s was one of rapid growth in both these fields.

A second factor was the steep rise in the production of apparatus for the chemical engineering industry in the 1930s.[18] The major products manufactured, however, were such things as rotary crushing machines, hydraulic presses and pumps for handling the transport of chemicals within plants, in evaporation installations and in distilleries. Most of these were products already in use in the mechanical engineering industry. In these circumstances chemical apparatus and chemical plant were likely to be categorized as chemical machinery.[19] In the companies that

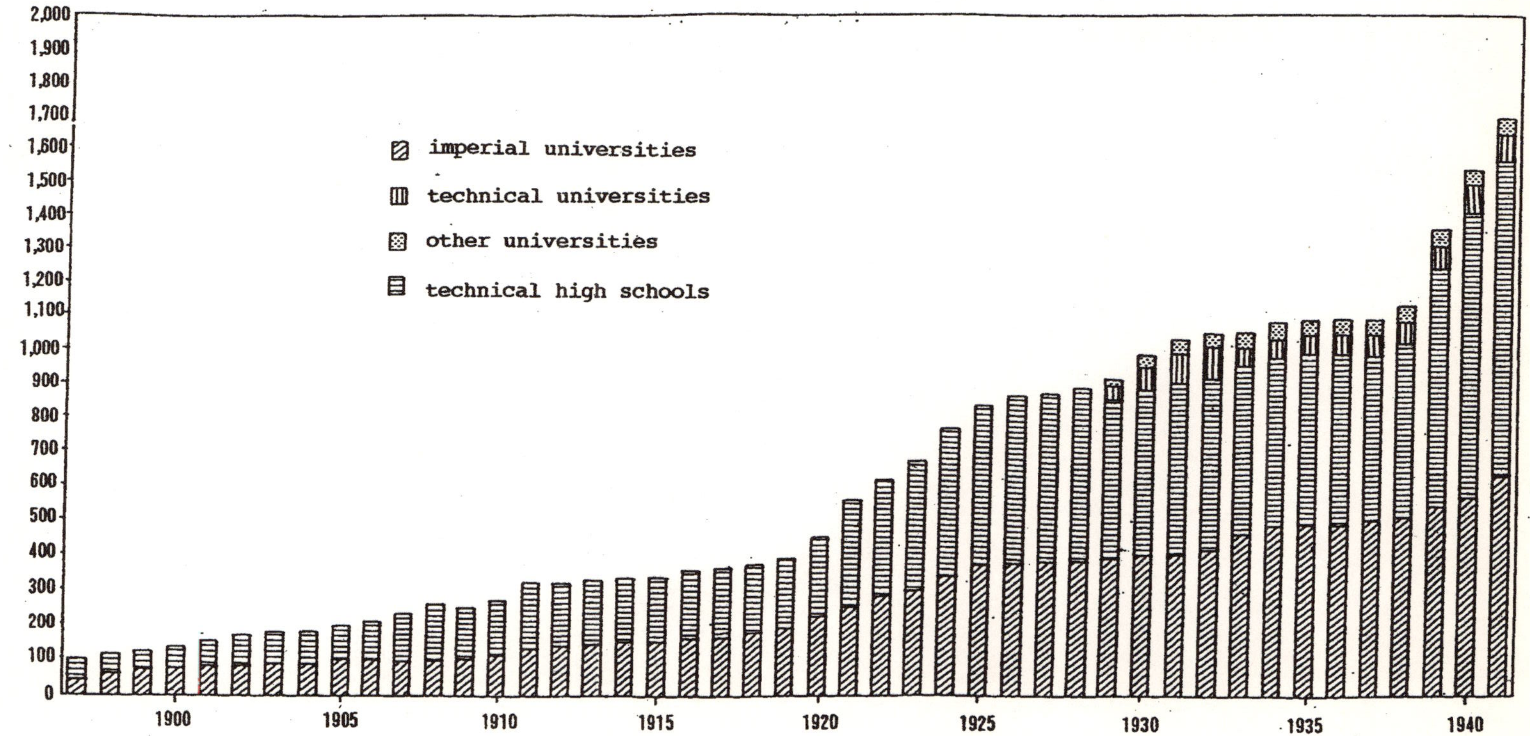

Figure 1 Professors and associate professors in science and engineering. The number of academic members during the period covered is taken from the *Annual Report of the Ministry of Education*.

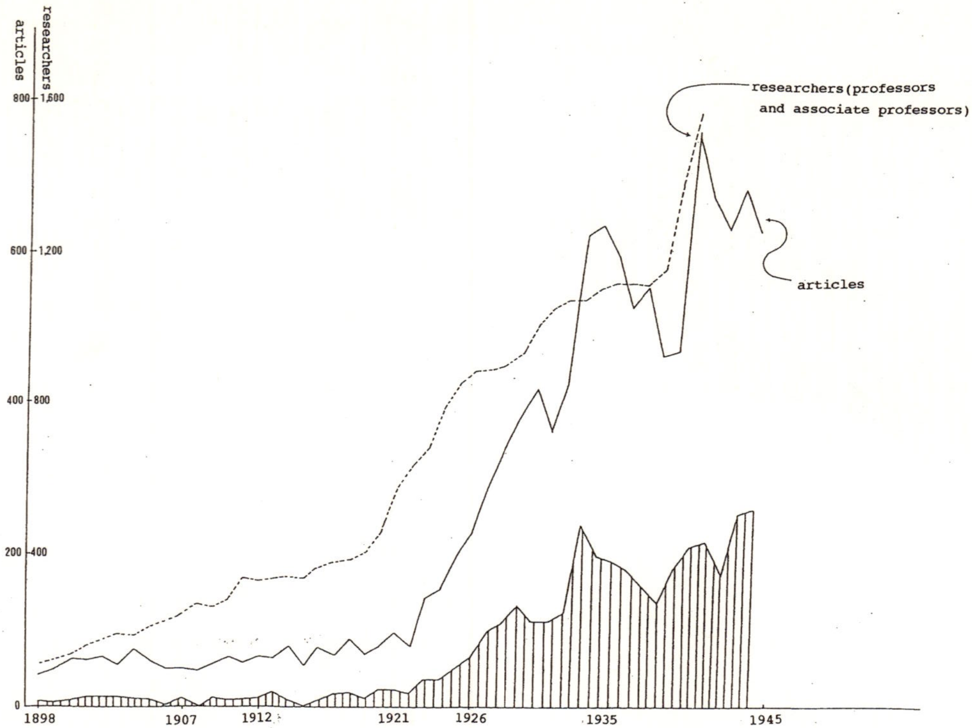

Figure 2 Articles in the *Journal of the Industrial Chemical Society of Japan* (Japan). The number of publications is taken from the section of *The Library Catalog of Kyoto Imperial University* dealing with engineering sciences.

manufactured these products the training of engineers and workers became very important since they were required to design and manufacture many types of products utilized by the chemical industry. In those days most design activity did not involve the development of new processes and plant but was involved instead with the modification of existing apparatus. Nevertheless, co-operation between various professionals engaged in this activity became more and more important. The way the Society of Chemical Engineering came to be named was influenced by these circumstances.

A third factor in the establishment of the Society was the growth of research in developing technologies capable of producing important chemicals and materials from indigenous natural resources such as alumina from Japanese clays, the liquefaction of coal to produce hydrocarbons, and ammonia from nitrogen, etc.[20] Accordingly, the design, development and construction of chemical apparatus assumed industrial significance. In the historical development of the chemical industry important apparatus had been imported from advanced countries. However, during the 1930s and 1940s the Japanese government strengthened the state-of-war regime, in being since the Japanese government began hostilities against China in 1931. Consequently, plant for the mass production of basic heavy chemicals had to be developed and constructed in keeping with the physical and chemical nature of local raw materials which, not unnaturally, often differed from the materials utilized in the developed countries. The Tsukijima Mechanical Engineering Company, for example, which had produced mainly plant for sugar production, began to construct oil refineries and aluminium plants: in the early 1930s few companies specialized in plant engineering

The importance of plant engineering becomes clear when one considers the development of alumina production from clay.[21] One usual process involved in the development was the Bayer process, which utilized bauxite as a raw material. The clay used in Japan contained less composites of alumina than bauxite. It was therefore a major problem to produce a high-quality product from poorer raw material if one were to compete economically with the Bayer process. No one in Japan succeeded in developing plant capable of processing native clays on a commercial scale, although some success was achieved at laboratory level. Generally speaking, Japan lagged in its research on the special problems encountered in upgrading plant from laboratory to commercial scale insofar as problems specific to engineering in the chemical industry were concerned. Immediately after its establishment the Society of Chemical Engineering convened several meetings devoted to the problem of developmental research at the intermediate level.[22]

The fourth factor concerned improvements in the training and education of chemical engineers. Students who graduated from colleges of applied chemistry and industrial chemistry came increasingly to take up work in mechanical engineering companies, designing and constructing chemical apparatus. Likewise, many students who had specialized in mechanical engineering took up employment with chemical companies. As a result a problem arose in universities and technical high schools as to how to teach mechanical subjects to students majoring in applied chemistry and vice versa. In the journal of the Society we find articles on such problems as how to train students taking courses in chemical machinery construction at technical high schools (1937),[23] and, in an article on the education of engineers in the chemical industry, on how co-operation was to be achieved between industrial chemists and mechanical engineers (1938).[24]

Among these articles a problem relating to the 'confrontation between chemists and mechanical engineers' was posed as follows:

> A mechanical engineer said: '... plants designed by chemists are difficult to construct because they don't understand mechanical engineering. We should tackle all these tasks ourselves once the size of the apparatus has been decided on by the chemists.' Against this some chemists responded: 'it is impossible to do so, because even an apparatus for agitating material should be designed and constructed by taking account of the chemical conditions in the reaction process'. Hearing these arguments, S. Uchida remarked that 'these arguments are like those in late 19th century America. The USA has since then established the quite new science of chemical engineering.'[25]

It is certainly the case that Uchida was correct in his assessment, and indeed his judgment is essential in gaining an understanding of how academic chemical engineering evolved. However, it is not clear whether the kind of chemical engineering that Uchida urged would have been of help in solving problems involving the issue of co-operation between chemists and mechanical engineers that arose in practice. In these circumstances the Society set up several schemes aimed at making materials available to help in the design and construction of chemical plant. Among these materials were a technical dictionary and a catalogue of process sheets. A process sheet shows the total connection of individual chemical processes involved in a chemical plant, and was useful in designing and constructing a plant.

CHEMICAL ENGINEERING TERMINOLOGY

The last point to be discussed here has to do with the confusion that existed in the 1930s relating to apparatus, plant, instrumentation and machinery which surfaced in discussion at the time of the establishment of the Society. In 1935 a meeting relating to the setting up of the Society was organized on the theme of chemical instrumentation development. Typical of the confusion that arose are the arguments that follow:[26] machines form part of chemical apparatus and therefore distilleries and evaporating plant should be deemed to be machinery.[27] Boilers should also be deemed machinery because they were produced by mechanical engineering companies. And so on. In his contribution to the debate[28] Professor Aoki of the Tokyo

Imperial University introduced a definition of chemical apparatus as used by Mataré, although this definition was not understood by all the participants. As for Mataré, his purpose was to draw attention to the need to distinguish the apparatus (the tool) from that which was being operated upon (the workpiece):

> During the process of work the tool is always active, whereas the workpiece is generally passive. Sometimes the workpiece imparts power to the tool, but this is not an essential feature of the apparatus. This happens in the same sense in which the stirring apparatus in producing spirits is no essential part of the workpiece. Around the apparatus, or inside it, something happens in which the apparatus, strictly speaking, takes no part. Only the tool acts because of forces which by natural law are part of it.[29]

Gaining a clear understanding was really dependent on the level of development that the engineering of industrial plant had reached. In those days in many cases the various parts of a chemical plant were often treated as products of the mechanical engineering industry and not as components of the total chemical process involved. The importance of the notion of chemical processes failed to attract sufficient attention among engineers to be widely discussed.

In all this we should take into account the critical importance of unit operations. Unit operations as a central concept of chemical engineering, as we have seen, may be defined in terms of the precise physico-chemical phenomena involved in each separate chemical process. According to R. Aris the notion of chemical engineering grew gradually as a means of bridging the gap between physics on the one hand and mechanical engineering and metallurgy on the other.[30] In other words, the notion of unit operations involved a 'trans-disciplinary' outlook *vis-à-vis* the individual disciplines in the early development of chemical engineering. In basic research on unit operations physico-chemical phenomena were to be considered as part of a process. In other research, corresponding to the direct demands of the industry, chemical apparatus and chemical plant were objects to be studied in an interdisciplinary way in the context of the chemical industry.[31] An understanding of the functions and structures of chemical apparatus was essential if co-operation between industrial chemists and mechanical engineers was to be achieved. Further, this kind of co-operation was a necessary part of the empirical basis of engineering science. In Japan the gap between practical needs and the various fields of scientific research resulted in confusion concerning use of appropriate terminology.

CONCLUSION

It was essential to bridge the gap between practice and theory in engineering if the engineering sciences as a whole were to develop. For chemical engineering to progress before the introduction of unit operations in pre-war Japan it was necessary for the idea to be seized that the apparatus used in chemical engineering was different in its nature from that used in mechanical engineering. This had to be learnt by the industrial chemists and mechanical engineers involved. Apparatus, plant and processes in the chemical industry had to be objects of interdisciplinary co-operation in research, in training and education, in publications and in the production of products in the field of chemical engineering. All this was part of the necessary framework within which chemical engineering could develop. The arguments that arose over the naming of the Society reflected the confusion then prevailing.

Notes and References

1. Unit operations are basic operations common to chemical plants, for example, distillation, drying, evaporation and so on, which terms originated chiefly as the application of principles of physics. After the introduction of the concept of unit operations by A.D. Little in 1915, William H. Walker, Warren K. Lewis and William H. McAdam developed this concept as a key notion of the American system of chemical engineering in the text *Principles of Chemical Engineering* (1923). The authors emphasized the fundamental character of unit operations. In the 1930s physico-chemical phenomena began to receive attention in chemical engineering.

2. N. Shimao, 'Shimomura Kotaro-sensei' ('Professor Kotaro Shimomura'), *Doshisha-jiho*, 1977, 53: 4.

3. The title of the Society's journal was changed from *Kagaku-kikai* (*Chemical Machinery*) to *Kagaku-kogaku* (*Chemical Engineering*) in 1953, and the name of the Society itself was changed to Kagaku-kogaku Kyokai (*Society of Chemical Engineering*) in 1956.

4. The earliest publication using the words 'chemical machinery' in its title is Eikichi Sato's *Kagakukogyo-kikai* (*Machinery for Chemical Industries*) (Tokyo, 1916).

5. The earliest Japanese student at MIT was Iwazo Suzuki. He enrolled in the Department of Chemical Engineering in 1908. See G. Jimbo, 'Kyokai no ayumiwo chushintoshita kagaku-kogaku no rekishi' ('The History of Chemical Engineering as it Affected the Development of the Society'), *Kagakukogaku*, 1986, 50: 167.

6. The first meeting of the Steering Committee for the establishment of the Society was held in October 1935. The meeting of the General Assembly leading to the foundation of the Society was held in September 1936. Membership totalled 1329 in March 1938, and 4506 in April 1940, before the end of World War II. The history of the Society has been recounted by Shun'ichi Uchida, *Enjinia eno michi* (*My Career as an Engineer*) (Tokyo, 1980); and Genji Jimbo *et al.*, 'The History of Chemical Engineering in Japan', in William F. Hurter (ed.), *History of Chemical Engineering* (Washington, DC, 1980).

7. See Olaf A. Hougen, 'Seven Decades of Chemical Engineering', *Chemical Engineering Progress*, 1977, 89–104; and W.F. Hurter (ed.), *op. cit.* (6).

8. Lewis K. Warren, 'Evolution of Unit Operations', *Chemical Engineering Progress Symposium Series*, 1955, 55, 2: 2.

9. Jean-Claude Guédon, 'Conceptual and Institutional Obstacles to the Emergence of Unit Operations in Europe', in W.F. Hurter (ed.), *op. cit.* (6).

10. Uchida, who studied chemical engineering at MIT, remarked: 'Chemical Engineering can be applied to every chemical industry, and should be considered as a basic science. Its advance has been derived from chemistry, physics and mathematics.' *Kagakukogyoron* (*An Outline of the Chemical Industry*) (Tokyo, 1933), 83. Saburo Kamei, who also studied at MIT, defined chemical engineering as Uchida had done but adopted a terminology relating to chemical machinery.

11. Professor Nobutaka Tomada of Tokyo Imperial University stated: 'Some machines utilized in the mechanical engineering industry and in other applications have been used in the chemical industry with little modification. The technology used here, as in chemical engineering, has been regarded as mechanical engineering applied to unit operations. From now on the technology of machinery used in chemical operations or chemical engineering should be developed on the basis of industrial chemistry.' 'Kagakukogaku to Kogyokagaku no Teikei') ('Cooperation between Chemical Engineering and the Chemical Industry'), (*Kogyokagaku zasshi*) (*Journal of Industrial Chemistry*), 1932, 35: 110. He published his *Kagaku-kikai-gaku* (*Outline of Chemical Machinery*) in 1941 in Tokyo.

12. Guèdon, *op. cit.* (9): 62.

13. Ernst Bartholomé, stressing a distinction in the conceptual framework of chemical engineering between the USA and Germany, remarked: 'Während insbesondere in den angelsächsischen Ländern Chemical Engineers vollkommen getrennt von den Chemikern als ein Zweig des Ingenieurwesens ausgebildet werden, ist es nach deutscher Auffassung günstig, für die Bearbeitung der auf der Chemie und physikalischen Chemie beruhenden Seiten des Chemical Engineerings Mitarbeiter zu haben, deren Ausbildung zumindest in den Anfängen mit der Ausbildung der anderen Fachrichtungen der Chemie übereinstimmt.' 'Entwicklung der technischen Chemie in Deutschland', *Chemie-Ingenieur-Technik*, 1976, 48: 913.

14. Guèdon, *op. cit.* (9), 46.

15. William C. Peck described the first notion of unit operations in Little's report to the President of MIT as follows: 'any chemical process, on whatever scale conducted, may be resolved into a coordinate series of what may be termed "unit operations", as pulverizing, drying, roasting, crystallizing, filtering, evaporating, electrolyzing and so on'. 'Early Chemical Engineering', *Chemistry and Industry*, June 1973, 2: 514.

16. Bartholomé, *op. cit.* (13), 916.

17. The following discussion shows a distinction between both groups – industrial chemists and mechanical engineers. The mechanical engineering group urged: 'the problems experienced by industrial chemists in chemical companies concern the utilization of construction materials used in plants. For example, chemists should examine the materials and the design of the plant before mechanical engineers complete the design work. Therefore cooperation between both sets of professionals is important.' Replying to this argument, Uchida remonstrated that 'chemical engineering is neither chemistry nor mechanical engineering'. *Kogyokagaku zasshi*, *op. cit.* (11), 110.

18. S. Kamei, 'The Statistics of the Chemical Industries: The Chemical Machinery Industry', *Kagakukikai Kyokai Nenpo* (*Annual Report of the Society of Chemical Machinery*), 1937, 1, 229–332; M. Kawanishi, 'Kagakukogyo oyobi Kagakukikaikogyo no Tenbou' ('A Perspective of the Chemical Industry and the Chemical Machinery Industry'), *ibid.*, 247–332. According to the statistics in this article, products in 1935 totalled 21.7 million (in Japanese yen) compared to 14.3 million in 1933.

19. Uchida urged that apparatus or plant should be called machinery because more attention was paid to moving parts than to static parts of apparatus at the design stage, S. Uchida, *Kagakukougyo to sono Sochi* (*The Chemical Industry and its Apparatus*) (Tokyo, 1935), p. 11.

20. Yuji Jido, 'The Development of the Process of Alumina Production from Clays and the Production Technologies Involved', *Ritsumeikan Business Management*, 1990, 28: 61–89 (in Japanese); Anthony N. Stranges, 'Synthetic Fuel Production in Pre-War and World War Two Japan: A Case Study in Technological Failure', *Annals of Science*, 1993, 50, 229–65.

21. Y. Jido, *op. cit.* (20), 67.

22. Discussion meetings on the following topics were held by the Society: 'Chemical Machinery in Germany' (1938) 'Measurement Instruments in Chemical Machinery' (1939); 'The Development of the Chemical Machinery Industry and the Chemical Industry under the State-Of-War Regime' (1939); 'The Promotion of Developmental Research and its Application to Industry' and 'The Promotion of Developmental Research at Semi-Commercial Level' (1940); 'The Problems of High-Pressure Devices' (1942).

23. *Kagakukikai*, 1937, 1: 109.

24. *Op. cit.* (23), 1938, 2: 11.

25. *Ibid.*, p. 25.

26. Report of a Conference on the Development of Chemical Machinery, *Kagakukogyojiho*, March 1935, Vol. 8, 9: 10–15.

27. *Ibid.*, p. 12.

28. *Ibid.*, p. 13.

29. Franz Mataré, *Die Arbeitmittel: Maschine, Apparat, Werkzeug ... Eine Abhandlung ihren Einfluss auf Industriebetrieb unter eingehender Berücksichtung des Apparatwesens* (Berlin, 1913), p. 34: 'Während des zweckerfüllenden Vorganges verhält sich hier der Arbeitgegenstand stets aktiv, das Arbeitsmittel dagegen bleibt prinzipiell passiv. Es hat allerdings dem Arbeitsgegenstand, wie wir sahen, mitunter auch Kraftwirkungen mitzuteilen, doch stellt dies kein wesentliches Merkmal des Apparatbegriffes dar, wie ja, um auf das oben angeführte Beispiel zu verweisen, das Rührwerk der Branntweinblase durchaus kein unlässlicher Bestandteil des Arbeitsmittels ist. Um den Apparat oder in seinem Innern spielt sich ein Vorgang ab, an dem er selbst streng genommen durchaus unbeteiligt bleibt. Nur der Arbeitsgegenstand agiert vermöge von Kraften, die ihm von Natur aus innewohnen.'

30. Rutherford Aris, 'Academic Chemical Engineering from a Historical Persepective', *Industrial and Engineering Chemistry Fundamentals*, 1977, 16, 1: 1.

31. In Germany the Fachgruppe für chemisches Apparatwesen was established in 1918 based on the former organization of DECHEMA. Cf. Max Buchner, 'Denkschrift über die Ziele und Aufgabe der Fachgruppe für chemisches Apparatwesen ... Abteilung für chemisches Grossapparate ...', *Festschrift 50 Jahre DECHEMA* (Frankfurt, 1970), 21–33.

Wilhelm Ludwig von Eschwege (1777–1855), a German Engineer of Mining and Metallurgy in Portugal and Brazil

FRIEDRICH TOUSSAINT

In 1822 the German poet Goethe, living at the court of Weimar, wrote in his diary 'Baron Eschwege has come from Brazil and shows jewels, metals and types of rock. His Highness bought a considerable number of pieces.'[1] Eschwege was then 44 years old and had returned from Brazil after 11 or 12 years' absence from Europe. They discussed Goethe's preferred theme, the genesis of diamonds in Brazilian rocks, a subject on which Eschwege had just had a book published in Weimar.[2] Here in Weimar lived his cousin and friend of his youth, Sophie von Baumbach, to whom he had become engaged. She was then 35 years of age and a lady-in-waiting at court.

Who was this Baron Eschwege (Figure 1), who had access to the Duke's court and to Goethe and equally to Alexander von Humboldt? Wilhelm Ludwig von Eschwege was born on 15 November 1777 on the manor of Aue near Eschwege in Lower Hesse.[3] The family was of the old Hessian nobility and his father a District Administrator. He was educated by a private tutor, and wished to become a soldier, especially since the learning of languages was difficult for him. His parents, however, were against this.

At the age of 15, in 1793, Eschwege was sent to the grammar school in Eisenach. This was at the time of the Coalition Wars with France, and he observed the marching through of numerous armies, especially the beaten French. In 1796 he completed his grammar school studies, afterwards proceeding to Göttingen University. For a time he studied contemporaneously with the famous mathematician Karl Friedrich Gauss, who was born in the same year (and who also died in the same year) as Eschwege. Complying with the wishes of his parents, Eschwege studied law but soon changed to public administration, preferring to prepare himself for a more practical career. To this end he followed courses in natural history, architecture, technology, commerce and forestry. His father died at the age

Baron Wilhelm Ludwig von Eschwege
Nach einer Zeichnung von Krumholz 1845

Figure 1 Portrait of Eschwege by Ferdinand Krumholz (1810–1878), a German painter who lived from 1844 to 1848 in Portugal and in Brazil from 1848 to 1853. The Eschwege portrait was probably done in Lisbon since Krumholz was known to Ferdinand of Coburg and was painter at the Portuguese Court.

of only 52 during Eschwege's time at university. His guardian became Freiherr von Baumbach, later his father-in-law. In 1799 Eschwege transferred to Marburg University since Göttingen did not offer certain specialities such as mineralogy, geology, mining and metallurgy which at Marburg were part of the course in public administration. In 1800 he successfully passed his examinations at the age of 23, and had then to look for employment.

In Brazilian publications it is mostly said that Eschwege had attended the Freiberg Academy of Mining. This error possibly arises from the fact that many well-known Brazilians had attended the Freiberg School, as for example Eschwege's first superior in Portugal, Bonifacio de Andrade, and his later adversary in Brazil, Manoel Ferreira da Câmara as well as Colonel Mursa, later director of the iron plant at Ipanema, Capitanía (province) of São Paulo.

In the Portuguese translation of 1979 of his principal work, the *Pluto*

Brasiliensis[4] (Figure 2), the editor, Mario Guimarães Ferri of the University of São Paulo, notes correctly that Eschwege had studied at Göttingen but five pages later says that he completed his professional training at the famous school of Freiberg! Following completion of his studies in 1801 Eschwege was employed in an unpaid capacity by the Hessian Government at the copper mine of Richelsdorf. When he decided that he could derive no further benefit at Richelsdorf he went to Clausthal to continue his mining studies.

The following spring he was asked by the Hessian Government whether he was willing to accept employment as director of a metallurgical plant in either Portugal or Brazil. Since he assumed that in such a post abroad he could expect quicker promotion than in the much more unpromising conditions at home, he accepted, as also because the terms of his remuneration were better: two thalers (about 600 milreis), most likely monthly, plus free accommodation with garden and fuel, the choice of an increase in salary after ten years or a pension for life on half salary, and certainly free travel from Germany and back. His departure for Portugal was somewhat delayed, so he used this time to improve his knowledge at the Rothe Hütte ironworks in the Harz mountains. In June 1803 he finally arrived in Portugal.

Eschwege's activities in Portugal are described in detail by himself in an article published some time after his return to Europe in 1835.[5] This work includes a broad description of mining and metallurgy in Portugal, as well as a history of mining in the kingdom, together with its technological development. Sometimes, however, his remarks on the state of affairs in the Kingdom of Portugal are somewhat critical, similar to those he was to make later – candour which won him enemies as well as friends.

At the time of Eschwege's arrival in Portugal mining and metallurgy there were in a state of decline despite earlier attempts at renewal by the Marques de Pombal (famous for having rebuilt Lisbon after the devastating earthquake of 1755). He arrived in the summer of 1803 at his new place of work, the ironworks of Foz d'Alge (mouth of the river Alge) south of Coimbra, accompanied by two colleagues and a servant. The colleagues were Martin Stiefel from Schmalkalden Thuringia (where a blast furnace plant was working) and Ludwig Wilhelm Varnhagen, whom he knew from university days. (Varnhagen later built and managed the first really productive, though unprofitable, ironworks with blast furnaces in Brazil.) Eschwege was greatly disappointed by what greeted him at Foz d'Alge: the technical state of the ironworks was as disastrous as the living conditions for the staff were unpleasant. Since he did not speak any Portuguese, he needed an interpreter.

Eschwege's superior in Portugal was the chairman of the Portuguese mining inspectorate, José Bonifacio de Andrade. De Andrade had been born in Santos, Brazil, at that time belonging still to the crown lands of Portugal. He had studied mining and metallurgy in Freiberg, Saxony, and thus spoke perfect German. Andrade later became famous, not so much as a scientist but as a politician. He taught the second emperor of Brazil, Dom

PLUTO
BRASILIENSIS.

Eine Reihe von Abhandlungen über Brasiliens Gold-, Diamanten- und anderen mineralischen Reichthum, über die Geschichte seiner Entdeckung, über das Vorkommen seiner Lagerstätten, des Betriebs, der Ausbeute und die darauf bezügliche Gesetzgebung u. s. w.

von

W. L. von ESCHWEGE,

Königl. Portugies. Ingenieur-Oberst und Oberberghauptmann u. s. w.

Nebst lithographirten Karten und Zeichnungen.

Berlin,
G. Reimer.
1833.

Figure 2 Frontispiece of *Pluto Brasiliensis.*

Pedro II, and played an important part in the struggle for Brazilian independence; he was even called 'patriarca da independencia brasileira'.

In his article in Karsten's archive[6] Eschwege describes very vividly his exhausting journey through northern Portugal to the ironworks at Foz d'Alge. On his way, seeing only totally unforested mountains, he asked a peasant where wood could be obtained for the ironworks. The answer was: 'Senhores, all these mountains which you see supply such a quantity of fuel that the furnaces are not able to consume it ... the "Cepa" [roots of the bilberry] are now over 30 years old, so they can be cropped.' From time to time the bilberry bushes were set alight in order to increase the strength of the roots.[7] Today this area looks completely different. All the mountains and valleys are covered with woodland having been reafforested with eucalyptus (Figure 3). After finding, with some difficulty, the exact location of the former ironworks, I visited the site in the summer of 1998. The river Zêzere, now dammed, forms a very long and narrow lake down to São Pedro de Tomar, nearly 40 km downstream. As a consequence the

Figure 3 Present-day view of the surroundings of the Ferraría de Foz d'Alge (photo taken by the author in 1998).

Figure 4 An earlier view of the plant. Photo taken at a moment of low water (Archives of the Câmara Municipal de Figueró dos Vinhos).

mouth of the small affluent of the Rio Zêzere, the Alge, enters the Zêzere higher up the valley, with the result that the plant is now closer to the Zêzere than previously. Some years ago the storage lake was emptied for maintenance work and some pictures were taken at that time of the ruins of the ironworks (Figure 4). Today little can be seen below the water, just the upper part of the throat of the furnaces.

Eschwege became seriously ill after his arrival but succeeded finally in getting the blast furnace into production. Since the works was overstaffed, he was sent around Portugal in order to inspect the state of Portuguese mining. His observations were published later in several articles in Germany. Since mining and metallurgy in Portugal were run by the army Eschwege was appointed an officer in the corps of engineers. At this time the Napoleonic invasion of Spain and Portugal took place. Whether Eschwege participated in the war against Napoleon is not clear but his promotion to captain was presumably by reason of his service in the army during the Peninsular War. In the course of his work at the iron plant he carried out many experiments in furnace operation as well as in the working of the finery. He succeeded in reducing charcoal consumption and began the work of reafforestation. He planted at that time pines (*pinus maritima* and *pinus sylvestris*), not what is planted today – eucalyptus, at that time not yet known in Europe.[8] The main output of the works at Foz d'Alge was guns and ammunition.

In March 1808 the royal court left Lisbon for Brazil in order to avoid captivity under Napoleon (João, the son of Queen Maria, became regent because of the insanity of the queen, and was proclaimed João VI, King of Portugal, in Rio de Janeiro in 1816 when Maria died). In 1809 Eschwege was summoned to follow the royal family into exile, something he did with great pleasure, since he had always wished to work in the New World. His first visit on arrival was to the court, where he was welcomed and immediately promoted to Major in recognition of his meritorious service in Portugal. His first task was the establishment of a mineralogical collection, following which he was offered the post of professor at the military academy. This post he declined since he preferred to be engaged in practical work in mining and metallurgy in the interior of the huge country. Together with the responsible minister he worked out a plan for his future work. The following programme was the outcome:

- inspection of the gold district in Minas Geraes, improvement of gold-washing;
- inspection of the diamond district;
- construction of an ironworks;
- exploration of the area of the Rio Doce and the protection of that area against the savage Botocudo Indians; and finally,
- study of the possibility of opening up the Rio Doce for navigation.

Today the main railway of the world's largest iron ore mining company, the Companhia do Vale do Rio Doce, runs here. Eschwege was to live in the town of Ouro Preto (formerly Villa Rica), at that time capital of the Capitanía Geral of Minas Geraes (General Mines) (Figure 5). Now began a period of 11 years of hard and effective work, with many discoveries and the acquisition of new knowledge. This was the century of great discoveries. Alexander von Humboldt had completed his expedition through South America only six years earlier. When in 1817 the future empress of Brazil, Leopoldine von Habsburg, arrived in Rio de Janeiro from Vienna[9] as bride of the crown prince, she was accompanied by several scientists: the botanist and mineralogist Johann Emanuel Pohl, the zoologist Johann Natterer, the landscape painter Thomas Ender and the Bavarian scientists J.B. von Spix and K.P. von Martius. All of them called on Eschwege since he was already well known and all European visitors felt it necessary to ask for his advice and avail themselves of his knowledge. The famous French naturalist, Auguste de St Hilaire, visted him together with the Russian Consul, von Langsdorff, before the start of St Hilaire's expedition through the interior of Brazil. This was the century of the real discovery of South America, and most of the explorers of Brazil asked Eschwege for his advice. Together with Spix he published maps of the country.

In 1811 Eschwege finally set out for Minas Geraes, a journey of some 270 miles which took him about two weeks (today by plane, every hour, slightly more than half an hour). Already in the same year he started work on the construction of an ironworks at Congonhas do Campo, he himself taking 20 per cent of the shares, thus assuming also a share of the risk (Figure 6). The desire to build this plant (a relatively small works) arose from his wish to anticipate two huge government-owned projects:

Figure 5 View (by Rugendas) of the town of Vila Rica (now Ouro Prêto), Eschwege's residence.

- In Minas Geraes a former student of Freiberg, the chief inspector of the diamond district, Manoel Ferreira da Camara, had since 1805 been planning a large ironworks with several blast furnaces at the Morro do Pilar.
- Meanwhile, in the province of São Paulo, at Ipanema, his old friend, Varnhagen, was also building a plant with blast furnaces after an earlier failed attempt to build an ironworks.

In his *Pluto Brasiliensis* Eschwege discusses in detail and with great acumen his quite modern economic ideas as to the possible basis of a large Brazilian iron industry. He devotes a whole chapter to this subject: 'Kann eine grosse Eisenhütte in Brasilien bestehen?' ('Is a large-scale iron smelter viable in Brazil?'). His conclusion was that it was not sustainable. The reasons were quite simple: though salaries for local workers were lower and materials cheaper in the interior of the country than on the coast, salaries for foreign experts were much higher than in Europe. This was not to mention transportation costs from the plant site in the interior to the coast

Figure 6 View of part of Eschwege's iron plant at Congonhas do Campo (the trompes) (photo taken by the author in 1985).

where the centres of consumption were located. The cost of shipping metal across the Atlantic was lower than the cost of moving iron by packhorse from the only areas where iron ore deposits occurred and where wood fuel was abundant.

In short, the production of iron to satisfy immediate local needs would be advantageous and profitable but not for export out of the area of production. In these regions the high cost of transport to the coast indicated the need for small-scale plant. A plant with blast furnace would be much too large. This was why Eschwege opted for such small-scale production and decided to build bloomery furnaces (Figure 7). Some of these arguments remained valid until recently. The course of events proved his point: his plant was always profitable while Camara's was a technical and economic failure. So also was Varnhagen's works at Ipanema which, though technically quite feasible, never paid dividends, through no fault of Varnhagen.

Eschwege's ideas were adopted throughout Minas Geraes, many such small local plants starting production with much success and producing iron and steel tools for agriculture, mining and so forth. His economic and technical success explains why in Brazil it is frequently asserted that he was the first to produce iron in Brazil. This is not in fact correct since Camara's blast furnace was tapped (for only a very short time) before Eschwege's bloomery furnaces came into production. The first technically functioning blast furnace was at Ipanema, though also not profitable.

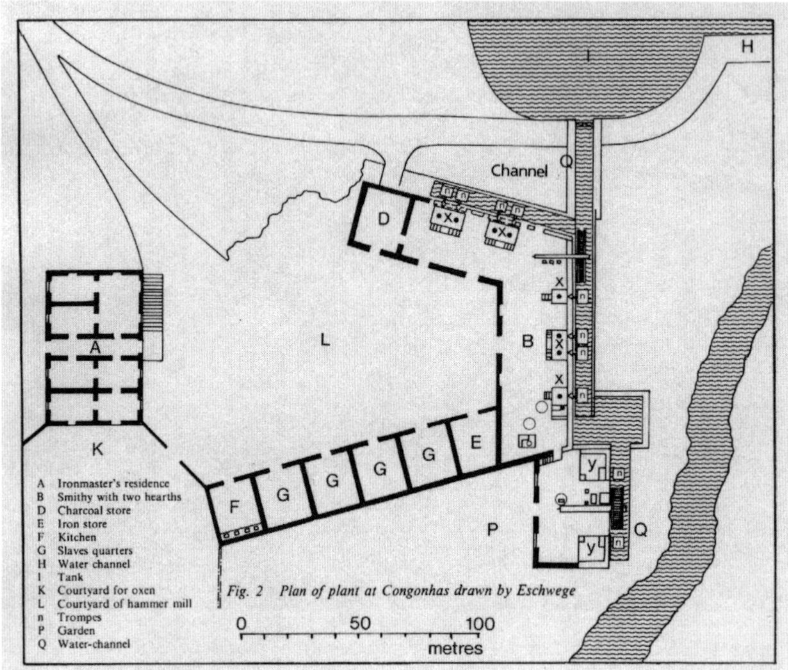

Figure 7 Plan of Eschwege's plant at Congonhas, redrawn from *Pluto Brasiliensis*.

Eschwege's plant is still visible at Congonhas do Campo, though partially in ruins. He named his ironworks *Fabrica Patriotica* which was certainly an act of homage to his host country. The name is still in use today. The iron ore mine in the same location is at present owned by the German Thyssen company. In his *Pluto Brasiliensis* Eschwege includes a plan of his plant. Interesting is the fact that at Congonhas he used trompes instead of conventional blowing apparatus. It is not known where he picked up knowledge of trompes: they were not much used in Germany at that time. Graham Hollister-Short was kind enough to acquaint me with Henning Calvör's book on mining machinery in the Harz area.[10] Calvör describes trompes for use in mining (ventilation) and in copper workng. The oldest description of trompes in Germany is to be found in Leupold's *Theatrum Machinarum Generale*.[11] Leupold refers to the Italian origin of these devices, described earlier by Giambattista della Porta in 1601.[12] Eschwege could have come across trompes in the Harz, though they were to be found in use mainly in the Alps and the Pyrenees, especially in Catalan forges, *forjas catalanas*. A detailed scientific description of them is to be found in Venturi[13] who, however, also notes that they were mainly in use in the Alps and the Pyrenees. So also does Karsten in 1816.[14] Trompes are very advantageous in cases where there is a small quantity of water but with a reasonable head. But it is interesting to know that the trompe was in extremely widespread use in Brazil (Minas Geraes) during the whole of the nineteenth century.[15]

Da Costa Sena and Ferrand make frequent reference to the Pyrenees and the Alpine region but never to Eschwege. Apparently Eschwege learnt this trompe technique in Brazil and was obviously very pleased with it since he introduced it into Portugal some years after his return to Europe in 1821. This confirms, contrary to what has been alleged, that he did not disapprove of foreign ideas; he explicitly says that these should be accepted and learnt from when they prove useful in special circumstances.

Eschwege's major publication was his *Pluto Brasiliensis*, in which are described most of his activities in Brazil. It was published in Berlin only in 1833. The book not only comprises reports of his personal activities but is also a reference book on everything to do with geology, mining and metallurgy in Brazil in the first half of the nineteenth century. He deals systematically with all metals. Most of the first volume is dedicated to gold (the most important mining product of Brazil at that time), its history, production figures and technical details of mining and metallurgical extraction.

The second volume starts with diamonds, the second most important product in terms of financial contribution to the public purse. The technique of washing diamonds was considerably improved by Eschwege and he presents valuable statistical information on production and trade, giving a detailed description of all types of diamonds found in Brazil (109 types of different colour, purity and aspect). He even deals with the problem of trafficking in these stones.

A separate chapter describes all other mining products: silver, copper, tin, quicksilver, platinum, cobalt, manganese, bismuth, lead (galena). Non-metallic products such as saltpetre, common salt, mineralized water sources, coal, sulphur and alum are also discussed. About coal Eschwege says that there are some deposits in the northern province of Bahia and quite considerable deposits in the south in the province of Rio Grande do Sul (but of very poor quality). This coal is geologically young and closer to lignite (brown coal) than to older coals. It would not be appropriate for use in iron metallurgy (it should be remembered that at that time in Germany mineral coal was in modest use only in Gleiwitz in Upper Silesia). Eschwege expected to find better quality coal as the country was more fully explored and more densely populated. The finding of better-quality coal deposits would be, he believed, of great importance since the areas of huge iron ore deposits were poor in wood fuel and thus needed mineral coal! (This is still a problem today for the Brazilian iron and steel industry which has to import coking coal to smelt the ore obtained from the world's largest iron ore deposit.)

The mining of lead-bearing ores kept Eschwege busy for quite some time. In northern Minas Geraes he discovered deposits of galena. In that area he also discovered two rivers, unknown up to that time, which he called the Verra and the Fulda (after the rivers Werra and Fulda in his home country, Hesse).

Several chapters of *Pluto Brasiliensis* deal with iron. In these Eschwege describes the ancient history of iron working (in Brazil):

- The 'Real Fábrica do Morro do Pilar' in the Province of Minas Geraes (a detailed history of the financial and technical failure of the blast furnace plant built by Manoel Ferreira da Camara;
- The 'Real Fábrica de Ferro de São João do Ipanema' (a history of the blast furnace plant at Ipanema in the province of São Paulo, managed by Eschwege's friend Varnhagen, including all financial reports);
- The 'Fabrica de Ferro do Prata em Congonhas do Campo' (his own plant) with an appendix on smelting experiments conducted by Englishmen at Gongo Soco; and a chapter also on native and meteoric iron.

This important work finishes with three chapters on the general economic position. In these he discusses whether a large-scale ironworks could survive in Brazil, what the effect of the suppression of slave trade would be on mining, and what measures were necessary for the development of the mineral industry generally.

In short, Eschwege wrote a book of extraordinary importance for Brazil. His knowledge of Brazilian geology and mineralogy is impressive, but his *Pluto* is far from being the sum of his achievements. His other publications on Brazil contain quite important material. Many of these articles had been published in Europe during his sojourn overseas, which meant that at that time he was already recognized as an important scholar in Europe. Much was published after his return from the huge cache of material he brought back with him.

Many maps, geographical and geological, were draughted by Eschwege, several of them jointly with Spix. Many of these continued in use until the beginning of the last century. A mineral found in the Rio Doce area was named after him – Eschwegeit. He became known in Brazil as 'O pai da geologia Brasileira' (the father of Brazilian geology).[16]

After a stay of nearly 12 years Eschwege returned to Europe in 1821. The reason for this was simply the new situation created by the proclamation of independence of Brazil under its first emperor, Dom Pedro I. Eschwege still felt loyalty to the Portuguese crown. He saw no useful role for himself in Brazil under the new regime (although his old friend Bonifacio de Andrade was much involved in the struggle for independence and would have supported him had he stayed on). Eschwege was extremely loyal, and revolutions were not to his liking. He presented himself immediately to King Dom João VI in Lisbon to offer his services. Emperor Dom Pedro I had given him leave for two years (he had been promoted in the meantime to Colonel). In Lisbon he was nominated an ordinary member of the Royal Academy of Lisbon (a learned society) to which he presented the first complete statistical and geographical report on the province of Minas Geraes. He left Lisbon for England on 21 October 1821 on the English packet boat *Lisbon*, reaching Falmouth seven days later. From Falmouth he travelled to Oxford and Windsor and then to London where he stayed until 18 November. From London he travelled to Paris where he visited Alexander von Humboldt who was living there at that time, busy with the publication of his masterpiece: his research on South America, *Voyage aux régions équinoctiales et du nouveau continent fait en 1799 ... et 1804*, published in Paris between 1814 and 1825.

From Paris Eschwege travelled via Kassel to Weimar where he met Goethe and the Archduke of Weimar, and certainly his future bride, Sophie von Baumbach. He went to Vienna where he presented a letter to the emperor of Austria, Francis I, from his daughter, Leopoldine, now empress of Brazil.

In March 1823 Eschwege married Sophie von Baumbach, and in June he arrived in Lisbon where he became Inspetor-Geral das Minas (Chief Intendant of Mining). A period of political unrest now ensued in Portugal which ended with his dismissal as Chief Intendant shortly after he had asked for two years' leave from the service of João VI and had returned to Germany. There he finished and published his *Beiträge zur Gebirgskunde Brasiliens* and his *Pluto Brasiliensis* in 1832 and 1833 respectively. In the meantime the situation in Portugal improved. The Brazilian emperor Dom Pedro I abdicated in favour of his son and returned to Portugal. Eschwege asked for a prolongation of his leave which he obtained on condition that he would return from Germany to be installed again as Chief Intendant of Mining, a condition that was not to be fulfilled.

Instead he was chosen to rebuild a former monastery on a mountain top above Sintra, where the old royal palace was located, and transform it into a summer residence for the royal couple. This was the last phase of

Figure 8 Castle of La Peña in Sintra (photo taken by the author in 1998).

Eschwege's life: he became an architect! The palace, which can be visited today, is surrounded by a beautiful park, laid out by a German forest superintendent. The castle, perhaps not very much to the modern taste, consists of a mixture of mauresque and manolinian styles, which were themselves already a mixture of the Gothic and the Byzantine (Figure 8). At weekends the castle is an attraction for the people of Lisbon since it is somewhat of a 'fairy castle'.

In 1850, aged 73, Eschwege finally returned to Germany. He died there in 1855 at the age of 78. He had been 47 years in the service of the Portuguese crown. He had arrived in Portugal at the age of 26 as an inexperienced, young, but ambitious man, and left as a famous scientist, decorated with the Brazilian Order of Christ, among many other decorations. The *Revista Militar*, which he inaugurated in Lisbon, published a laudatory obituary, and King Ferdinand of Portugal, in a letter to Eschwege's widow, called him a true friend whom he would never forget.

Notes and References

1. Johann Wolfgang von Goethe, 'Tag- und Jahreshefte 1822', in *Sämtliche Werke* (Zürich, 1977), Vol. 11, 951–2.
2. W. von Eschwege, *Geognostisches Gemälde von Brasilien und wahrscheinliches Muttergestein der Diamanten* (Weimar, 1822).
3. Friedrich Sommer, *Wilhelm Ludwig von Eschwege 1775–1855. Lebensbild* ... (Stuttgart, 1928), p. 7.
4. W. von Eschwege, *Pluto Brasiliensis* (Berlin, 1833); Brazilian edn., edited by M.G. Ferri, 2 vols. (trans. D. de Figueredo Murta) (São Paulo, 1979).
5. W. von Eschwege, 'Bemerkungen über den Bergbau und Hüttenbetrieb in Portugal', in C.J.B. Karsten (ed.), *Archiv für Mineralogie, Geognosie, Bergbau und Hüttenkunde* (Berlin, 1835), 8: 185–225.
6. *Op. cit.* (5), note 4, 190–92.
7. *Ibid.*, note 4, 192.
8. *Ibid.*, note 4, 198–9.
9. It may be of some interest to note that the main colours of the Brazilian flag (green and yellow) are the colours of the Portuguese House of Braganza (green) and the House of Habsburg (yellow).
10. Henning Calvör, *Historisch-chronologische Nachricht und theoretische und praktische Beschreibung des Maschinenwesens und der Hülfsmittel bey dem Bergbau auf dem Oberharze* (Brunswick, 1763), 7–10.
11. Jacob Leupold, *Theatrum Machinarum Generale, Schauplatz des Grundes mechanischer Wissenschaften* (Leipzig, 1724), § 331, 108.
12. *Ibid.*, § 340, 109; Giambatista della Porta, *Pneumaticorum Libri III* (Naples, 1601), cap. viii, 58–9.
13. J.B. Venturi, 'Untersuchungen und Erfahrungen über die Seitenmitteilungen der Bewegung in flüssigen Körpern, *Gilbert's Annalen der Physik* (Halle, 1800), III, 130–42.
14. C.J.B. Karsten, *Handbuch der Eisenhüttenkunde* (Halle, 1816), 477–90.
15. Joaquim Candido Da Costa Sena, 'Viagem de estudios no centro da Provincia de Minas', *Anais da Escola de Minas de Ouro Prêto*, 1881, 1: 106–43; Paul Ferrand, 'A industria do Ferro no Brasil (Provincia de Minas Geraes)', *Anais da Escola de Minas de Ouro Prêto*, 1885, 4: 122–39.
16. Dr Johann Baptist von Spix (1781–1826), Bavarian naturalist, who was a member of the entourage of Archduchess Leopoldine von Habsburg when she travelled to Brazil for her wedding with Prince Pedro, later Emperor Pedro I.

List of Eschwege's Publications on Brazil and Portugal (Excluding Numbers of His Publications in Portuguese)

1. 'Extracto de uma memoria sobre a decadência das minas de ouro da Capitanía de Minas Geraes e sobre vários objetos montanísticos. Academia Real de Sciencias de Lisboa', *Memorias*, 1811.
2. 'Brief über Brasilien', *Neue allgemeine geographische Ephemeriden*, F.J.J. Bertuch (ed.), 1815.
3. 'Mineralogische Nachrichten über Brasilien', *Jahrbuch der Berg- und Hüttenleute* (Nuremberg, 1815).
4. Idées générales sur la constitution géologique du Brésil', *Annales des Mines* (Paris, 1817).
5. 'Physikalische und bergmännische Nachrichten aus Brasilien', *Annalen der Physik*, L.W. Gilbert (ed.) (Leipzig), 1818, Vol. 60.
6. 'Vorkommen des elastischen Sandsteins', *Annalen der Physik*, L.W. Gilbert (ed.) (Leipzig), 1818, Vol. 60.
7. *Journal von Brasilien* (Weimar, 1818).
8. 'Observations sur la manière de voyager dans l'intérieur du Brésil', *Nouvelles Annales des Voyages de Géologie et d'Histoire* (Paris), 1819, Vol. 3.
9. Über einige merkwürdige brasilianische Gebirgsformationen', *Annalen der Physik*, L.W. Gilbert (ed.) (Leipzig), 1820, Vol. 62.
10. *Geognostisches Gemälde von Brasilien und wahrscheinliches Muttergestein der Diamanten* (Weimar, 1822).
11. 'Geognotische Beobachtungen über einen Theil der Capitanie von São Paulo', *Taschenbuch für die gesammte Mineralogie*, K.C. von Leonhard (ed.) (Frankfurt am Main), 1822.
12. 'Voyage de Rio de Janeiro au Comarca d'Ilha Grande fait en 1810', *Nouvelles Annales des Voyages de Géologie et d'Histoire* (Paris), 1823, Vol. 20.
13. 'Note on the Geognosy of Brazil', *Edinburgh Philosophical Journal*, July 1823, Vol. IX.
14. 'Esquisse géognostique du Brésil, suivi d'une dissertation sur la gangue originaire du diamant', *Annales des Mines* (Paris, 1823), Vol. VIII.
15. 'Noticias e reflexões estatísticas a respeito da provincia de Minas Geraes', *Memorias da Academia Real de Sciencias de Lisboa*, 1825, Vol. IX.
16. *Brasilien, die Neue Welt* (Brunswick, 1830).
17. *Beiträge zur Gebirgskunde Brasiliens* (Berlin, 1832).
18. 'Bosquejo geognostico do Brasil, com uma dissertação sobre a matriz dos diamantes', *Geologia Elementar de Nereo Boubée* (Rio de Janeiro, 1846).
19. *Odologia dos engenheiros constructores ou guia para construcção e conservacção das Estrades em Portugal e no Brasil* (Lisbon, 1843).

Symposium: The Current State of the History of Technology in Britain

Introduction

GRAHAM HOLLISTER-SHORT

It has been grievously obvious for many years that in Britain the history of technology as a discipline has lacked any meaningful institutional existence. Whatever the quantum of other observable activity has been or is now, it cannot alter this fact: it is still Hamlet without the Prince. Given the importance of the subject, which I take it few would wish to deny, this has always been a highly unsatisfactory state of affairs. But why should things have turned out like this? In the 1960s it was possible to be reasonably optimistic. Two chairs in the subject had been created *de novo* in London and Manchester, and no one then could have predicted that each of the centres, instead of growing, would later experience implosion. Apart from other factors, questions of timing were undoubtedly involved. As to timing, the 1980s, when these events were played out, certainly coincided with what I think of as the onset of a 20-year Ice Age in British academic life. Job opportunities in the university sector had by then virtually ceased to exist. Expansion had abruptly stopped, while such had been the mass recruitment of the 1960s that few vacancies arising from retirement could be expected before the late 1990s. Movement in the university sector, in a word, had seized up to such an extent as to call for initiatives to ease the gridlock. I recall from my own experience that the British Academy began offering research fellowships in the early 1980s (at the rate of three a year to begin with) as a means of introducing some flexibility into the situation, while at the same time the Department of Education and Science began making funds available for – telling phrase – 'new blood' appointments in the various disciplines. Even so, I knew of many new excellent PhDs in the Department of History of Science and Technology at Imperial College who were never able to obtain permanent positions despite staying in contention, as it were, living precariously from year to year by taking on stopgap and short-term research projects. Before this became the norm the last of them to get lucky enough to obtain a university post did so shortly after he had obtained his PhD – but that was in 1975. These recollections, I realize, may well seem like the disclosure of a kind of *vie intime* but do not, I think, distort the situation in which history of technology was then placed, and indeed in which it is placed still. I grieve for the talent lost with those men and women. What is more, it means that were the situation to change for the better in the future, they will not be available as senior scholars with what might have been international reputations able to attract high-calibre recruits to the subject. None of the contributors to this symposium touches on this aspect of the pre-history of the current state of the history of technology, although I think it is germane to the discussion.

But this and other considerations would probably not have led me to decide to devote such a large section of the present volume to a debate on the current state of the discipline had it not been for reading, a few years back, Lutz Engelskirchen's paper, 'Establishing the History of Technology as an Academic Discipline in the Federal Republic of Germany after 1945'. This was published in Volume 4 of *ICON* the journal of the International Committee for the History of Technology, in 1998. Perhaps not too many readers of this volume will have read Engelskirchen's survey. Those who have not might well be inclined to suppose that whatever might have happened in Germany was not likely to have much or any direct relevance to the question under examination here. Maybe so, except that I believe the paper has heuristic value for anyone attempting to understand what has taken place in the UK in the last 40 years, any future account of which period could well be entitled 'Failing to Establish the History of Technology ...' At the very least the *ICON* paper throws an oblique light on the aetiology of that failure.

In Western Germany, at the time when its sovereignty was restored in 1955, it was strongly felt, as elsewhere in Europe, that the divorce that had taken place between the humanities and the sciences (and indeed the growing divisions within the sciences themselves) was dangerous and called for some response. And further, that if technology *was* an integral part of man's cultural endowment, if not indeed a prime mover in the successive phases of its formation, as Fernand Braudel believed, and if, as was obvious, technology had created, and was visibly continuing to create, immense and rapid changes in the human condition, then it was important that it become the object of intensive examination:

> Nowhere was this more obvious than in mankind's newfound ability to annihilate itself with nuclear weapons. The study of technology's relationship to society was seen as an absolute necessity if technology was to be handled in a responsible manner. (Engelskirchen, *op. cit.*, p. 150).

Similar considerations were also exercising minds in the UK at this time, that is, in the late 1950s. Sir Eric Ashby's essay of 1958, *Technology and the Academics* is still valuable for an understanding of the resistances that would be encountered:

> the close integration of the humanities and science [from which one might expect new combinations of thought] has proved too difficult for the universities to organise. (Ashby, *op. cit.*, p. 74).

In expanding this idea he wrote:

> Technology in universities could be made to play a far more important role ... It could become the cement between science and humanism ... for technology is inseparable from humanism. The

> technologist is up to his neck in human problems, whether he likes it or not ... the social consequences of his work are therefore an integral part of his profession. (*op. cit.*, p. 82).

From this, study of the history of technology naturally follows, although Ashby acknowledged that the provision of specially designed courses in the humanities relevant to technology would present some difficulty. The *Essay* reads now like an epitaph.

I say 'epitaph' because whereas in England such action as was taken arose from the determination of a very few well-placed individuals, in Germany the Federal Republic possessed a body – the Wissenschaftsrat – which had the power to make decisions relating to the shaping of higher education throughout the German state. The Wissenschaftsrat was convinced that new bridging subjects were needed 'to facilitate communication' between the different worlds of research, especially between the humanities and engineering. The history of technology, as one of these bridging sciences, was to be introduced not only into the traditional universities (which then, as in England now, had not perceived engineering sciences as part of the cultural inheritance) but into the technical universities (Technische Hochschulen) as well. Although, as readers will see from Hans-Joachim Braun's contribution relating to the current situation, there are 'still significant deficits', they will also note that the history of technology is now taught at 22 universities and technical universities and at three polytechnical institutes. The number of professorships exceeds the above figures if one includes those whose titles are combined with other fields of study, as for instance economic history.

Before the time of the Wissenschaftsrat's action, the study of the history of technology in Germany and England had proceeded in not dissimilar ways. In Germany the leading figure from before World War I was Conrad Matschoss, 'Altmeister der deutscher Technikgeschichte'. From the appearance of Volume 1 in 1909 until his death in 1941 Matschoss edited the *Beiträge für Technikgeschichte*, published by the Verein Deutscher Ingenieure, the last volume of which, Volume 30, appeared posthumously in 1942. In England at almost the same time H.W. Dickinson played a similar leading role in developing scholarly research in the history of technology and also in setting up the Newcomen Society for the Study of the History of Engineering and Technology. The first volume of the Society's *Transactions* appeared in 1921 but it is to the presidential address which opened Volume 4 of 1923–24 by a distinguished engineer, L. St L. Pendred, that I wish to refer. It is entitled 'The Value of Technological History'. In it he sought to show that history of technology had

> those qualities which broaden the mind and sharpen the intellect and that it is as *essentially human* [my italics] as any other kind of history. (*Transactions*, *op. cit.*, p. 7).

But the small company, mostly engineers, who composed the membership remained a small company down to the time when, largely as a result of the initiative of the Rector of Imperial College, Sir Patrick Linstead, the decision was taken to press ahead (in 1959) with the creation of a new chair in the history of science and technology at Imperial College. The press release relating to this of 6 December 1961 spoke of the field of study as having been made intentionally wide so that (as it was hoped) it would

> embrace both the genesis of basic ideas and tools and the recent history of science and technology, the latter particularly because of the great acceleration in applying scientific discovery with important social and political results.

Interestingly enough, it was ICI (Imperial Chemical Industries), which had earlier subsidized the publication of the five-volume *Oxford History of Technology* (1954–58), which agreed now to help fund the chair for the first seven years. But similar thoughts were in the air further north. At the very time that the search continued to find an occupant for the chair at Imperial College, a similar initiative was taking shape at the then Manchester College of Science and Technology (now UMIST, University of Manchester Institute of Science and Technology) by Professor B.W. Bowden. Unaware of what was taking place at Imperial College, Bowden wrote to Professor P.M.S. Blackett of Imperial College on 10 May 1961, informing him that

> we are trying in this College to establish an entirely new chair in the history of science and technology. It is a subject which is almost new to the English academic world, important though it undoubtedly is.

Bowden went on to ask whether Blackett could suggest suitable names. So here were two institutions now fishing in the same waters. And there, über Höhe und Tiefe, the parallel with Germany ends. If a common feeling existed across western Europe, it was going to be crucially a matter of the strength or weakness of institutional infrastructures which would determine how far things were changed on the ground so that the grinding down of the cake of custom might begin.

But how do the contributors to the symposium see the situation as it is now? They divide quite clearly into two camps: a minority who, while not complacent, certainly do not share my own pessimistic view of the situation, and those who in one way or another are as dissatisfied with it as I am. In the first camp both David Edgerton and Graeme Gooday, although their positions differ in some respects, agree that while there are extraordinarily few posts in history of technology in Britain, it is possible to point to a considerable output of scholarly papers on the subject. David Edgerton points out that in the period 1994–1999 ten articles have been published each year by British authors in various journals devoted to the history of technology. He notes that

> perhaps those who believe there ought to be or is a British tradition in the history of technology should review this literature

although he observes earlier (and correctly, I believe) that the history of technology could [not] draw on a long and intellectually interesting tradition of study within Britain as the history of science [had been able to do].

He sees the emergence of SCOT (Social Construction of Technology) in the 1980s as an unwelcome apparition, as a bid (a failed bid, he believes) to revive the moribund condition of the discipline here in Britain.

Gooday points to the considerable number of those who, while they might not describe themselves as historians of technology, 'participate in this subject under other job descriptions'. In a footnote to his paper he lists some 18 names. I know the work of a good many of those historians, and it is often impressive. It is when they are tempted to range outside the limits of the bridgeheads they have established on this or that aspect of the broader history of technology that doubt creeps in. One of the 18 has stated in a recently published book that the medieval European post-mill was a sophisticated response to the design limitations of the earlier 'primitive' (sic) tower mill. This is in fact the exact opposite of what is commonly taken to have been the direction of development. But it is when Dr Gooday lines up actual historians of technology in the cross-hairs of his sights that I think he is less than fair. He offers a five-point critique, and I would like to comment briefly on two of his points. The second of them is entitled 'History of technology is more than just the history of artefacts'. Indeed it is, but one cannot do without such histories. The period in the history of technology in which I am particularly interested – early modern Europe – is over much of its extent still a kind of trackless Matto Grosso. My own research has largely been focused on the development of mining technology. Now although mining was even in Renaissance Europe undoubtedly a vitally important industry, there was, as I discovered, little enough in the literature, whether in the German, French or English languages, that could throw much light on the mechanical technology miners then employed. Reconstructing the stages through which primitive mine pumping machinery was in the course of the sixteenth century brought to such perfection that it continued to serve with little modification for the best part of the next three centuries was an essential first step to understanding what lay beyond its merely physical form. If the actual technology of the period can then be given the black box treatment, there has at least to be a black box – that much can be taken for granted. Even so, the value of such artefactual research for an understanding of the nature of creativity in the material world should not be underestimated. There is no reason to believe that creativity in technology is any different in its nature from creativity in any other field.

Dr Gooday's fifth point, which has to do with the market-place and the competitive forces which he insists historians of technology must recognize and accept rather than 'bewailing the philistinism of the masses', I find a little difficult to understand if for no other reason than that the masses seem to be hooked on the history of technology anyway. British television in the last several years has carried substantial numbers of programmes which, whatever flags they may have sailed under, were mainline, and sometimes very good, history of technology, given the constraints of the medium.

Hugh Torrens, like Rupert Hall and Angus Buchanan, is very pessimistic. He writes below that Alex Roland, a past President of SHOT, the Society for the History of Technology, offered evidence in his presidential address of 1997 that in the US the TV history channel, following a poll in 1995, reported that 'the item of greatest interest to the public is the history of science and technology'; so no problem there either. BBC sound radio has also been a rich source of good-quality material. It is in academe that the problem is located.

Hall and Buchanan find little in the present situation to praise. They both contrast it unfavourably with the situation some 40 years ago. In 1963 Rupert Hall became the first occupant of the newly created Chair in the History of Science and Technology at Imperial College. Shortly afterwards, in 1964, Angus Buchanan set up the Centre for the History of Technology at the University of Bath. The discipline seemed then to be well set to grow in the next decade as history of science had grown earlier when, between 1957 and 1961, five universities established teaching posts, joining Oxford and Cambridge where history of science had achieved recognition somewhat earlier. Their experiences have left them well placed to throw much light on the causes and consequences of what they see as history of technology's long-term decline. Is there a way out of the indifference and apathy seen by Rupert Hall as the real cause of the problem? Angus Buchanan suggests that 'a princely endowment' would be one way of turning the situation round. For that to succeed one would need high-calibre appointees trained in the history of technology itself, the Catch 22 here being that the long declension of the subject means that such experts and expertise are very hard to find, for reasons which I have mentioned earlier. Nevertheless, a large transfusion of money (Wellcome money, but isn't it always?) has certainly worked well enough in raising history of medicine above the academic horizon. Why should not the history of technology likewise benefit from a Maecenas?

Hugh Torrens, like Hall and Buchanan, finds much wanting in the contemporary scene. Noting the long history of mandarin-like academic distaste for such non-purely-cerebral and earthy activities as actually require grappling with the material world and, worse, the absence of any significant disciplinary base at present for the history of technology in British higher education, he puts great emphasis on the need to

> address the current antipathy to technology and its history in this country, and try better to reveal the real importance of technology as the ultimate creative activity.

As he remarked in a personal communication, The Royal Society, The British Association for the Advancement of Science, the organization Public Understanding of Science and the Save British Science movement are all working energetically for the cause of science and its history. There is no corresponding effort being made for technology and its history.

Ian Inkster, who has greatly assisted me in various ways in the setting up of this symposium, also sees the situation as unsatisfactory but is concerned to show, so to speak, that if one throws history of technology out of the academic door, it will come back through the window. How is this? The answer, put as briefly as possible, is that work in the history of technology and the perspectives opened up by its historians (not in Britain – or rarely so – unfortunately) have provided those one would think of as mainline historians of global and large-scale world history with an extremely valuable tool kit. He writes:

> Put at its most direct, we now suggest that major interpretations of global or very general history rest on both the findings of historians of technology and on interpretive perspectives which place technological imperatives and innovation at or near the centre of discourse.

Further, he believes that general histories will not long survive their own theses if the springs of new knowledge welling from the history of technology are cut off by its slow disappearance or demise. Since technology is the product of the imagination as well as of the material needs of humankind, it goes to the heart of what man does and what by doing he becomes.

***

I hope readers will enjoy the six views regarding the health of the history of technology in the UK and will find Hans-Joachim Braun's account of how things have gone in Germany revealing also. The subject is too important for anyone to think of those participating in this debate as engaging in some latter-day Big-Endian versus Little-Endian ideological or terminological dispute. Aristotle believed politics to be the master science because to understand the polis one had to understand everything else. I believe that for our time technology occupies an even more powerful place in the world and exhibits ever more strongly the potency to propel social structures into solution, the very thing that led Marx in his time to write that its power was such as to cause 'all that is solid to melt into air'. Such power should be studied at the highest levels and, whatever the future outcome, should certainly not be ignored.

Reflections on the History of Technology in Britain[1]

DAVID EDGERTON

Nearly a decade ago, prompted by an argument that in Britain the history of technology was going down the tubes, I wrote a short article for the *Newsletter* of the British Society for the History of Science claiming that the history of technology, understood as a focus for research rather than a discipline, was not only healthy but growing.[2] Perhaps I was too given to discounting, even inverting, the whingeing of sectional interests to understand what was going on, but I did find a great deal of activity, under headings of history of science, business history, economic history and others, that was contributing to our understanding of the history of technology, and British technology in particular. I still think my positive judgement was broadly right. I write as someone who has taught in the successors to two departments which would be at the centre of any account of the early academic institutional development of the history of technology in Britain: the Departments of the History of Science *and Technology* created in the early 1960s at UMIST (Manchester) and at Imperial College, London.[3] Both departments went through a difficult period following the retirements of their creators (Rupert Hall and the late Donald Cardwell) in the early 1980s. Both have re-emerged, indirectly, as Centres for the History of Science, *Technology and Medicine* at the University of Manchester (rather than UMIST) and at Imperial College, in the context of a London-wide programme.[4] Thus I could hardly turn down Graham Hollister-Short's request for some personal autobiographical reflections on the state of the subject, even though I disagree with his premise that the subject has been in a sorry state, and am conscious of the dangers of confusing biography with history.

Historians of technology are of necessity interdisciplinary and they come from many backgrounds. Many of the contributors to this volume have first degrees in history and postgraduate degrees that could not be construed as history of technology in a narrow sense. I have been described as a business historian, an economic historian, a military historian, a Marxist historian, as well as a historian of science and of technology. I have a first degree in chemistry, took an option in the history and philosophy of science, and did a dissertation (in effect an MA by research) on the history of industrial research.[5] For my PhD, which I started in 1981, I worked on British political and economic history around World War II, under the supervision of a historian/sociologist of science (Gary Werskey), in the Department of Social and Economic Studies at Imperial College; my immediate colleagues were industrial sociologists and economists.

I don't recall being introduced to the history of technology as a subject at Oxford, though I don't suppose either Margaret Gowing or Allan Chapman made too much of a distinction between science and technology. Under Allan Chapman's supervision I worked on a standard theme of US histories of technology: the history of industrial innovation and research laboratories, under the influence of David Noble's *America by Design*.[6] I don't suppose I had any idea that British history of technology existed at all. Even at Imperial I had only the vaguest sense that there had been or was any history of technology as a specialist area of study in Britain.[7] From my perspective the Business History Unit at the London School of Economics was at that time the main centre for the study of the history of British technology.[8] My first published paper appeared in *Business History* in an issue devoted to technology.[9]

In 1984 I went to the Department of Science and Technology Policy (formerly Liberal Studies in Science) at the University of Manchester (which included a number of historians/sociologists of science[10]) to teach the 'economics of science and technology'. I taught courses with a substantial historical content, and I had to mug up rather quickly on the history and the theory: I relied for intellectual orientation on Nathan Rosenberg's collections of essays and on the political theorist Jon Elster's *Explaining Technical Change* (1983). Although I was to retain an interest in the economics of technical change and science policy, in 1985 I went to UMIST, at a time when John Pickstone and the historians of medicine moved to the University. I taught history of technology to undergraduate civil and mechanical engineers and history of science to undergraduate chemists. I found it was much easier to put together a history of science survey course than one in the history of technology. The UMIST history of technology textbooks, if I may call them that, by Cardwell and Pacey, were something quite different from textbooks in other subjects.[11] Altogether more stimulating was the seminal collection of essays, by Donald MacKenzie and Judy Wacjman, *The Social Shaping of Technology* (1985). This reader brought together a range of different traditions of thinking about the history of technology very successfully into one book, pointing to lots of ways in which the study of technology in history might go.[12]

At UMIST there was no prospect of developing a centre for teaching research in the history of technology, and I transferred to Manchester University where, under John Pickstone, a Centre for the History of Science, Technology and Medicine was being created. The Centre developed a wide range of courses in the history of technology. A good deal of research was done on twentieth-century British technology. We had at various times Geoffrey Bowker and Geoffrey Tweedale working on computing, Colin Divall on engineering and Richard Coopey on the history of the Ministry of Technology. This was the context in which I wrote my note in the BSHS *Newsletter*. From my point of view, lamentations were inappropriate.

In my own research my aim has been to put British technology and science in their place in the history of twentieth-century Britain, not merely to fill a gap, but in order to demonstrate that taking account of science and technology meant writing a different kind of British history. It seemed to me that the study of the history of technology on a systematic basis was probably best done alongside the study of the history of science and medicine. Although colleagues then and since have taken the view that the history of science, technology and medicine should be integrated into mainstream history *departments*, my hardly original view is that specialist departments should exist, as a nucleus of expertise and training, and as providers of teaching programmes for scientists, engineers and medics. Without such an institutional structure there is a great danger of losing our subjects altogether. In any case in some respects history taken as a whole was becoming less receptive to the history of technology, if perhaps more so to parts of the history of science and the history of medicine. Economic history, itself declining in popularity, was turning increasingly towards cliometrics, which, while yielding very interesting results in the study of technology in particular, was orienting itself round the language of economics rather than the material of the real economy. Social history was turning away from labour history and its related concerns with the labour process (an exceedingly important stimulus to the study of technology) to more cultural concerns (though the concern with knowledge was useful). And, of course, much of straight history was unbearably dull compacting of papers found in the Public Record Office (particularly in some of the fields I was interested in). There were of course very good positive reasons for combining the study of the history of science, technology and medicine, especially for the nineteenth and twentieth centuries. Indeed much British history of the physical sciences was very strongly pushing what has been usefully called the 'technological conception of science', sociologists of science were becoming very interested in technology, and there were unexplored but obviously interesting comparisons and analogies to be made between technology and medicine.

At the very beginning of 1993 I moved to Imperial College. The programme there might easily have died completely had it not been for Jim Secord. The teaching of history of technology at master's level was in abeyance, and separate from the history of science and medicine. With the creation of a new centre at Imperial College, we first integrated courses in the history of technology into a single package of courses to form a number of MSc programmes (including one in the History of Technology and another in the History of Science and Technology), in the joint MSc run with the historians of science at UCL (University College, London) and the Wellcome Unit for the History of Medicine. We now have a single MSc programme with many options. Until recently one could identify particular options in the history of technology (History of Technology Core Course; Technology in the Modern Era, Technology in the

Twentieth Century), but we have increasingly integrated the histories of science, technology and medicine at the level of options. What is particularly interesting is that it is not always the case that the 'historian of technology' teaches the technology bits. A 'historian of medicine' might well teach on handlooms, and a 'historian of science' on V2s (or indeed a whole undergraduate course in the history of technology). A 'historian of technology' might be found talking about the history of the history of science. Furthermore, students are encouraged to write essays that cover science, technology and medicine. The net result is that many more students are exposed to the study of technology than ever before, while there are almost certainly more students than ever before specializing (in effect) in the history of technology, though with greater intellectual rigour than would be possible by studying the history of technology alone. At undergraduate level we have a course in the History of Technology available to all science and engineering undergraduates, as part of a large package of courses. It is worth noting that, given the choice, most engineers will prefer to take courses other than History of Technology, just as most scientists take courses other than History of Science.[13] They prefer Philosophy, Politics, European History.[14]

Our integrated approach to the study of the histories of science, technology and medicine brings out peculiarities of history of technology very clearly. For example, collective works of reference and synthetic surveys by British authors have been systematically worse than comparable efforts for science and for medicine; indeed of a quite different intellectual register. Also striking is that there isn't a 'history of technology' literature which covers the sorts of topics a general history of technology course should touch on; and again and again we find the need to use readings from a wide variety of disciplines (and here I am not referring just to British literature).[15] I am arguing in effect that while we do need (and have) specialist training in the history of technology, that training cannot be adequately provided from within the boundaries of 'history of technology'.

For all its benefits these particular difficulties are inherent in such integrationist strategies. The history of technology could never command the resources the Wellcome trust began to pour into the history of medicine. Nor could it draw on a long and intellectually interesting tradition of study within Britain as could the history of science. More than either it needed contact with the more general historical disciplines, and some social scientific ones. Furthermore, the demise of the joint ESRC/SERC[16] committee had a great impact on PhD and MSc funding in this area: it has proved difficult to ensure that ESRC recognized its responsibilities for the history of technology. There were also intellectual dangers, illustrated by the SCOT programme. In the mid-1980s British sociologists and historians of science made a bid to rejuvenate the study of technology and its history by applying the lessons of the sociology of scientific knowledge. I had little sympathy for the programme (or its supposed opposition).[17] What I particularly disliked about SCOT was its lack of understanding of the range of intellectual resources that could be and had been brought to bear on technology.[18] It seemed to me silly to imply that anyone serious had maintained that technology was not 'socially shaped' and it was obvious that there were lots of different ways of thinking about technology in history. SCOT threatened to give us little more than methodological statements prefacing routine case studies of innovation, rather than helping us understand technology in particular historical contexts, for example the industrial revolution, or British technology since the late nineteenth century. In the latter case the problem was not the 'technological determinism' seen by many historians or sociologists of technology as a key interpretative problem or historiographical 'Whiggism' or presentist progressivism, but rather 'inverted Whiggism' or 'declinism': the tendency to find seeds of failure in the past to explain what is taken to be decline.

I fear that 'declinism' has been a feature not only of histories of British technology, but also of histories of British history of technology, often it seems using just the same sort of explanations and images. Graham Hollister-Short's question was why the history of technology failed to thrive in Britain. This question needs redefinition. While it is the case that programmes established in the 1960s and 1970s were not able to build sustainable research programmes or produce graduate students to continue and expand the work of the pioneering generation in the academy (in contrast to the history of science)[19] there is a bigger picture not of decline but of strong growth in studies of the history of technology in Britain. The history of technology was undertaken, increasingly so, within the contexts of history of science, sociology of science, and economic, social, labour and business history. Certainly we now find many historians with, to put it at its lowest, very strong interests in the history of technology in many departments concerned mainly with the history of science which had no such people twenty or thirty years ago. Furthermore, it is very encouraging to see particular mediaevalists, archaeologists and sinologists, not to mention economic and social historians, in many institutional contexts, making wonderful contributions to the study of technology. Think too of the place of research in the history of technology in the Science Museum today. And even if one looks at history of technology fairly narrowly defined there is probably more activity in Britain than is readily acknowledged, and almost certainly more than in the 1980s. In *Technology and Culture* there were eight articles by British-based authors between 1994 and 1999, eight in *History and Technology,* eight in the *British Journal for the History of Science*, eight in *ICON*, 1995–1998, and about 20 in *History of Technology,* 1994–1998.[20] This amounts to an annual British production of around ten articles.

If the field has not declined, but rather strengthened, that does not of course mean that it is as strong as it should be. The question we were asked was: how could we locate history of technology in the mainstream of historical study in the UK? Part of the answer is to get the history of technology into the mainstream of the history of science and medicine, and there has been some success here, though there is still some way to go. In terms of the wider historical profession it seems to me that it is up to historians of technology to show that we have something to say to the generality of historians, and also to such particularly relevant subgroups as military historians, business historians, economic historians, environmental historians, and so on. This must be much more than a claim that technology is pervasive: it must be a specific historical claim which addresses a particular historical question. Elsewhere I have argued that much, probably most, history of technology, could not engage with such historical questions since it was concerned not with the history of technology, but rather with the history of invention, innovation, and so on.[21] That is not to say that histories of invention will not be of interest to historians, only that they will be of interest for different reasons. But whatever the best way forward, if we are to build the systematic study of the history of technology in Britain, an excessively negative account of the recent development of the subject is unhelpful in understanding not only what has been done, but also what is to be done.

Notes and references

1. With thanks to Graham Hollister-Short and Andrew Warwick.
2. No. 38 June 1992.
3. Both were elite technical schools which formed part of the Universities of Manchester and London respectively. They were (I think) the only two history of science departments with history of technology in their titles – older departments were of the History *and Philosophy of Science*, later ones were more varied but still tended to exclude technology (think of Science Studies, Edinburgh; Liberal Studies in Science, Manchester, and History and Social Studies of Science, Sussex); the only Centre for the History of Technology was at Bath (1964).
4. The department at UMIST was closed down, following the transfer of all but one of the staff to the University of Manchester.
5. I took courses with Margaret Gowing on the history of nineteenth- and twentieth-century British science, and the philosophy of science with Rom Harre; my tutors were (fond memories) Allan Chapman and a very young Peter Lipton.
6. I met Robert Bud at this time; he too had worked on the history of industrial research.
7. An intellectually very able engineer and social scientist who was to become a close friend had come from India to do a PhD in the history of technology but quickly transferred his interests.
8. Jonathan Liebenau, a Penn graduate, was there as were two Imperial historians of science, now working on technological topics: John Hendry and Mari Williams. It is also worth noting that both David Jeremy and Geoffrey Tweedale were long associated with the Unit. The Director of the Unit, Leslie Hannah, was working on the history of British electricity supply. It is interesting to compare his treatment of the topic, and in particular of the British story, with that of Thomas Hughes.
9. The other contributors were Mari Williams, John Hendry and Jonathan Liebenau, who edited the special issue.
10. Including Jonathan Harwood, Edward Yoxen, Philip Gummett and Geoff Price.
11. D.S.L. Cardwell, *Technology, Science and History* (1972); Arnold Pacey, *The Maze of Ingenuity* (1974), *The Culture of Technology* (1983). I found Pacey useful, Cardwell less so.
12. The second edition (1999) contains less history than the original.
13. I had thought this was an Imperial peculiarity until I attended a conference at ETH (Eidgenössische Technische Hochschule) Zurich on the non-technical training of engineers. The MIT experience was identical.
14. History of Science and History of Technology have had about 40–50 students each in recent years, compared to 70–100 for Politics and for European History.

15. To take another comparison: at Imperial College I teach two undergraduate courses: one in the history of nineteenth- and twentieth-century technology, the other on European History 1870–1989. It is the latter which attracts more students (science and engineering students, that is). Although the former is closer to my heart in many ways, the latter is more rewarding to teach. I can enjoy and invite my students to study carefully books by Eric Hobsbawn, Norman Davies and Mark Mazower, knowing their efforts will be rewarded. For technology there are no British equivalents at all, and in fact no suitable US texts either. Ruth Schwartz Cowan's *More Work for Mother* is always popular, and William McNeill (not of course a 'historian of technology') on technology and war goes down well. Histories of US technology by Pursell, Schwartz Cowan and Hughes are not very suitable for the British/international students we teach.

16. Economic and Social Research Council and the Science and Engineering Research Council. The latter no longer exists in this form.

17. The debate between two British protagonists, Angus Buchanan and John Law, in the pages of *Technology and Culture* in 1991 seemed to me to be fruitless.

18. 'Tilting at Paper Tigers', essay review of Donald MacKenzie's, *Inventing Accuracy, British Journal for the History of Science*, Vol. 26 (1993), pp. 67–75.

19. Which did expand very strongly, especially in Cambridge, though not, comparatively speaking, in places like Imperial or UMIST where the history of technology was strongest.

20. In reverse chronological order the authors are: *Technology and Culture,* Nahum Campbell-Kelly, Tann, Collins and Kusch, Clavering, Zeitlin, Divall, Donnelly. Compare with 1980–84, when there was but one article from Britain. In *History and Technology* they are Edgerton, Marks, Gooday, Hughes, Bulkeley, James, Nahum, Weindling.

21. 'De l'Innovation aux Usages: dix thèses éclectiques sur l'histoire des techniques', *Annales HSS*, July–October 1998, Nos 4–5, pp. 815–837. Thanks to John Krige it has also appeared in English as 'From Innovation to Use: Ten (Eclectic) Theses on the History of Technology'. *History and Technology*, Vol. 16 (1999), pp. 1–26.

The Flourishing of History of Technology in the United Kingdom: A Critique of Antiquarian Complaints of 'Neglect'

GRAEME GOODAY

INTRODUCTION

I am sceptical about complaints that the history of technology is unduly neglected in the United Kingdom. While it is undoubtedly true that fewer professionals are employed as 'historians of technology' than in many other specialist fields of history, it does not follow from this that insufficient institutional or public attention is paid to the subject.[1] In this paper I shall suggest five points that such complainants might have overlooked, and argue that there are questionable assumptions in the narrow vision of the history of technology that they implicitly or explicitly adopt.[2] I shall be at pains to spell out what I take these flawed assumptions to be, and propose a more inclusive interdisciplinary vision of the subject which presents the history of technology in Britain as being in a relatively healthy condition. As one who writes on this subject within the academic discipline of history and philosophy of science, I shall emphasize that history of technology can benefit greatly from its intersections with other fields in the same way that history of science has so manifestly done during the last three decades.

The first assumption I challenge is that the history of technology is simply the history of engineering, and hence that the engineer is the only interesting human subject of study; such a narrow origins-centred approach overlooks the important 'after-life' of technologies in the hands of its users and *a fortiori* the often fascinating historical literature that it has inspired. The next position I criticize is that which takes too narrowly artefact-centred an approach to what constitutes 'technology': this position fails to acknowledge that 'technology' encompasses much more than mere hardware, and thus neglects much valuable historical writing that adopts a broader construal of the subject. Then there are those who take too narrow a view of what kinds of methodology can legitimately be pursued in writing the history of technology: in assuming that only *their* kind of specialist historiography can be rigorous or informative, some scholars unhelpfully exclude approaches fruitfully complementary to their own. Next, there are those who unhelpfully restrict their attention to apparently benign and beneficial technologies; a less ingenuous approach would also accept an obligation to consider technologies with morally problematic consequences, and include within their purview, for example, the many historical accounts of technologically generated risks, especially of pollution. Finally, there are those who are unduly optimistic about the amount of attention that could possibly be accorded to history of technology in a scholarly market place that is teeming with many other varieties of interesting historical writing. Given such inevitable competition, I suggest that it is only by reaching out to wider audiences and taking seriously their multifarious and often highly critical views of technology, that historians of technology can hope to win more attention for their work from the reading public.

In what follows I shall cash out this critique, and cite examples from the research of British historians to show that a vigorous network of scholars in a range of disciplines does currently nurture the history of technology and addresses the subject to a wide range of different audiences.

HISTORY OF TECHNOLOGY IS MORE THAN JUST THE HISTORY OF ENGINEERING

Not a little history of technology is written by those who once worked or still work in the engineering profession. Unsurprisingly their technically informed writing is marked by a strong emphasis on the role of engineers in conceiving, creating and refining new forms of technology. It is evidently important to document the role of engineers in the early life of any technology, and such writing can readily be found in early issues of *History of Technology* and it still predominates now in the *Transactions of the Newcomen Society*.[3] While those who write such histories might find an audience for their work among the community of (ex)engineers, they are doubtless disappointed that their work does not attract a wide popular following. There are three reasons why this might indeed be so: one is that such highly technical writing often presupposes a level of specialist understanding that is beyond the grasp of those who have not received a training in engineering, and is thus found by them to be unapproachable. Second these stories are so factually detailed and preoccupied with matters of heroic priority and technical ingenuity that they do not offer the kind of broad contextual narrative that appeals to the aesthetic sensibilities of non-engineering audiences. Most importantly, however, such engineer-centred accounts fail to meet the principle maintained by both scholars and non-specialist readers alike that while writing the history of engineering is an essential part of the subject, it is not by itself *sufficient* to constitute the full history of technology. It is simply not enough to compose a detailed exposition of the conception and creation of a technology and end one's narrative at that point. Why should this be so?

Let us consider the claim that, having explained how a technology was designed and made by engineers, there is little of interest to say about its subsequent life in the hands of owners and users. Analogies are helpful here in identifying the absurdity of this origins-centred position. Could anyone write a biography of a human subject just by confining attention to the processes of procreation and birth? Can we capture the importance of a hundred pound note just by looking at the process of its design and printing? Would an investigation of the fission processes alone in a nuclear reactor by itself enable us to grasp the awesome global significance of plutonium? Unless one subscribes to the most implausibly hard-bitten form of physical determinism, the answer in all cases is patently 'no', since what happens later on in the career of anything generally has at least some significance in shaping its development and wider human value. Indeed, it is now accepted as a truism amongst most historians of technology that the fate of a technology lies in the hands of those outside the engineering community. Discretionary humans can choose to buy it or not buy it; to use it or not use it at all; to implement usages recommended by the maker or develop their own; to demand changes by the maker, implement their own changes, or to accept it as it is; to reverse engineer it or leave its operations mysterious; to upgrade it, preserve it, cannibalize it, or discard it as newer models arise. Following the wisdom of anthropologists and archaeologists, as well as historians of material culture, we can note that useful artefacts do not cease to change when they arrive in the hands of their users, and their history should therefore not cease there either.

Studies that take up the above-mentioned 'post-engineering' themes should, accordingly, be included within any sensible survey of history of technology. Arnold Pacey's writings on the history of technology are perhaps the most informative and wide-ranging on this topic. While showing full respect for the creativity of engineers, he focuses attention on the way that ordinary users have often *adapted* technologies to their national cultures and/or local resources to ensure that their operations are viable within existing constraints.[4] The historical process of interchange of technologies between cultures is, as Pacey shows, one of the most fascinating to document. Feminist critics of technology, especially as deployed in the so-called 'developing nations', have shown that an important part of the history of technology has been the rejection of technologies that Western engineers have developed on the grounds of their cultural, practical or economic inappropriateness. For example, Patricia Stamp has noted that women in Africa generally rejected solar stoves developed by resourceful British engineers in the 1960s since rural African families do not eat their main meal during daylight hours. The *inappropriateness* of such technologies for their intended users is surely a crucial part of their history, but is alas characteristically overlooked by those who fetishize the role of the engineer as master of all he [*sic*] touches and surveys.[5] Taking this critical line still further we can note that engineer-centred histories not only fail to see the sovereignty of the consumer in determining the fate of an artefact, for they can also fail to see that non-engineers can also design and make technologies. As Adrian Wilson has noted in his highly original book *The Making of Man-Midwifery*, it was late seventeenth- and early eighteenth-century surgeons who developed the forceps out of the 'hook' and 'crotchet' used to deliver dead babies. Wilson then goes on to show that it was expectant mothers who demanded the use of the forceps by these incipient 'male-midwives' once it became clear that using them might help to avoid death in difficult deliveries; the mere creation of the forceps was not in itself a sufficient condition of their widespread deployment in gynaecology, nor was the enthusiasm of surgeons for their new technology. For several reasons then, the history of the midwifery forceps cannot be told merely as a story of engineering innovation.[6]

Studies of technological history that broaden their scope of engagement beyond the actions of past engineers are very important and constitute an increasingly popular area of history of technology for both writers and readers. If engineering historians consider that such fertile scholarship does not constitute a counter-example to their vision of a sadly under-acknowledged field they should present a coherent argument to explain *why* they believe the actions of end-users do not matter to the history of technology. That such traditional historians have not done so is perhaps telling evidence of their culpable ignorance of this wider literature.

HISTORY OF TECHNOLOGY IS MORE THAN JUST THE HISTORY OF ARTEFACTS

Some historians of technology focus on the development of particular technological artefacts and thus effectively narrow down the history of technology to the history of the *material products* of technological activity. Insofar as such studies focus on the material artefact in use, they clearly capture something valuable about the irreducibly physical presence of much technology throughout history. Yet it is all too tempting to focus on the materiality of many technologies and fail to notice that there is much more to technology than its mere physical form. To press the point, one can argue that documenting the genesis of an artefact as an embodiment of virtuosic or aesthetic technique is a form of technological connoisseurship that belongs in the history of art or antiquarian studies, not the history of technology *per se*. Histories of technology must, after all, surely be oriented towards the *use* of objects – whether to sustain the careers of engineers or meet the needs of their owners and operators – not the passive admiration of objects in themselves. Indeed, focusing thus on the 'usefulness' of technology rather than on its materiality enables one to account for the way in which the history of technology can also coherently be the history of non-material artefacts too. It can thus informatively include the history of patents, as Christine McLeod has shown,[7] and the history of software and computer games – which as Martin Campbell-Kelly has noted are now central issues in the recent history of computing.[8] The future historian of virtual reality technology will doubtless have much more to say about the development of its 'virtual' rather than material components.

Applying this sort of critical scrutiny to our understanding of what *counts* as technology, we soon encounter further reasons for broadening our conception of the history of technology to include more than just chronicles of artefacts (virtual or material). As the philosopher of technology Carl Mitcham has pointed out in *Thinking Through Technology*,[9] there are three other major ways of understanding what constitutes technology – technology as 'knowledge,' 'activity', and '[extension of] volition' – and thus three other kinds of subject that can be addressed in the history of technology. In a fully matured history of technology these issues are not easily separated, of course. Any thoroughgoing account necessarily integrates discussions of how knowledgeable use of artefacts in certain kinds of activities has extended human capacities and powers in particular selected directions.

Considering technology as knowledge is to focus upon the issue of skill or 'knowing how to' use a technological artefact to a given end: the extent to which an artefact or machine can effectively serve a useful purpose is surely dependent on the technological knowledge of its operators. Unless its operators have the pertinent knowledge, artefacts cannot so readily serve a technological purpose: *in extremis* a piece of material hardware can be reduced to the status of recalcitrantly inert heterogeneous lump if nobody knows how to operate or mend it. My own first writing on the history of technology was largely in this area of the history of technological skill. It concerned the development of electrical engineering education in Britain in the 1880s, and explained how novice electrical engineers learnt to design and test lighting systems in order to minimize the human and commercial risks of system breakdown.[10] It is in a similar vein that we should consider the importance of Ben Marsden's work on W.J.M. Rankine's development of the notion of 'engineering science' in mid-nineteenth-century Glasgow[11] and Gerrylyn Roberts' discussion of the importance of specialist training for women at home who first used the 'All-Electric Kitchen' in the 1920s.[12]

Considering technology as 'activity' is to examine the continuous ongoing processes that are constitutive of technological usage. The premise of this sort of historical writing is that the historically important problems of technological hardware do not cease once it has been made, for important issues arise in regard to its maintenance and adaptation to new contexts. This genre of historical writing can thus focus on engineers' work on technological maintenance and development as is embodied in Jim Donnelly's studies of the development of chemical industrial installations in the late nineteenth and early twentieth centuries.[13] Alternatively, it can investigate users' everyday implementation of a technology, and how they found new ways of deploying it. In this mould we can find Asa Briggs' study of *Victorian Things*[14] and Iwan Morus' recent study of how new electrical technologies in the first half of the nineteenth century were put to an extraordinary variety of uses by the electrical showmen and practitioners of London.[15]

Looking at technology as 'volition' is to study the ways in which technology is deployed as an extension of human capacities in order to accomplish new goals, or to enhance the fulfilment of existing goals. Thus there is a significant body of literature in the history of technology concerned with how technological hardware has been designed and used with the specific goal of extending human power. Allan Chapman has written a very helpful account of how the telegraph system was developed in order to extend the scope of timekeeping so as to attain national synchronization of railway timetables – very useful for the administrative concerns of both governments and railway companies.[16] Jon Agar's forthcoming work on governmental technologization of the civil service will show how its historical development has been very closely integrated with the emergence of technologies specifically designed for information management.[17] To some extent we can study technologies from the point of view of the ways in which they extended the power of ordinary users; thus, for example, new contraceptive and reproductive technologies have sometimes been portrayed as enabling women especially to achieve liberation by overcoming biological contraints. This view has been strongly criticized from a number of directions, notably Ann Oakley who observes that the history of choices about which reproductive technologies have been made for whom has been very much dominated by male decision-makers.[18] A similarly dystopian but also plausible rendition of 'technology as power' has concerned the way technology has been used to extend the power of large and already powerful institutions, particularly governments and corporations. This is exemplified in histories of nuclear energy by Tony Hall and Walter Patterson,[19] and of the media by James Curran and Jean Seaton.[20] In their accounts, the adoption of large-scale technologies in the public sphere has often severely constrained the liberties of individuals, and has posed major threats to the democratic nature of social interaction. Unsurprisingly such concerns largely do not feature in the account of traditional technophile historians who lament the persistent reluctance of humans to adapt to the wonderful empowerment that new technologies are often alleged to have brought.

HISTORY OF TECHNOLOGY NEEDS MORE THAN JUST ONE PARTICULAR 'HISTORICAL' APPROACH

A recurrent complaint one hears is that Britain does not have a community of historians of technology of the size or visibility of that in the USA. Looking at the membership profile of the Society for History of Technology, it is clear that the USA dominates the field by whatever criterion one measures scholarly activity. This is perhaps partly because technology is widely recognized in the USA to have played a crucial role in the establishment of the country as it exists today – while the same could not quite so clearly be said of British views of technology in the making of their own national identity. There is a further matter here, however, which concerns the explicit identification of scholars as historians of technology in the USA, and this has much to do with the way that their field was formed as reaction to the hostile exclusivity of historians of science. Following the History of Science Society's famous refusal in 1957 to publish articles on the history of technology in its journal *ISIS*, Melvyn Kranzberg and his allies were inspired to launch the Society for the History of Technology and its associated journal *Technology and Culture*. Nobody could deny that this unhappy snub from HSS resulted in sufficient righteous indignation for history of technology to thrive in a climate of scholarly competition *against* history of science.[21]

This problem has not arisen in the UK, since history of science has always been recognized as a broadly inclusive category that has nurtured much history of technology. However, it has also meant that, unlike in the USA, British practitioners who work in the large overlapping domains of history of science and technology are not obliged to identify themselves exclusively as either historians of science or historians of technology. This is not unreasonable, since during the last 150 years a large proportion of the history of technology – however conceived – has surely had some form of connection to science; moreover many contemporary readers seem unable or unwilling to differentiate between science and technology. Since such scholars in the overlapping domain often publish promiscuously in a wide range of journals, their work might be less familiar to those who resiliently refuse to look beyond periodicals dedicated to history of technology. Ironically, for example, one can now find articles on technology in *ISIS*, such as Will Ashworth's prize-winning piece on the extension of techniques developed in Bentham's dockyard technologies to mathematics education.[22] Further examples of history of technology flourishing within history of science journals can readily be found, particularly in the *Annals of Science* and *History of Science*.[23] Such scholarship highlights the role of science in developing technologies as well as the importance of technologies in science in stimulating the development of theories, notably thermodynamics, and as tools of scientific investigation. In the context of such writing it is impossible to separate the history of science and the history of technology.

It is a basic claim of this paper that, in order to find the most interesting work in the history of technology, one must often look across a wide range of publications. One can look to more traditional domains of economic history and military history to find important new work being done by Ian Inkster[24] and David Edgerton[25] as well as Donald Mackenzie's more radical study of the history of Inter-Continental Ballistic Missiles.[26] Transport history, especially the history of railways, is a particularly fruitful domain of study as is evidenced by work at the Institute of Railway Studies in York.[27] The history of computing is starting to show a real growth in the UK, as is epitomized in the work of Martin Campbell Kelly and Jon Agar.[28] Then there are the more exotic and radical areas of scholarship to consider. However unappealingly 'fashionable' one might find work in the history of material design, the work of Adrian Forty shows, for example, how fruitful this can be at capturing the historical interest of the many relatively uncomplicated and uncontroversial technologies.[29]

Institutional historians are good at showing how there are often intimate interconnections between professional organizations and the emergence of new fields of technological endeavour.[30] Scholars in the field of communication studies have produced histories that place the development of media technologies in the appropriate context of the evolution of broadcasting.[31] Feminist studies have also been very forceful in revealing the complex links between gender and technology in both the home[32] and elsewhere. Cynthia Cockburn's work has been especially helpful in this regard in interpreting the complex gender relations in the history of printing.[33]

In the field of medical history one finds a healthy interest in the domain of technology: e.g. Chris Lawrence writing on the history of medical technologies such as the thermometer and sphygmomanometer.[34] Ann Oakley's feminist critique of diagnostic antenatal technologies is essential reading to understand how women have not been the primary beneficiaries of such technologies – indeed such technologies have been used to overlook or override their testimony about their own bodies.[35] If there is such vigorous interest in technology in other specialist historical sub-disciplines, what then do the historians of technology have to worry about – unless they find their vanity piqued by the fact that their particular speciality is not the dominant form? Surely they should cherish these alternative modes of writing about technology and embrace their fertile suggestions from alternative viewpoints, and not complain that such research is not necessarily published in history of technology journals or perhaps argued in ways that match their prior conceptions of what ought to be included under the heading of 'history of technology?'

HISTORY OF TECHNOLOGY SHOULD BE ABOUT MORE THAN JUST BENIGN VICTORIOUS TECHNOLOGIES

In conventional or 'traditional' histories of technology much is made of the link between technologies and human progress, with the more or less explicit assumption that the history of technology is primarily the history of 'good' technologies that have improved human welfare. This is apparent, for example, in Buchanan's *The Power of the Machine* in which he focuses on the technologies of power generation, and the application of new sources of power to manufacturing, transport and construction, as does Donald Cardwell in his capacious overview of the subject.[36] Such applications, Buchanan plausibly contends, have transformed the conditions of many people and made life easier and more pleasant for at least some in the world. He is, however, reluctant to address the exceptionally difficult topic of nuclear power beyond noting that it has not brought unequivocal benefits to human kind, and notes in passing that its future prospects might be better.[37] In so doing, however, Buchanan underplays the crucial historical point to which other scholars have devoted rather more attention. This is that Britain did not originally acquire nuclear power in order to extend its energy resources, but rather to generate plutonium for its nuclear weapons in the early stages of the Cold War.

Although later in this book, Buchanan's generally irenic style shifts towards the important dilemma faced by users of technology – that the increased comforts offered by technology are offset by the potential damage to society – he locates this destructiveness in a rather indefinite future.[38] What Buchanan thus avoids narrating in any detail is the messy and often depressing *history* of nuclear power and its intimate connections to nuclear weaponry which one might naturally think to have been the prerogative of the historian of technology. Towards the end of Buchanan's book, he addresses the topic of the *actual* damage caused by the application of technologies in the past – both nuclear and non-nuclear – as something of an afterthought for future contemplation. Judging by the contents of their writings, however, Buchanan and Cardwell seem reluctant to accept the notion that they should give a balanced treatment of the benefits and harms associated with technological usage instead of devoting mere asides and codas to these sorts of issues.

Fortunately, however, there are historians and historically inclined scholars in other fields who are determined to draw attention to the social problems associated with technology as an integral part of the history of technology – irrespective of how inconvenient technophile scholars might find this to be. There has been some scholarship on nuclear power, on the environmental pollution from power generation, pesticides and insecticides, radioactive contamination from nuclear power plants, the loss of liberty and openness in nuclear politics. As indicated above, such themes have been taken up by Walter Patterson and Tony Hall outside the immediate disciplinary confines of history of technology in ways that place due emphasis on the political presuppositions of nuclear technology while also sparing readers exasperating and diversionary micro-details of the hardware concerned. If specialist historians of technology had been more concerned to write their accounts with due reference to the history of human power rather than to mechanical power, such awkward and troublesome subjects could have been handled well within their expertise and might have benefited from their informed commentary on the issues. Sadly they have not been so concerned.

More important still from the point of view of the ordinary person is the problem of unemployment created by the deployment of labour-saving and more specifically labour-replacing technology in the industrial sphere. Tellingly, Buchanan does not discuss this point – not even mentioning the Luddites who raised this vexed issue in the early nineteenth century. We have to rely instead on the writings of labour historians such as Maxine Berg[39] properly informed by political and economic theory to tell us this important history of technology as experienced by the working and formerly working population. On this theme of technologies as producer of unhappy consequences we can note that histories of environmental pollution point very strongly to an aspect of history of technology which technophile historians seem strikingly reluctant to recognize.[40] The assumption seemingly held by the latter that histories of pollution are somehow separate from the history of technology shows a deeply impoverished understanding of the role of technology in past societies and an irresponsible failure to uphold their professional obligations not to mislead their readers about the broader import of their subject.

HISTORY OF TECHNOLOGY SHOULD APPEAL TO THE SENSIBILITIES OF A WIDE AUDIENCE

A final point to be addressed is the level of interest in the history of technology that can reasonably be expected from both academics and the ordinary reading public. This notional baseline of the largest feasible size of readership must be borne in mind when considering whether or not more people ought – for whatever reasons – to be paying more attention to the history of technology. Judging by the volumes that dominate the shelves in most bookshops, it is not hard to discern that many potential readers of history of technology are preoccupied with more obviously glamorous historical studies of art, architecture, fashion, politics, empire and warfare – not to mention the evergreen subject of biography. It is very hard to see that history of technology of whatever form could displace more than a small fraction of this audience over to its texts. Obviously to be persuaded that history of technology is worth reading, some attempt has to be made to appeal to the sensibilities of readers, and address issues pertaining to technology that actually concern them.

Particularly *ineffective* ways of reaching out to a wider audience for historians of technology are as follows. First one can make the solipsistic assumption that historians of technology produce the only sort of historical writing that is important and lament that the reading public simply has not noticed this yet, but hope in optimism that it inevitably will; this is absurd for the reasons already given above. Or one can assume that if only the public were to wake up and pay attention, it would find intensely detailed antiquarian accounts of ancient, obscure and defunct technologies as utterly fascinating as their authors do; this patronizing and rather bizarre view overlooks the fact that many ordinary readers are already wide awake and perfectly aware how dreary it can be to read dry technical treatises written by elderly men obsessed with the most arcane forms of mechanical minutiae. A third equally impotent approach to increasing the readership of history of technology is to announce that technology is the most important topic in modern civilization and that everyone of necessity must therefore avail themselves of the most recent writings on the history of this subject or suffer (often unspecified) dire consequences. This is clearly nonsensical since there are many topics at least as historically important as technology, and it is not remotely clear that neglecting to read the history of these or of any other topics will of itself precipitate a major world crisis. If historians of technology write with these sorts of ridiculous tactics and assumption, its potential readers will quite understandably and justifiably neglect it.

History of technology instead has to be written in a manner which addresses the complex ways in which technology matters to its readerships especially in their daily experience as more-or-less skilled users of hardware, and participants in highly technologized working and living environments. Clearly these histories must focus on the way in which past technologies matter in terms of human needs and welfare, and not present material artefacts as either existing in a self-contained universe of their own or owing their continued existence purely to the brilliant intellects of engineers. Technology must not be presented as intrinsically important but rather in instrumental ways in relation to its importance in other spheres, specifically science, economics, health, the environment and employment. It must also address the ways in which new technologies in the past have affected their users, particularly their autonomy, values, income and standard of living, and this means specifically understanding their ambivalence towards technology as a bringer of loss of freedom, increase of stress and enforced dependence on the reliability of suppliers and maintenance organizations. The material hardware itself can hardly be responsible for these woes. Histories of technology must address the ways in which the interests of ordinary hardware users have often been subordinated to those of profit and industrial power.

CONCLUSION

I have argued in this paper that before they make sweeping judgements about the alleged neglect of history of technology, historians should think carefully about what they include under the heading of 'technology', what they count as 'history,' why this subject should matter to the rest of the world, and what they can do to attract the interest of a wider audience. Rather than spending time bewailing the lack of interest in their subject or attacking the apparent philistinism of the indifferent masses, I contend that if historians want more attention paid to their work, they should take rather more responsibility for solving whatever problem is revealed by considered reflection. In short, if it is that they want more attention paid to *their* particular kind of work, I suggest that they should take more trouble to write the kind of history that speaks in some meaningful way to contemporary audiences who have much else competing for their attention. Perhaps before writing more history of technology, they should stop to read enough of what historical scholars in overlapping and adjacent fields have written about technology and its past. That way they might be inspired to cease their cries of 'neglect' and actually *notice* that much good historical work is in fact being done in the UK at the moment.

That it is not all being done under the rubric of 'history of technology' should not be a particularly relevant consideration. The extremely wide range of issues that is raised by the historical study of technology requires a very wide range of forms of expertise on the part of scholars who investigate it to do these issues justice. Just as history of science now benefits enormously from contributions from those who work in cultural history, sociology, anthropology, politics, post-colonial studies and so forth, so can history of technology draw usefully upon those who work in adjacent fields. Although this might seem to some to be a form of unwarranted 'balkanization' of the subject, historians of technology should make a virtue of the necessity of the division of specialized labour – unless they can become 'Renaissance men' and master every approach to the subject currently pursued by a range of different scholars.

Then again, if the Jeremiahs insist that the criterion of a healthy field of history of technology is the existence of a large(r) number of specialist university appointments, it is up to them to show that the fertility and importance of the field merit it sufficient prestige for much greater institutional funding to follow and to attract younger scholars to join this field *qua* history of technology. They will only have themselves to blame if they do not accomplish this goal.

Notes and References

1. There are only a handful of university posts in Britain specifically dedicated to history of technology and named as such, namely at Imperial College, London, and the University of Manchester. There are, however, many more who participate in this subject under other job descriptions, and those best known to me are: Jon Agar (University of Manchester), William Ashworth (University of Liverpool), Maxine Berg (University of Warwick), Martin Campbell-Kelly (University of Warwick), Colin Divall (University of York/National Railway Museum), Jim Donnelly (University of Leeds), Robert Fox (University of Oxford), Ian Inkster (Nottingham Trent University), Sean Johnston (Dumfries College, University of Glasgow), Frank James (Royal Institution), Christine MacLeod (University of Bristol), Donald Mackenzie (University of Edinburgh), Ben Marsden (University of Aberdeen), Richard Noakes (University of Leeds), Iwan Morus (Queens' College, Belfast), Arnold Pacey (Open University), Jennifer Tann (University of Birmingham), Crosbie Smith (University of Kent at Canterbury).

2. For the claim that the core of the history of technology is the study of the sources of power, see R. Angus Buchanan, *The Power of the Machine: The Impact of Technology from 1700 to the Present* (London, 1992/1994), p. 4.

3. See for example, volume 68 of the *Transactions of the Newcomen Society* for 1996–97.

4. Arnold Pacey, *The Culture of Technology* (Oxford, 1983); *The Maze of Ingenuity* (London, 1974); *Technology In World Civilization* (Oxford, 1990).

5. Patricia Stamp, *Technology, Gender and Power in Africa* (Ottawa, 1989).

6. Adrian Wilson, *The Making of Man-Midwifery* (London, 1995).

7. Christine MacLeod, *Inventing the Industrial Revolution: The English Patent System, 1660–1800* (Cambridge, 1998).

8. Bill Aspray and Martin Campbell-Kelly, *Computer: A History of the Information Machine* (New York, 1996).

9. Carl Mitcham, *Thinking through Technology: The Path between Engineering and Philosophy* (Chicago, 1994), pp. 161–266.

10. Graeme Gooday, 'Teaching Telegraphy and Electrotechnics in the Physics Laboratory: William Ayrton and the Creation of an Academic Space for Electrical Engineering 1873–84', *History of Technology*, 1991, 13: 73–114.

11. Ben Marsden, 'Engineering Science in Glasgow – Economy, Efficiency and Measurement as Prime Movers in the Differentiation of an Academic Discipline', *British Journal for the History of Science*, 1992, 25: 319–346.

12. Gerrylyn Roberts, 'Electrification', Chapter 3 of Colin Chant (ed.), *Science, Technology and Everyday Life* (London, 1989).

13. Jim Donnelly, 'Consultants, Managers, Testing Slaves – Changing Roles for Chemists in the British Alkali Industry, 1850–1920', *Technology and Culture*, 1994, 35: 100–128.

14. Asa Briggs, *Victorian Things* (London, 1988/90).

15. Iwan Morus, *Frankenstein's Children: Electricity, Exhibition, and Experiment in Early Nineteenth-century London* (Princeton, 1998).

16. Allan Chapman, 'Standard Time for All: The Electric Telegraph, Airy, and the Greenwich Time Service', in F.A.J.L. James (ed.), *Semaphores to Shortwaves* (London, 1998), 40–59.

17. Jon Agar, *The Government Machine* (forthcoming, MIT Press).

18. Anne Oakley, *The Captured Womb: A History of the Medical Care of Pregnant Women* (Oxford, 1984).

19. Tony Hall, *Nuclear Politics: The History of Nuclear Power in Britain* (Harmondsworth, 1986); Walter Patterson, *Going Critical* (London, 1985).

20. James Curran and Jean Seaton, *Power without Responsibility: The Press and Broadcasting in Britain*, 5th edn. (London, 1997).

21. John Staudennmaier, s.j., *Technology's Storytellers: Reweaving the Human Fabric* (London/ Cambridge MA, 1985), pp. 1–2.

22. William J. Ashworth, 'Memory, Efficiency, and Synthetic Analysis: Charles Babbage, John Herschel and the Industrial Mind,' *ISIS*, 1996, 87: 629–53.

23. Ben Marsden, 'Blowing Hot and Cold: Reports and Retorts on the Status of the Air-Engine as Success or Failure, 1830–1855'. *History Of Science*, 1998, 36: 373–420.

24. Ian Inkster, *Technology and Industrialisation: Historical Case Studies and International Perspectives* (Aldershot, 1998).

25. David Edgerton, *England and the Aeroplane: An Essay on a Militant and Technological Nation* (London, 1991); *Science, Technology, and the British Industrial 'Decline', 1870–1970* (Cambridge, 1996).

26. Donald MacKenzie, *Inventing Accuracy: A Historical Sociology of Nuclear Missile Guidance* (Cambridge, MA, and London, 1990).

27. Colin Divall, 'Learning from America', *Railroad History: The Diesel Revolution* (Westford, Massachusetts, 2000), 124–42.

28. See examples above, *op. cit.* (8) and *op. cit.* (17).

29. Adrian Forty, *Objects of Desire: Design and Society since 1750* (London, 1986).

30. Colin Divall, James F. Donnelly and Sean Johnston, 'Professional Identity and Organization in a Technical Occupation: The Emergence of Chemical Engineering in Britain, *c.*1915–30', in *Contemporary British History*, 1999, 13: 56–81.

31. Andrew Crisell, *An Introductory History of British Broadcasting* (London, 1997).

32. Christina Hardyment, *From Microwave to Mangle: The Mechanization of Household Work* (Cambridge, 1988).

33. Cynthia Cockburn, *Brothers: Male Dominance and Technological Change*, 2nd edn. (London, 1991).

34. Chris Lawrence, Incommunicable Knowledge: Science, Technology and the Clinical Art in Britain, 1850–1914', *Journal of Contemporary History*, 1985, 20: 503–520.

35. Ann Oakley, *From Here to Maternity: Becoming a Mother* (Harmondsworth, 1986). See additional examples from Wilson and Oakley cited above, *op. cit.* (6) and *op. cit.* (18).

36. Donald Cardwell, *The Fontana History of Technology* (London, 1994).

37. Buchanan, *op. cit.* (2), pp. 78–79.

38. *Ibid.*, pp. 236–7.

39. Maxine Berg, *The Age of Manufactures, 1700–1820: Industry, Innovation and Work in Britain*, 2nd edn. (London, 1994).

40. See the examples in Hannah Brady (ed.), *Dirty Words: Essays on the History and Culture of Pollution* (London, 1990).

Where Is the History of Technology?

RUPERT HALL

Recently a letter-writer in *The Times* has deplored Britain's lack of interest in the teaching of technology as an academic subject. The writer, remarking 'we seem likely to repeat the past hundred years neglect of technical education' cites evidence to support the contention 'that engineering, science and technology are being relegated, ignored even, at a time when influence in support of these disciplines is necessary'; influence, that is, in lofty spaces where advisers of government operate.[1] Lamentation over Britain's technical incompetence is indeed more than a century old, and a great many governmental measures offered in answer to it, from the foundation of the Royal College of Chemistry in 1845 and the Education Act of 1870, to the White Paper of 1993,[2] have never been found adequate. There were, of course, two distinct issues here: one being the effectiveness of practical *training*, for which Britain relied heavily on the apprenticeship system – which operated at two levels: I shall not pursue this issue – and the other being the provision of science-based technical *education*. With the demise of apprenticeship and the increasing importance of the technical school, or college, or university, only the second issue remains for discussion; yet the question is still open: has technical education in practice effectively replaced and improved upon the older British method?

That is not a question I am equipped to discuss. The *general* deficiency of British education in mathematical and scientific subjects as well as technology has, however, been asserted very often; as an example I refer to a lecture delivered to the British Academy by Dr S.J. Prais in 1993.[3] This used sampling techniques to indicate (amid much else) that whereas university graduates were as well represented in the British workforce (1988–91) as in those of other European countries – indeed in some cases the comparison was markedly in favour of Britain – the proportion of those with lower qualifications was much less in Britain while that of wholly unqualified British workers was much higher (64 per cent against an average of 34 per cent in other European countries considered). Again, international tests of the mathematical abilities of 13-year-old children administered in 1963–64 showed that those of the USA performed least well with an average score of 16 per cent, France and Britain came next (18, 19) the Netherlands and Germany higher with 24 and 25 while Japan did best with an average score of 31 per cent. The introduction of comprehensive education since that time seems to have reduced the level of British competence further.[4] Of another arithmetical test Dr Prais writes: 'Only the top fifth of all English pupils, with 62 per cent correct, were able to approach the attainments of the German lowest two-fifths of all pupils.'[5]

Again, the vicissitudes of English educational policy are not my concern. I seek rather to start from what I take to be the correct (if highly regrettable) position that in Britain mass education has produced no easy familiarity in the general population with mathematical, scientific or technical subjects. Let us suppose that there is an adequate group of university entrants in mathematics and science – though this too might be questioned. This cannot at any time constitute more than a small part of the whole population.

I wish now to turn to a further point made by the letter-writer I mentioned before, Mr M.M. Charlish, who revives the arguments of Samuel Smiles (1862) that kings, queens and statesmen fill too many of the pages of history and biography: 'Surely some niche ought to be found for the mechanic without whose skill and labour, society as it is, could not exist.' This too has indeed often been echoed since. I remember it as often voiced by that learned scholar Dr Charles Singer, who himself wrote extensively on the history of medicine, science and technology, in all of which fields he carried out pioneering research during the early decades of the century.[6] I do not myself believe that many students of history in this century have been fed on a diet of J.R. Green's *History of England* or even on such an immortal classic as Gibbon's *Decline and Fall of the Roman Empire*. Rather they absorb a mixture in which political and constitutional history are no more than aspects competing with economic, social, industrial, military, religious and sexual history. They may in fact learn a good deal about 'the mechanic' and his conditions of life as they were in Smiles's day and in centuries before and since. True, they learn nothing (in most institutions) about the history of science and technology. The United States has advanced on a broader front than we have in this respect, for with us the border lands between the historian and the scientist or the historian and the engineer are little explored. Are we to conclude that Britain cares nothing for the history of technology? It is not difficult to think of counter-arguments. The National Museum of Science and Industry is still a popular place of resort for young people, and many other old industrial sites or rural museums in virtually every county attract steady crowds of visitors (at any rate in summer), whether they be large like Beamish and Ironbridge or perhaps only a single engine, mine or works. Probably nowhere else in the world are there so many well-preserved and well-displayed monuments to the mining, manufactures and transport of the past as there are in Britain. Television has helped by presenting to the nation a number of programmes (of varied quality), of which the most recent was that guided by Mr Fred Dibnah, a sound practical man of long experience with engines. It cannot be denied that people do like to see windmills turn, pistons slide, hear whistles pierce the ear, and so on. I know of one stretch of railway in South Wales that never existed in the heroic days of coal and iron but was recently built, in most attractive country, to attract summer tourist traffic. The steam locomotive is an unfailing attraction, at Didcot and York or wherever a stretch of abandoned line can be revived.

All this is excellent. It demonstrates daily that the past of technology can be both fascinating and instructive. It enhances the public awareness of the fact that time is an infinite dimension, not merely 'today'. But it is not mere academic snobbishness to suppose that this is not the history of technology. The visitor can see flashbacks of certain episodes without acquiring any sense of continuity. It is true that the best museum displays can awaken some understanding of cause and effect, as well as wonder and surprise, but this and every form of analysis and criticism is better conveyed by the written word. The visual and experiential approach is no doubt the best first step in appreciation of the past, as it may be a first step also to an actual 'hands-on' mastery of ancient tools and processes, but it is for most people an entertainment beyond which few advance to serious historical study.

The history of technology exists in many forms: much can certainly be communicated by vision (and even by smell) which cannot so well be conveyed by words. Biography of the 'heroic' model has been a strong element since the days of Samuel Smiles; analytical history as applied in the history of metals by Cyril Stanley Smith or in the history of medieval building by the use of stress-analysis is a promising modern form; there have been many 'linear' histories of particular types of structures or machines, at all levels – ships, bridges, aeroplanes, dams, canals, almost endless works on cars and locomotives – though there has been less work on productive machines (such as looms, presses, lathes and other machine tools) and even less on extractive industry. A good deal of this writing is straightforwardly descriptive with little attempt at analysis, and to find out how structures or machines were actually built one often has to open a fresh investigation. Some parts of descriptive history of technology are records of what is unique or very unusual: if this did not exist, experimental devices and innovations in design might well be lost to sight for ever. I have been trying to define what I call the 'hard' history of technology, because buildings rather than architects, engines rather than engineers, things rather than people are its subject. It has been much developed by British scholars and writers: the excellent *Transactions* of the Newcomen Society, the works of H.W. Dickinson, Rhys Jenkins and Charles Singer and many more from the earlier decades of this century are of this type, as indeed were those of German pioneers in this field (such as, after the very early work of Johann Beckmann (1739–1812), Ludwig Darmstaedter, F.M. Feldhaus, Albert Neuburger) and various French and Italian scholars. The 'soft' history of technology, concerned with the lives of people either as workers in agriculture and industry or as 'consumers' affected by them, has been strongly represented in the American journal *Technology and Culture* after being explored earlier by social and economic historians. I mention the writings of J.U. Nef as an older example of such studies merging sociology, economics and technology: in his large work on the early British coal industry and elsewhere Nef examined the effects of changes in fuel on the development of civilization; *War and Economic Progress* (1950) investigated the particular relation (positive or negative) between economic growth and the incidence of wars.

'Hard' and 'soft' history of technology may alike be presented in a more accessible, perhaps mainly visual, vehicle or at a fully professional level, such as defines the 'academic' historiography of technology. I place the word 'academic' in inverted commas because in fact universities have contributed little to the writing of the history of technology in this country and, I suspect, in most other countries with the exception of Germany and the USA. France has for two hundred years possessed in the *École Polytechnique* one of the world's greatest scientific and technical universities, but this institution has done nothing (to the best of my knowledge) for the historical aspects of its studies. In the United States academic interest in the history of technology, developing later than that in the history of science (already present in the 1920s), has expanded considerably in the last generation, greater emphasis being laid on the 'soft' history of technology than has been the case with the European development, taking place over a longer time. I do not of course mean that there was no earlier work in the USA of the 'harder' variety; let two examples suffice: W.B. Parsons's *Engineers and Engineering in the Renaissance* (1939) and A.P. Usher's *History of Mechanical Inventions* (1929). I have already noted works of the 'hard' variety published by German academics in the early part of this century, many of which served as tools and models for scholars in other countries; it remains to record the great expansion of professorships in the history of technology occurring in (West) Germany since World War II. It is well known that German technical education has been highly organized and of excellent quality for many generations; now Germany is the only country in Europe (and perhaps in the whole world) of which it may be said that the history of technology is well established there as an academic discipline. The Germans began strongly from the first years of this century with the opening of the Deutsches Museum in Munich (1906) and the foundation of a journal *(Beiträge für Technikgeschichte*, 1909). A first Chair in the history of technology was created at the Charlottenburg technical university in 1936 for Conrad Matschoss, founder of the *Beiträge*. Changes in the principles and structure of scientific and technical education introduced in the early 1960s with the object of increasing *Verklammerung* (mutual understanding) between arts and sciences within the universities encouraged the foundations of new professorships of history of technology; four new Chairs and a lectureship were created in that decade and many more have followed since. Most were at technical universities. And the history of technology became, as it has also done in the USA in more recent years, a discipline dominated by trained historians rather than engineers:

> Many of the historians from the Deutsches Museum and the VDI (Society of German Engineers) saw their life's work endangered, and were understandably unhappy about a development which saw the study of the history of technology increasingly divorced from what they had always taken it to be.[7]

In other words, upon the 'hard' foundations of the history of technology in Germany there had arisen a 'soft' superstructure – no less worthy, but different. From the above it is clear that the statement: 'British universities have not fostered the history of technology as have those of Germany and the USA' requires interpretation and qualification. The 'history of technology' that has flourished in these countries is not by any means the same historiography as that practised in these countries before 1960 and in Britain to the present day. But I have no wish to embark upon such a discussion or to insist upon one or another definition of what the history of technology really is or should be. However, it is certainly true that while the Newcomen Society continues to be as active as it has been for the last 70 years, while books on engineering history and analogous topics continue to be published in Britain, while interest in 'industrial archaeology' continues to be strong, and while both economic and social historians continue to address themselves to certain themes related to the history of technology, so that the discipline certainly exists here as it has for many years, it is nevertheless true that it has failed to establish itself in British universities. Indeed, we have in the last two decades lost the professorships of the history of science and technology once operating in two of our universities (London and Manchester) and regained only one, and though teaching of the history of technology continues it can hardly be said to thrive. To draw an obvious comparison, the study of medical history (supported by the Wellcome Trust) is in a much more healthy condition: quite the opposite of the situation 20 years ago.

Why has the academic history of technology, and its teaching as a discipline, regressed in Britain from a promising start in the 1960s? It is not the case, I believe, that professional engineers and technologists are less interested now in the history of their subject than their predecessors. Nor, I believe, could this be said of practitioners of the very wide disciplines of economic and social history; perhaps rather the reverse is true. Continuing public interest in industrial archaeology and the support given to many ventures in conservation from the *Great Britain* enterprise downwards in scale suggest that the audience for academic history of technology *outside* the universities may be considerably larger and more effective than it was 40 or 50 years ago. On the other hand, it is certainly true that all *smaller* university subjects have suffered greatly from the increased pressure on finances that has screwed down steadily over that period. It is well known that many universities rely heavily on the fees paid by foreign students; it is easy to guess that most of these favour courses most likely to be profitable in career terms. 'Training' has waxed while 'education' has waned. Institutions place their bets on well-tried runners.

It may be generally supposed that if an engineer or technologist is trained to acquit himself or herself ably with respect to the professional tasks undertaken, nothing more can be expected or should be offered to him or her. Whether they have a knowledge of the culture of their country, whether they know anything of the history of their discipline, whether they have any sense of the social responsibility placed upon those who undertake engineering works or manufacture – these may be seen as matters of no concern to the teaching institution. I suppose that it was precisely because they did not adopt these views, because they wanted to produce well-rounded and socially conscious engineers and technologists, that the Germans decided to plan for *Verklammerung*. I do not know what opinions prevail now but I am confident that 20 or 30 years ago many who were responsible for the education of such professionals in Britain thought much like their German colleagues and sought to innovate by providing appropriate instruction and encouragement. However, many interests compete for the small number of hours to be spared from strictly professional teaching. Hence the history of technology is one interest competing with many others – social studies and law, for example, and perhaps most powerful of all, various forms of study concerning the environment. Nor, surely, can anyone doubt that professional competence ought to be accompanied by concern for the human and environmental consequences of engineering and industry. When we appreciate that appropriate historical studies can illuminate the need for such concern, it is not surprising that our German colleagues have developed the 'soft' history of technology.

'Soft' history of technology is clearly 'relevant' (to use a popular term) and the wonder is rather that there has not been more activity in this area in this country, as in Germany and the USA. Being a historian rather than an engineer I am not qualified to judge whether the 'hard' history of technology is 'relevant' in the same social sense, though I am confident that it is essential for a full historical account of the development of Western society. The technical and scientific factors conditioning – often limiting – the transport and manufactures, structures and warfare and so on at *any* state of the world demand to be understood. 'Soft' history of technology deals rather with the effects than the nature of such conditioning.

Yet I feel that something more fundamental has held back the history of technology in this country: its deep anti-intellectualism. Britons value the practical, and distrust and mock theory. So we have argued that apprenticeship is better than college learning: experiment, trial-and-error, suck-it-and-see are better than reliance on theories. Once, some 50 years ago, walking beside the Serpentine in Hyde Park, I met a man towing hull-models through the water on a long stick. He told me he was testing possible hull-forms for cargo-ships. Was he mad, or was this the best that British naval design was then capable of? Do we still call technical designers 'draughtsmen' as though all that was needed for their work was a T-square and French curves, enough to make the pretty drawings with which architects beguile their clients? 'Technology' was still at that time, or only a little before, an alien word, rather dubious, while the word 'technic' or 'technique' – even its spelling is uncertain – has never, in spite of the influential example of Lewis Mumford, been fully naturalized in the English language: technique is often akin to trickery. The *Shorter Oxford Dictionary* tells us that it may mean 'a trick or knack' or, specifically, 'skill in sexual relations'! Therefore something out of place in the building of bridges, ships or motor cars. If courses teaching technology are unpopular in schools, and perhaps even the demand for university places is not as keen as it might be, is it to be wondered at that the history of this subject is of even less value? A musician would look a fool if he lacked the most elementary knowledge of Monteverdi and Mozart, Bach and Brahms – though what they wrote is even now called 'ancient' music – because the works of these composers are still very much alive; Roman roads and bridges, seventeenth-century canals like London's New River, are still with us but how few of us care how such works were planned, levelled and carried out? So these remain very dusty, antiquarian issues. I believe that if we possessed a stronger intellectual interest in the man-made world around us (of which the English landscape, for an example, is still no insignificant part), if we were curious about the towns, structures, artefacts everywhere to be seen, if we were puzzled to know why our lives are so different from those of our grandparents, everyone of us would wish to explore the history of how things were done and made. But we are dull and apathetic and easily titillated by trivialities.

Notes and References

1. Letter from Mr M.M. Charlish, *The Times*, 4 September 1999, p. 23.
2. *Realising Our Potential*, Cm. 250, H.M.S.O. 1993.
3. S.J. Prais, 'Economic Performance and Education: The Nature of Britain's Deficiencies', *Proc. Brit. Academy*, 84, 1993: 151–207.
4. *Ibid.*, pp. 171, 176. In a test made in 1964 more than half the British pupils correctly answered an easy question on simple proportionality; fewer than one-third did so when the test was repeated in 1981. In Japan the percentage of correct answers was 61.
5. *Ibid.*, p. 183.
6. In history of technology with *The Earliest Chemical Industry* [the trade in alum], Folio Society (London 1948). He initiated *The History of Technology* in five volumes, published by the Clarendon Press (Oxford, 1954–1958).
7. Lutz Engelskirchen, 'Establishing History of Technology as an Academic Discipline in the Federal Republic of Germany after 1945', *ICON*, 4, 1998, p. 153.

Reflections on the Decline of the History of Technology in Britain

R. ANGUS BUCHANAN

ABSTRACT

There is a consensus among British historians of technology that all is not well with their discipline. This impression is derived in part from the contrast with other advanced nations, and particularly the United States of America, in which it has grown to be a recognized and well-resourced subject of university study. But in Britain, despite some promising indications of activity in the 1960s and 1970s, the history of technology has not achieved the critical mass necessary to ensure its healthy survival in the highly competitive world of late-twentieth-century academic politics. There are several reasons for this: the subject has never been made welcome by other historical studies, which have been suspicious of its use of physical evidence; it has failed to distinguish itself sufficiently from the history of science, which acquired academic articulation before the history of technology; and it has an uneasy relationship with the social sciences, especially the 'social constructionists', who threaten its historical identity. The result has been that the initiatives of 20 years ago have not been sustained once university economies and assessment exercises began to take effect. Without a princely endowment which would enable at least one outstanding unit in the history of technology to be created, it seems likely that the most hopeful prospect for the future is for individual practitioners to continue to strive for scholarly excellence, developing the literature of the subject, stimulating discussion and controversy on important themes, and seeking to demonstrate the essential importance of the history of technology to our understanding of modern society.

REFLECTIONS ON THE DECLINE OF THE HISTORY OF TECHNOLOGY IN BRITAIN

The editor has invited me to consider why the academic study of the history of technology has failed to become a well-established discipline in British universities. Such an invitation presents a temptation to indulge in autobiographical reflection, and I promise that I shall make no attempt to resist the temptation. But first it is important to define the subject of our enquiry. The history of technology is a branch of historical studies which has only comparatively recently been identified as a distinct entity deserving recognition alongside political, social, economic and other specialized aspects of the discipline. It is concerned with the capacity of humankind for making and doing things, and with the artefacts thus produced. Such artefacts may be studied and interpreted in order to discover the motives and achievements of those who made them, and so to acquire an understanding of human creativity and of the potentiality of invention. It should hardly be necessary to add that such an understanding, applied to the sources of power, to industrial productivity, and to the development of transport and the means of communication, is of inestimable value to the processes of education. Without some approximation to such understanding, indeed, it is difficult to see how any sound assessment of human history can be made, because all the successes and failures of men and women in modifying the human environment have derived, in a word, from technology.

Yet despite its manifest importance both as a historical study and as an aspect of the education of any citizen of the modern world, the history of technology has not received, at least in Britain, the attention which it deserves. There are several obvious reasons for this. First is the disadvantage of being a comparative latecomer in the academic spectrum, with other disciplines reluctant to allow it full recognition commensurate with that which they receive themselves. This is particularly the case among other historical studies, which have shown little readiness to make it welcome. Traditional general histories continue conspicuously to ignore it: a recent addition to the New Oxford series of general histories of Britain, on the mid-Victorian period, rates I.K. Brunel one reference – as a patron of art – and takes the phenomenal nineteenth-century transformation of Britain by the railways and the steam engine virtually for granted, requiring no special explanation.[1] Such myopia about the impact of technology is in no way untypical: a long catalogue could be compiled of similar indifference. There have been some very significant exceptions, especially among economic and social historians, with writers such as Sir John Clapham, A.E. Musson and Eric Robinson, and J.R. Harris giving careful attention to the role of technology in industrialization.[2] But even such comparatively enlightened authors have normally been uneasy about the minutiae of specific techniques, and have taken care to remain firmly within the framework of documentary analysis.

This points to a second disadvantage of the history of technology as a historical study: being concerned with artefacts, its ambit includes the consideration of non-documentary evidence. The function of a steam engine or a printing machine may be reduced to documentary form, provided that there is some prior understanding of the artefact itself, and the history of technology has been willing to incorporate this physical dimension even when, as in the case of much modern 'black box' technology, it is not possible to present a simple, easily comprehensible, account of the technology under review. Historians are temperamentally ill at ease with such physical evidence, and so historians of technology have been tempted to look elsewhere, for alternative disciplinary linkages. Geography and archaeology have both been considered, and insofar as they are familiar with field work and physical evidence these disciplines can give some encouragement to historians of technology. Archaeologists, in particular, have begun to see industrial archaeology – which comprises, practically speaking, the physical evidence of the history of technology, especially since the Industrial Revolution – as a period division within their own range of specialisms, and this promises to be a very helpful liaison. Of course, it could be argued that archaeology is itself part of history, being concerned with precisely that sort of physical history which traditional historiography has, to its own loss, ignored. But that would be ungrateful towards those archaeologists who have been most helpful in finding a home for industrial archaeology, and in any event it does not help the history of technology.

Another disadvantage of the history of technology has been that it has developed under the shadow of the history of science. The two disciplines share some of the same subject matter, insofar as the history of technology is frequently, although not invariably, concerned with the application of scientific discoveries. But the students of scientific innovations became academically articulate before the students of innovative artefacts, so that the latter acquired a dependency relationship on the former which tended to keep them in a perpetually subordinate position. The history of science, moreover, has stressed its relationship with the philosophy of science rather than with more practical aspects of the subject. It has also indulged in a long debate between 'internalist' and 'externalist' approaches to the discipline which is of little interest to historians of technology because technology is of its essence a social phenomenon, making sense only in the context of social needs and human aspirations. Not until the publication of the first five volumes of the Oxford *History of Technology* in the 1950s was the scope of the history of technology as a separate and different study sketched out, and curiously that was written largely by historians of science, although they had come to recognize the independent methodology of the history of technology.[3] The relationship has remained an uneasy one, however, even or perhaps especially in those university departments which have tried to combine the two disciplines.

There have been other disadvantages which have made it difficult for the history of technology to establish a foothold in a competitive academic market. To some extent it has suffered at the hands of social scientists who have been interested in it as a field in which theories of 'social construction' can be projected, or business histories assembled. The assumption that the history of technology, like the history of science, is subject to the normal rules of social organization, is unobjectionable, but it is galling to be instructed about this necessary relationship by social scientists who frequently show little knowledge of history or sensitivity for historical methodology. This intervention may be useful, insofar as it introduces perspectives to the interpretation of historical events, but in the process it threatens to undermine the approach which is basic to all historical research of beginning with the evidence rather than the generalization. Similarly, the objectives of business histories may be illuminating in regard to some technically innovative enterprises, but they rarely add significantly to technological history. In the United States, on the other hand, where the history of technology is much stronger than in Britain, the sociological and business analysis of historical phenomena has been regarded with more favourable expectations.[4]

The strength of the history of technology in the United States and elsewhere underscores the poor performance of the discipline in Britain. In the USA there is a thriving society, the Society for the History of Technology (SHOT), with an excellent journal, *Technology and Culture*, which enjoys an international circulation and maintains a lively discussion of current issues in the field. There are many Chairs in the History of Technology at American universities, with corresponding staff support, and a healthy 'market' of academic posts for new entrants to the discipline. Much the same can be said of the status of the history of technology in Germany, France, the Netherlands and Sweden, for although the coverage is uneven each has acknowledged centres of excellence in the discipline, with prominent and respected academics in charge. Elsewhere in Europe, and in Japan and Australasia, there are some outstanding practitioners in the history of technology, often well placed in leading universities. In the less developed and developing countries support for academic enterprises in general has a lower priority, and the history of technology has only a haphazard presence.[5]

It is bizarre that Britain should fall into the category of less developed and developing countries as far as recognition of the history of technology is concerned, but such seems to be the case. This is all the more remarkable because 20 years ago the situation appeared to be quite promising. Donald Cardwell was a presence to be reckoned with at UMIST, within the University of Manchester, with able lieutenants in Richard Hills and Arnold Pacey. Rupert and Marie Boas Hall, although primarily historians of science, provided a welcome for the history of technology in the person of Norman Smith at Imperial College in the University of London. At Cambridge, the huge reputation of Joseph Needham gave substance to the discipline, even though his immersion in Chinese science and technology tended to isolate him in the rather abstruse history of science practised there. A number of individual scholars managed to assert a presence for the history of technology within adjacent disciplines at other universities, and contributed to what seemed at the time an encouraging prospect.

Even then, however, there was an anomaly in the academic structure of the history of technology in the sense that its primary institution was not of an academic character. The Newcomen Society for the History of Engineering and Technology had carried the flag for the discipline since its foundation in 1920, and its *Transactions* had become one of the major repositories for serious research in the history of technology. But the

Society has been ambivalent about academic linkages and it has tended to stress the practical engineering aspects of the subject and their relationship with museum conservation. It has undergone several renewals of personnel and of enthusiasm and has continued to present good work in the history of technology over eight decades. Twenty years ago it reorganized itself for a more proactive role in promoting the history of technology, but even though it has doubtless become more efficient the results in terms of advancing the study of engineering and technological history have been disappointing. This is mainly because the Society has still not impinged significantly on British academic life, and the reason for this is probably the great preponderance of retired engineers among its members, with no career interests in the subject.[6]

Among other promising initiatives in the second half of the twentieth century has been the Centre for the History of Technology, Science and Society, which I established at the Bristol College of Science and Technology, shortly to become the University of Bath, in 1964. The cumbersome title, replacing the inelegant 'Centre for the Study of the History of Technology' with which it started, was the result of an unsuccessful attempt to accommodate the research interests of colleagues. It is thus best referred to as simply the 'Centre for the History of Technology'. It was the result of personal preoccupations which had something of a revelatory character for me. I had been recruited in 1960 to a new 'General Studies' Department in one of the ten designated 'colleges of advanced technology' (CATs), with the assignment to teach modern social and economic history to engineers and applied scientists. This was under an admirable scheme which required that one-tenth of the teaching in all courses leading to the new 'Diploma in Technology' should be in non-specialist subjects, and which opened a door for teaching in economics, history and social science. This scheme was abandoned when the CATs became fully fledged universities and proceeded to define their own degree requirements, which were almost invariably intensely specialized. But the scheme lasted long enough to oblige me to devise a way of making history palatable to my students, who were all practically minded technologists and scientists.[7]

My solution was to develop the history of technology as a 'bridge' discipline, and once I had begun to explore its possibilities I became fascinated by its research opportunities. Among other initiatives, I put on an Adult Class in the subject at the Bristol Folk House. It was not well attended, but through it I made some valuable contacts and began to come to grips with the literature. I made contact with Bristol City Museum and struck up a creative partnership with the lively young Curator in Technology, Neil Cossons. I joined the Newcomen Society and attended my first Summer Meeting. Newcomen friends urged me to find a research base for George Watkins, a Bristolian who had assembled an astonishing collection of photographs of stationary steam engines. And through the Council for British Archaeology (CBA), which had recently appointed an 'Industrial Panel' with a 'Survey officer for Industrial

Monuments' in the person of Rex Wailes, retired engineer and past-President of the Newcomen Society, I was encouraged to launch the 'National Record of Industrial Monuments' (NRIM), using record cards prepared by the CBA.[8]

The combined effect of these stimuli was to lead me into the creation of the Centre for the History of Technology, and it has operated continuously under my direction since 1964. This is quite a respectable spell for an academic institution, and it has passed through several phases. To begin with, it was heavily involved in industrial archaeology, both nationally through the CBA and the NRIM, and locally through the study of the surviving artefacts of technological history in the Bristol region. Then the long association with George Watkins, who served for 20 years as a research officer and visiting fellow, ensured that steam technology remained prominent in the activities of the Centre. This was gradually replaced by a concern with the history of the engineering profession and with engineering biography, particularly with Brunel studies. In recent years, the Centre has promoted an important initiative in developing the study of the history of gunpowder technology. All this has taken place alongside vigorous participation in various institutional activities, creating new bodies such as the Bristol Industrial Archaeological Society (BIAS) and the Association for Industrial Archaeology (AIA), and taking a formative role in the Newcomen Society, the Royal Commission on the Historical Monuments of England (RCHME), the National Trust, and the International Committee for the History of Technology (ICOHTEC).

ICOHTEC has been particularly important in acquiring an international perspective on the history of technology. I was present at the establishment of the Committee, at the International Congress for the History of Science, Technology and Medicine, held in Paris in the summer of 1968. I was selected as the British representative on the Executive Committee, and subsequently served as Secretary General (1981–1993) and President (1993–1997). In 1970 I attended the first ICOHTEC Symposium, held at Pont-à-Mousson in France, as one of the three British representatives: the other two were Donald Cardwell and Frank Greenaway. In retrospect, it seems that our representation did reasonable justice to the mixture of interests in the history of technology in Britain: Cardwell was a physicist-turned-historian of science who had a keen appreciation of the significance of scientific initiatives in technological development; Greenaway was a leading authority in chemical engineering at the Science Museum; and I represented an attempt to expand historical sensitivity in the understanding of technological innovations. Together, we presented a mixture of robust possibilities, although this was probably appreciated more by ICOHTEC than by British academic life. Over the years, with the latest of 27 very successful international symposia having been held in Prague in 2000, the British participation has been persistently small but active: apart from myself, the most regular British attenders have been Graham Hollister-Short and Alex Keller, who have been the first two editors of *ICON*, the ICOHTEC journal. The Committee has been sustained by a very substantial North American participation, and by strong support from Germany, France and other European countries, with contributions from Australia and Japan. Despite modest but determined advertising of its activities, ICOHTEC has failed to generate a broad base of support in Britain. Even the Symposium held in Bath in 1994, which was very successful in most respects, did not significantly modify this lack of British support. It may be regarded, indeed, as accurately reflecting the restricted role of the history of technology in Britain, in contrast with the lively activities in the other advanced industrial countries.[9]

It should be clear from this account that the history of technology as a respected field of study has had, to put it mildly, a very checkered history in Britain. Nevertheless, there were good reasons for optimism until the early 1980s, with a few active individuals making a small but determined contribution towards the development of the subject, and with its success as an international discipline. But by the end of that decade, the existing university chairs in the subject had been vacated and the retiring professors had not been replaced. And even though two new professors in the history of technology were created in the 1990s at Sheffield Hallam and at Bath, these were both personal appointments and they also have now retired and not been replaced. The recent appointment of David Edgerton as Professor of the History of Science and Technology at Imperial College, running a seminar and research programme [now suspended, ed.] in association with the Science Museum next door in South Kensington, has remedied a serious omission but it still constitutes the exception which proves the rule. That rule may be plainly stated: for 20 years, the history of technology has been in decline in Britain. From a start in the 1950s and 1960s which was admittedly low in quantity but of promising quality, the discipline has by 2000 dwindled almost to vanishing point in British universities.[10]

How can this decline be explained? It could be argued that it merely represents a phenomenon which has been a persistent feature of British universities since 1980: the impact of government economies, brought about by changes in funding and the institution of widespread quality assurance and research assessment procedures, has been particularly severe on small departments. Whereas large departments have managed to survive by shedding a few marginal posts and adopting the new game plan with a show of zeal, small sections and departments have been vulnerable to the loss of key posts and have then been denied the resources to replace them. In some cases alliances have been struck between departments which had hitherto found each other uncongenial in order to achieve the critical mass necessary for mutual survival. Several history of science units appear to have come to such a satisfactory arrangement, and so are weathering the storm of intrusive economies. With the history of technology, however, this has not happened. Units which were first weakened by the loss of key personnel through death or retirement were then offered alliances which inflicted a loss of identity or were shunted into the margin until natural causes removed them from the scene.

While I have no doubt about the substantial accuracy of this summary, there is, of course, much more to it than that. Other small disciplines, after all, have been satisfactorily assimilated into larger groups and ensured survival into the next round of academic infighting. The trouble with the history of technology has been that no appropriate academic accommodation has been forthcoming. This is partly because, as at Bath, there was no suitable history department to provide a home. Elsewhere, however, it has been due in part to the academic myopia about the history of technology which, as we have seen, has persisted in many traditional history departments. The distaste of most academics for the role of 'servicing' other departments, because it is regarded as low-level and repetitive teaching and detracts from research, has not helped here. Much history of technology has inevitably been offered at undergraduate level as a 'bridge' discipline to help in putting specialist studies into perspective, and such linkages have been particularly prone to economies. The history of technology, seen as a regular item in the academic syllabus, requires sensitivity and sympathy from both historians and technologists, and such empathy has been noticeably absent while other disciplines have been protecting their own corners.

Even though the quantitative weakness of the history of technology in British universities has been a formidable problem, it is probably true to say that historians of technology have not been as astute as they might have been in securing a continuing role for their discipline. They have not set out with a determination to storm the fortresses of entrenched ignorance and prejudice as they might have done. Such a course might not have won them friends, but it would have increased their influence. I recognize, for example, that I lack the ability to understand how any intelligent person could fail to understand the importance of the history of technology – a deficiency which leaves me expecting favourable rather than negative assessments, bewildered and frustrated by turn-downs from grant-awarding bodies, and generally vulnerable to more sure-footed academic operators. Other historians of technology have shown similar unreality in terms of being reluctant to take on their colleagues in the struggle for academic survival. I have observed colleagues in the social sciences almost literally hammering on the doors of grant-awarding bodies and assessment panels until they received what they wanted. But such activity was not for us. Whether we were being too gentlemanly, or behaving like prima donnas, or simply that we were all loners without much sense of group action, is not for me to judge: the fact remains, there were never enough of us to make a success of the history of technology as an independent academic discipline once the expansive years of the 1960s and 1970s were over.

Then it could be argued that some of us have pursued the wrong research projects. In particular, it has been suggested that the study of industrial archaeology has led too many of us to deploy too much energy on the conservation of industrial monuments. This, however, is to beg the more fundamental question of the relationship between the history of technology and industrial archaeology, because it has always seemed to me that the latter constitutes the physical evidence essential to the interpretation of the former, so that the two should be regarded as indivisible. On these grounds, it was a legitimate aspiration to hope that academic recognition for industrial archaeology would redound favourably on the history of technology. In the event, the converse has been closer to reality: the failure of industrial archaeology to win academic recognition, for reasons which we need not pursue here, has rendered much research effort by sympathetic historians of technology largely useless, at least in terms of academic status. I have to concede that in practice attempts by myself and others to incorporate industrial archaeology in the history of technology have not been conspicuously successful, and that the main achievements of industrial archaeology have been in conservation and museum studies. It seems likely now that industrial archaeology will find its natural academic home as a part of archaeological studies, so that historians of technology need not participate directly in it.[11]

So far as there is substance in this criticism regarding the choice of research projects, it applies most forcibly to the output of books on the history of technology because it could be said that we have failed to generate a thriving and controversial literature. In North America *Technology and Culture* has been very successful in keeping the history of technology in the collective scholarly mind, and has maintained a vigorous review policy which has encouraged scholars to take part in continuing debates on many aspects of the subject. The lack of such a 'platform' in Britain is probably the biggest weakness of the history of technology as an academically respected discipline. Notwithstanding the massive research of Joseph Needham and his team on Chinese technology; the flood of books about railways and canals; the continuing flow of worthy contributions from members of the Newcomen Society; and some passable textbooks on the subject; there is actually a shortage of monographs stimulating widespread discussion among historians of technology.[12]

If this analysis of what has gone wrong with the history of technology in Britain is correct, it suggests several courses of action which could arrest the decline. First of these, in the current climate of British universities, would be the investment of a princely endowment by an individual or an institution prepared to make, say, £5 million sterling available to create an Institute for the History of Technology at one of our leading universities. This would, at a stroke, endow a permanent Chair and secure a research staff which could promote undergraduate courses, undertake servicing of specialist departments and develop a variety of research projects around its members and research students. It could also take a leading part in the international representation of Britain in the history of technology. Failing any immediate prospect of such a robust solution, the handful of surviving historians of technology need to nourish scholarship in the subject, promoting discussion and controversy among a broad academic and public constituency, encouraging schools to present the subject to their students, and generating public participation wherever possible. Above all, they need to keep up a stream of stimulating books and articles which will bring the subject positively into the public eye and identify it with a lively literature and broad-minded practitioners. Here the journal *History of Technology* has played a particularly important role – as witness the present symposium – in encouraging the development of an 'invisible college' of historians of technology, eager to share insights and research experience.

I remain of a sufficiently optimistic disposition to believe that by such measures, the situation could even now be turned round. It is crucially important that the decline of the history of technology in Britain should be reversed, and that the national scandal of its present plight should be remedied. It is not resorting to a crude monocausal species of explanation to argue that, among other things, an understanding of the history of technology is essential to a comprehension of the ways in which our society has come into existence, and to gaining a clear view of the problems confronting it at present and a capacity to project likely developments into the future. These topics constitute an outstandingly important educational agenda in which the nature of invention and the social prerequisites for successful innovation, the means of technological diffusion and change, and the way they have reacted with political and social objectives over a wide range of objectives, are all essential features. The ability to tackle such an agenda is virtually unique to the history of technology, and we neglect it at our peril.

Notes and References

1. See K. Theodore Hoppen, *The Mid-Victorian Generation, 1846–1886* (Oxford, 1998): the solitary reference to I.K. Brunel is on p. 410.

2. See Sir John Clapham, *An Economic History of Modern Britain: The Early Railway Age, 1820–1850* (Cambridge, 1926); A.E. Musson and Eric Robinson, *Science and Technology in the Industrial Revolution* (Manchester, 1969); and John R. Harris, *Industrial Espionage and Technology Transfer: Britain and France in the Eighteenth Century* (Aldershot, 1998).

3. The editors of *A History of Technology*, published in five volumes by Oxford between 1954 and 1958 were Charles Singer, E.J. Holmyard, A.R. Hall, and Trevor I. Williams: they all approached the subject from a history of science perspective, although with some industrial experience, and selected their authors from a wide range of expertise.

4. The most prominent statement of sociological analysis in relation to the history of technology has been that of the 'social constructionists' Wieber E. Bijker, Thomas P. Hughes and Trevor J. Pinch (eds.), *The Social Construction of Technological Systems: New Directions in the Sociology and History of Technology* (Cambridge, MA, 1987). For my detailed critique of this view, see R.A. Buchanan, 'Theory and Narrative in the History of Technology', *Technology and Culture*, April 1991, Vol. 32, No. 2, Pt 1, 365–376. Thomas Hughes is, of course, an outstanding historian of technology, but most of the contributors to the volume are not historians.

5. Although SHOT is an American institution, it has a worldwide membership amongst historians of technology: it has held three of its annual conferences in Europe, and its book reviews and annual bibliographical review perform an international service.

6. The Newcomen Society was in the field a generation before SHOT, and it went international before SHOT was established, but its American Branch became the independent Newcomen Society of North America and concentrated on business histories. Meanwhile, despite the continued excellence of its *Transactions* it has abandoned any attempt to maintain a bibliographical review of the subject, and its membership has remained firmly engineering, practical, and only incidentally academic.

7. The creation of the CATs was recommended in a government White Paper on *Technical Education* in 1956 (Cmd. 9703), and ten of them were created in the next four years: eight of these subsequently became universities – Aston, Bath, Bradford, Brunel, City, Loughborough, Salford and Surrey. They were potentially good ground for the history of technology, but apart from a research presence at Bath it has not figured prominently in the curriculum of any of them.

8. The NRIM was a pioneering effort to achieve a comprehensive record of industrial monuments, but it encountered formidable operating difficulties and was eventually passed over to the Royal Commission on the Historical Monuments of England (RCHME), now part of English Heritage. The Record was assimilated into the National Monuments Record (NMR) held at Swindon.

9. ICOHTEC was established by an international group of scholars of whom the leading member was Melvin Kranzberg of the USA. As Kranzberg was also instrumental in the foundation of SHOT, there has always been a close affinity between the two organizations. The current President is Carroll Pursell of Case Western Reserve University, and the Secretary General is Professor Dr Hans-Joachim Braun of the Universität der Bundeswehr Hamburg, Germany.

10. The Chairs at Sheffield Hallam and Bath were held respectively by Francis Evans and myself. Other isolated historians of technology in academic posts at British universities in the 1990s were Michael Duffy at Sunderland, Ian Inkster at Trent, Colin Hempstead at Teesside, and David Jeremy at Manchester Metropolitan. Robert Fox is the second occupant (after Margaret Gowing) of the Chair of Professor of the History of Science at Oxford, but his interests are strongly concentrated on technological change.

11. This is happening at the University of Leicester, where Marilyn Palmer currently leads a small industrial archaeology section within the School of Archaeology and has promoted courses and research projects, also providing an academic base for *Industrial Archaeology Review*.

12. My own work, *The Power of the Machine: The Impact of Technology from 1700 to the Present*, (London, 1992), attempted to provide such a text and to define the subject area: such is the speed of technological change, however, that such works require frequent updating.

Some Thoughts on the History of Technology and Its Current Condition in Britain

H. S. TORRENS

> No longer can the history of technology be caricatured as the amateur pursuit of retired engineers stuck in the hagiographic mode ... The history of technology emerges from these volumes (*ICON*, volumes 1–3, 1995–1997) as among the most exciting sectors of interdisciplinary history today.[1]

INTRODUCTION

I write this soon after visiting two departments once active in the history of technology in British universities: those of the late Donald Cardwell at Manchester[2] and Angus Buchanan at Bath. This was after a visit to Germany where, as Hans-Joachim Braun (Secretary-General of ICOH-TEC) reminded me, there are now over 20 chairs in this field. In Britain there is now only one, David Edgerton's at Imperial College, in a partnership with the Science Museum (which, as Tom Rolt pointed out, is nine-tenths filled with technology[3]). Other individuals are, of course, active in this field in universities here, working in other, mainly History of Science, departments.

But Science (knowing) is not the same as Technology (doing), however interdependent they have become industrially.[4] Their separate histories should also clearly interrelate. Too often they do not, as the American Paul Lucier demonstrated for the history of geology. This is one science which has the closest possible connections with technology, through mining. Lucier has shown how little its few historians had been, or were now, concerned with practical, technical aspects.[5]

Britain seems to have a long history of antipathy to the technologist, as Martin Wiener tried to point out in his politically powerful book.[6] Wiener was much criticized, partly for not having proved that this attitude started around 1850, as he claimed, but I think he made the case nevertheless that an upper-class British elite had long been antagonistic to people capable of 'doing' things. Anecdotes of John Rennie senior (1761–1821) illuminate this (and the problems faced by the historian of technology!). In 1832 William Buckland, Oxford professor, lecturing on metals, reported to his students how

> Mr Rennie once worked at the anvil. Travelling afterwards with a nobleman in Scotland, ye carriage spring broke. Rennie got out and mended it himself with his coat off, at ye next forge, and his Noble Friend, very courteous hitherto, never spoke to him again.[7]

When Samuel Smiles reported the same incident, whether in 1861 or in 1905 (from an 1860 source), the broken item was now either a horseshoe or the axletree. The Vulcan at the forge Rennie used was now either 'not at home' or drunk – in either case forcing Rennie to do the repair himself. In one version it was still the nobleman who was so unhappy to have to travel with Rennie while in the other it was his fellow travellers. Disgust at having to mix with someone with dirt under his fingernails was the one uniform feature.[8]

We must redress the current antipathy to technology and its history in this country, and try better to reveal the real importance of technology as the ultimate creative activity. I tried to urge this in a paper given at Liverpool in 1997[9] at a second conference on 'Perceptions of Great Engineers'.[10] I intended there to describe reactions to the first conference of 1994. However, since that first volume of conference papers had not even been reviewed by the Newcomen Society (our British Society for the History of Engineering and Technology, which sponsored the meeting),[11] I spoke instead about 'Current Problems Facing the History of Technology', a topic I return to here, with personal thoughts on its condition in Britain. So slight a reaction as that mentioned above is due, I feel, to the absence of any significant disciplinary base for the history of technology in British higher education.

THE CULTURE OF HISTORY OF TECHNOLOGY

> If the study of technology is important, it is somewhat surprising, when one looks at the state of the discipline at university level, to find how widely the provision of teaching posts varies between western countries. At a rough estimate there are about 80 professorial posts in the USA, over 20 in Germany, four in France and one in the UK.[12]

The contrast between Britain and the USA, where *Technology and Culture* (*T&C*), published since 1959, has succeeded in placing technologies within their various historical cultures, is striking. Daily we face technological revolutions – whether that of the Internet or in transport – and so can see how profoundly technology impacts on everyone (unlike much science). More and more, technology divides humanity, with First and Third World countries drifting apart across a gulf that separates those having access to modern technology from those who have none.

Technology now both so controls *and* divides mankind that its importance can no longer be ignored or denied. The significance of the history of technology simply follows from this.[13] It has been taken up with enthusiasm by American academic historians, as the annual bibliographies issued as supplements to *T&C* demonstrate. Alex Roland reviewed the situation of this subject in the USA in 1997 and pointed out how much it owed to Melvin Kranzberg (1917–1995). Roland also reported a 1995 poll finding by the US TV *History Channel* that 'the item of greatest interest to the public [there] is the history of science and technology'.[14] This explains, to me at any rate, why Americans are such enthusiastic consumers of new technology, whether electronic or genetic, whereas the British are so much more ignorant and Luddite (especially in areas such as genetic modification). Americans have magazines like *American Heritage of Invention and Technology* to reinforce the vision of America as the enthusiastic home of technology. Having had the opportunity to teach history of technology in the USA (but never in the UK) has added to my frustration that this most important of subjects is being so neglected here. I can only agree with two writers to *The Times* in 1990. One noted how

> our education system is incestuous, run by academics teaching academic subjects with no knowledge of either the history or the current operation of technological wealth creating industries.

The other noted how

> the omission of more specific references [at school] to the impact of technology on history was a grave and fundamental weakness. It underlines the lack of real understanding about the impact of technology in [our] adult world.[15]

History of technology in Britain is supported first and foremost by the Newcomen Society, set up in 1920 at a meeting to honour the centenary of the death of the great steam engineer James Watt (1736–1819),[16] but which then chose to name itself after Thomas Newcomen (1664–1729). This was because Watt's name 'might not be entirely free from notes of discord inherited from the past'.[17] At the Society's 1980 Diamond Jubilee the then president commended 'one feature which I feel to have been most valuable to the development of the Society ... the high proportion of members who have been trained as engineers'.[18]

The contrast, for whatever reason, between those writing on the history of technology in the UK and the USA could not be greater. In 1980 about 50 per cent of Newcomenite authors were professional engineers, while only 10 per cent of those writing in *T&C* were engineers. In July 1991 it was confided to me by a former president of the Newcomen Society how 'engineers needed to do the history of technology' and how he did not want the Newcomen Society 'highjacked by academics or museum curators'.

As an academic but, as it happens, an amateur historian of technology, I much regret such adversarial attitudes. We should surely all be working together if this subject is as important as I believe it to be, and not treating one another as unwelcome intruders. Maybe this is only a reincarnation of the old blight of British engineering, its long fragmentation – into railway, canal, electrical, telegraph, civil and mechanical (etc.) segments. Similar fragmentation has certainly developed in the history of engineering here, with a particularly British diversion into the cul-de-sac (as I see it) of industrial archaeology, which has never delivered the promise of academic credibility once hoped for it. This is a failing it shares with the history of technology here, unlike the history of science, which *has* achieved a notable footing in some British universities. Such problems may explain, I feel, why the first volume of *Perceptions of Great Engineers* was never reviewed by the Newcomen Society. Its authors may have been simply been perceived as intruders/outsiders, when only insiders/engineers could/should have undertaken such technical history.

SOME PROBLEMS FACING THE HISTORY OF TECHNOLOGY IN BRITAIN

> It is extraordinary, and indeed shameful, that the nation in which the Industrial Revolution began can show such indifference to the foundation on which our entire culture now rests. We depend on technology to clothe, feed and amuse ourselves, as well as to earn our bread. Surely the history we teach ought to reflect this?[19]

Idolatry and Ancestor Worship

Apart from Isambard Kingdom Brunel (1806–1859), who has engendered his own idolatry,[20] the steam engineer James Watt (1736–1819) has been one of the few recipients of such attention. This has been encouraged by the remarkable survival of records concerning both his business and his person, along with so many of those of his business and personal associates. These supported the Scotto-centric view, which Robert Edward Anderson (1840–?) gave in the old *Dictionary of National Biography*. This was strewn with errors, some of which I have tried to reveal.[21] Others followed in his tracks, like the equally Scottish Andrew Carnegie biography of 1905.[22]

Such ancestor worship has led to a related condition which John Griffiths has named Wattolatry.[23] But he again reveals the problems facing the historian of technology. His book – sponsored by British Gas to celebrate 'the invention of gas lighting in 1792' by Watt's associate William Murdoch (1754–1839) – pointed out that Murdoch's first practical gas-making apparatus was not developed until 1794. Despite this, the Newcomen Society happily celebrated a visit in 1997 to Murdoch's Cornish house 'where gas lighting was first installed in 1792'.[24] Griffiths also discussed problems of the surviving historical record regarding contacts between Watt and Murdoch. Griffiths suggested reasons for the imbalance between the '400 to 500 items of direct correspondence' between them that he had expected, and the surviving tally of 'just 52 letters to Murdoch and only some 20 from him'. He gave two possible reasons for this imbalance, one frustrating (that these archives had been lost), the other sinister (that earlier 'Wattolators' had weeded them). Some of the problems with either explanation have been pointed out.[25] The difficulties in resolving this case are typical of the history of something 'done' rather than of something 'known' (and thus better recorded).

Objects can also be worshipped, like the famous Iron Bridge in Shropshire, symbol equally of a birthplace for the Industrial Revolution and of the Ironbridge Gorge Museum Trust's activities. This bridge is often claimed as the world's first iron bridge, a claim which is far from true. There can be only one first, but there are all sorts of iron and it is best to carefully call this 'the first known large-scale use of cast iron in bridge-building'. Earlier iron bridges in Britain have been documented but their impact on those in charge of our heritage has been minimal.

One such earlier bridge was the 72-foot-long wrought iron construction spoken of as erected 'a few days ago' in early January 1770 at Kirklees in Yorkshire. This no longer survives, and was merely described by Nortcliffe as 'perhaps the first iron bridge?'[26] The question mark proved prescient when another 'first' iron bridge was soon revealed to have been erected at Boroughbridge, Yorkshire, no later than 31 October 1769 by John Smith junior (1725–?1783).[27] These discoveries showed only that any claim made for such a structure as being the 'first iron bridge' should be carefully considered. Instead, we find a former president of the Newcomen Society telling members in 1997 that the Iron Bridge at Sunderland of 1796 was 'the second built in Britain'.[28] This ignored the two earlier bridges noted above *and* the Coalbrookdale Company's bridge, still partly surviving, erected at Trentham, Staffordshire, in 1794. This had been described in a paper of 1982 which noted that the major problem facing historians of British cast-iron bridges in the critical decade 1780–1790 was 'the lack of documentary records of the leading foundries of that period'.[29]

There Is No Such Academic Subject in Britain

What I have labelled 'Idolatory and Ancestor Worship' is characteristic of older writing in the UK (as elsewhere) on what we would now call 'History of Technology'. But whereas other countries, notably the United States and Germany, where the discipline of history of technology has been well established for 40 years or more, have produced an abundance of scholarly modern work against which the shortcomings of the older tradition are readily observable, the same cannot be said of the UK. This, it seems to me, can be traced in large part to the failure to properly fund and properly institutionalize the subject. There is as well the long tradition in Germany of the importance of technology, which is reflected in its universities. British car buyers today know all too well the significance of 'Vorsprung durch Technik' to Audi in helping to sell its German technology here through advertising. The first attempt to create a Chair of Technology in Britain was the short-lived crown appointment made at Edinburgh University in 1855. This was discussed by J.A. Wilson, the biographer of George Wilson (1818–1859), the chair's first and last occupant. She noted how, when the chair was first announced, it proved 'no small puzzle to the public ... what Technology meant' but that soon 'people [were] learning how to spell it'![30] Wilson's inaugural lecture of 1855 entitled *What Is Technology?* is worth reading today for the light it sheds on the history of technology.[31] Wilson's inaugural lecture and his occupancy of the Edinburgh chair were chosen by Robert Anderson as the topic of his presidential address in 1992 to the British Society for the History of Science.[32] He showed how much Wilson achieved in his four years in the post, especially in the development of museums of technology, and despite his regrettably short occupancy of the chair.

More modern British attitudes to this subject are illumined by Sir Eric Ashby's essay of 1958, *Technology and the Academics*. Here he noted how

> the mere technologist [is] tolerated in [our] universities, because the State and industry are willing to finance them. Tolerated, but not assimilated; for the traditional don is not yet willing to admit that technologists may have anything intrinsic to contribute to academic life. It is not yet taken for granted that a faculty of technology enriches a university intellectually as well as materially.[33]

Since Ashby wrote, the material advancement of our universities has become all that concerns the many who now so expensively manage them. In addition, our former polytechnics have been transmogrified into more conventional, and equally anti-technological, universities. How much worse the situation here is now than it was in 1958 was best brought home to me at my own university (Keele) when wholesale disposals of library books started there in the 1990s.[34] The opportunity was also taken to create space for 'much-needed new technology'. These disposals meant that holdings in the history of technology suffered worst because there was no faculty of technology at Keele which might have been asked for an opinion, or any engineering departments either. The thought that such material might be valuable to historians of technology appears not to have occurred to the Information Services managers who oversaw the disposals. As a result, long runs of the *Minutes and Proceedings of the Institution of Civil Engineers*,[35] and the *Mechanics Magazine* (all previously expensively re-bound) were discarded. Another item which also vanished was Mary Tasker's 1980 Historical Association pamphlet number 47, *Teaching the History of Technology*. New journals, founded to support the gadarene rush into print demanded by British Research Assessment Exercises, such as the new *Journal of Industrial History*, will fail to fill the ecological niche left empty by the history of technology here. They merely help to fragment its study further.

History of Technology Is Best Carried out by Engineers

Engineers are too often regarded in the UK as the only people capable of mastering the technicalities involved in researching the history of technology. These *are* obviously of vital importance, but too little attempt is then made in such research to place technologies in their proper historical or cultural contexts. I have failed to find any such 'stimulating' history being written in Britain as is provided by Rachel Maines.[36] This was history wonderfully placed in its context but which caused its author great problems when dealing with engineers![37] The attitude of engineer-only historians also explains, again if only to me, why we have such a problem in Britain with the Great Man Syndrome. This results when many historians use, and reuse, only secondary literature without full, and first, recourse to primary sources, whether documents, locations or artefacts/manufacts. Another result is too much of the Best/First from Britain syndrome. We duly honoured the computing pioneer, Charles Babbage (1791–1871), at a centenary meeting in 1991, but I failed to hear it pointed out that the first such 'difference engine' was German (1784), that Babbage was not so technologically ahead of his time, that he was not so badly treated by the British Government (their £17,000 investment producing rather little result) or that Swedish engineers succeeded with this technology from 1843, unlike Babbage himself.[38]

Despite their feeling that history of technology is best handled by those with engineering training, the attitude of our engineering bodies here to that history remains ambivalent. Unlike the Royal Society, which makes notable efforts to support the history of science in Britain, there seems no such regard for history within the Royal Academy of Engineering, our body of equivalent standing for engineering. But our Engineering Council was happy to use history in 1984 when it produced a memorable series of advertisements one of which asked the highly pertinent question: 'Why isn't there an Engineers' Corner in Westminster Abbey?'[39]

Museums and the Heritage Industry

This problem concerns the management of heritage in Britain. Unlike the United States, with its magazine *American Heritage of Invention and Technology*, we instead have *English Heritage*, which concentrates on landscape and buildings, only occasionally concerning itself with past industry.[40] The recent progress of the heritage industry here has lowered expectations concerning the truth, complexity and reality of the past. This is a subject admirably examined by Robert Hewison.[41] It sometimes seems to me that what C.F.G. Masterman said could *not* come to pass in 1922 has indeed now happened. He wrote:

> Lancashire may be largely destroyed by the manufactures of the East; coal may become increasingly expensive to work or be faced with impossibly low prices . . . or be largely replaced by oil . . . But . . . I do not think that we shall ever come to fulfil Mr. Bernard Shaw's prophecy of an England of lodging-house keepers and trained guides pointing out to visiting crowds the sites of the places where Shakespeare lived or Gladstone died.[42]

A particularly incisive analysis of the historical experience offered at the World Heritage Ironbridge Gorge Museum complex has been offered by Bob West.[43] I personally blame heritage management for the current indifference to factual errors on four plaques put up at various times to honour the English practical geologist and engineer, William Smith

(1769–1839). One, at his birthplace, Churchill, Oxfordshire, memorialized him controversially as 'The Father of British Geology', which Scots might dispute on behalf of James Hutton (1726–1797). Other plaques at Rugbourne Farm, High Littleton, and 29 Pulteney Street, Bath, Somerset, have wrong dates. The fourth, re-erected in about 1932 on Tucking Mill cottage, South Stoke, Somerset (having been previously – since 1889 – on a nearby but subsequently demolished mill) still remains in the wrong *parish*, as was pointed out over 25 years ago.[44]

Heritage management can so easily induce a lack of attention to such inconveniences. This in turn can fail to alert people to the fact that the history of technology may be complex, important or of real value. The Ironbridge Institute at the Ironbridge Gorge Museum Trust used to offer a series of short courses on the history of technology, and employed highly qualified people to teach it. But Heritage Management and Industrial Archaeology are the only courses now highlighted in their latest brochure. These subjects are seen as more important for postgraduate training than any analysis of history. Yet Industrial History courses had proudly launched the Institute in 1981,[45] although such courses were singled out for cuts in funding within months.[46]

All this has had significant consequences for other museums involved in preserving our industrial past in Britain and its representation at places such as the Beamish North of England Open Air Museum or the Black Country Museum. We can only hope that recent academic initiatives at the new (1995) Institute of Railway Studies at the National Railway Museum in York will bear fruit, despite earlier fractious debates between the Director of the Science Museum and railway historians.[47] The other Science Museum initative (if still in being) at Imperial College, London, with its chair in the history of science and technology (now the only such chair in the UK) must also be applauded. Nor should we forget the efforts being made by some professional engineering institutions here, like the Historial Group of the Institution of Electrical Engineers or the Panel for Historical Engineering Works of the Institution of Civil Engineers. But whoever tries, something more, surely, urgently needs to be done about the current state of the history of technology in Britain.

ACKNOWLEDGEMENTS

I owe grateful thanks for help, support and information to a number of people, among them in particular, David Bick (Newent), Angus Buchanan (Bath), Victoria Haworth (Newcastle), Graham Hollister-Short (London), Adrian Jarvis (Liverpool), Joe Marsh (Manchester), Alan Smith (London), Tony Woolrich (Bridgwater) and Nick Wyatt (London). Some of these kindly read and commented on an earlier version of these very personal thoughts. Their comments are gratefully acknowledged.

Notes and References

1. P.M. Rattansi, 'Review', *Times Higher Educational Supplement* (hereafter *THES*), 4 December, 1998: 32.

2. J.V. Pickstone, 'Obituary', *British Journal for the History of Science* (hereafter *BJHS*), 1999, 32: 485–8.

3. L.T.C. Rolt, *The Mechanicals* (London, 1967), 126.

4. On this topic I have found S. Dasgupta, *Technology and Creativity* (New York, 1996) very stimulating.

5. P. Lucier, 'A Plea for Applied Geology', *History of Science*, 1999, 37: 283–318.

6. M.J. Wiener, *English Culture and the Decline of the Industrial Spirit 1850–1980* (Cambridge, 1981).

7. Buckland, Lecture XIV, March 1832, J.E. Jackson lecture notes, British Geological Survey, MSS 1/635.

8. For the 1860 version see S. Smiles, *The Autobiography* (London, 1905), 253–4; for the 1861 version see S. Smiles, *Lives of the Engineers* (London, 1861), 281–2.

9. A. Jarvis and K. Smith (eds.), *Perceptions of Great Engineers II* (Liverpool, 1998), 5.

10. D. Smith (ed.), *Perceptions of Great Engineers: Fact and Fantasy* (London, 1994).

11. The only reviews known to me appeared in 1994 in *The Engineer* and in 1995 in *Notes and Records of the Royal Society, Trevithick Society Newsletter* and *The Journal of the Railway and Canal Historical Society.*

12. G. Hollister-Short, 'Editorial', *ICON*, 1998, 4: 3.

13. D. Landes, *The Wealth and Poverty of Nations* (London, 1998) is particularly stimulating.

14. A. Roland, 'What Hath Kranzberg Wrought?', *T&C*, 1997, 38: 698.

15. *The Times* 25 April and 11 September 1990.

16. J. Tann (ed.), Introduction to the second edition of H.W. Dickinson and R. Jenkins, *James Watt and the Steam Engine* (Ashbourne, 1981).

17. A. Titley, 'Beginnings of the Society', *Transactions of the Newcomen Society* (hereafter *TNS*), 1946, 22: 38.

18. D.H. Tew, 'Introduction', *TNS*, 1981, 51: 193.

19. A.P. Woolrich, *History Today*, April 1999, 49, No. 4: 60.

20. A. Vaughan. *Isambard Kingdom Brunel: Engineering Knight-Errant* (London, 1991).

21. D. Smith, *op. cit.* (10), 28–9.

22. A. Carnegie, *James Watt* (Edinburgh, 1905), a volume in the Famous Scots Series.

23. J. Griffiths, *The Third Man: The Life and Times of William Murdoch 1754–1839* (London, 1992), 32 and 345.

24. Flyer for visit on 14–17 August 1997.

25. H.S. Torrens, 'First-Class Third Man?', *Notes and Records of the Royal Society*, 1994, 48: 161–3. It is also likely that many of the contacts between them never reached paper.

26. D. Nortcliffe, 'The First Iron Bridge?', *Industrial Past*, 1980, 7(1): 14–17.

27. P. Smith, 'John Smith and England's First Iron Bridge', *Journal of the Railway and Canal Historical Society*, 1996, 3, No. 165: 178–85.

28. R.W. Rennison, 'The Sunderland Bridge of 1796 and what followed', *Newcomen Bulletin*, 1997, 167: 33–4.

29. H.S. Torrens and B. Trinder, 'The Iron Bridge at Trentham', *Industrial Archaeology Review*, 1982, 6: 45–55.

30. J.A. Wilson, *Memoir of George Wilson, M.D., F.R.S.E.* (Edinburgh, 1860), 414.

31. (Edinburgh and London, 1855), 1–26.

32. R. Anderson, 'What is Technology?: Education through Museums in the Mid-nineteenth Century', *BJHS*, 1992, 25: 169–84.

33. Subtitled *An Essay on Universities and the Scientific Revolution* (London, 1958), 66.

34. *THES*, 24–31 December 1999: 5, and *The Bookdealer*, No. 1429, 23 September 1999: 3–4.

35. A.P. Woolrich's letter to the *The Times*, 18 June 1998, recorded this disposal (but not its perpetrators).

36. R. Maines, *The Technology of Orgasm*, Johns Hopkins Studies in the History of Technology, 1998, New Series 24.

37. *Idem.*, 'A Sew-Sew Sex Life?', *THES*, 21 August 1998: 16, and C. King, 'Love Machines', 1999 on http://www.wired.com/wired/archive/7.01/vibrators.

38. See G. Tweedale, 1991, 'Review of M. Lindgren's 1990 Book, *Glory and Failure*', *BJHS*, 1991, 24: 202–3, and Anon, 'The Early Difference Engines', *Uppsala Newsletter History of Science*, 1987, No. 7: 6.

39. *The Times*, 28 November 1984.

40. *English Heritage*, No. 21, March 1993.

41. R. Hewison, *The Heritage Industry: Britain in a Climate of Decline* (London, 1987).

42. C.F.G. Masterman, *England after War* (London, 1922), 44–5.

43. B. West, 'The Making of the English Working Past', in R. Lumley (ed.), *The Museum Time Machine* (London, 1988), 36–62.

44. J.M. Eyles, 'William Smith's Home Near Bath: The Real Tucking Mill', *Journal of the Society for the Bibliography of Natural History*, 1974, 7: 29–34.

45. *The Times*, 3 July 1981.

46. *The Times*, 13 January 1982.

47. See the ill-tempered debate in *Steam Railway*, December 1993 and January 1994, and the letter from 'representatives of . . . the railway heritage industry', *The Times*, 15 April 1994.

Pursuing Big Books: Technological Change in Global History

IAN INKSTER

> The fact of Western technological precedence is there.
>
> David Landes, 1998, p. 514

> Technological change seems likely to take over from unaided human wickedness as the focus of fears of barbarism.
>
> Fernandez-Armesto, 1995, p. 493

INTRODUCTION

A major problem in teaching the history of technology to undergraduates lies in its characterization as peripheral to the 'main-line' or 'conventional' historical enterprise as has evolved in British universities since the war. This is a theme taken up in the present special issue of *History of Technology* by writers who have experienced the disappointing transition from the suitably harnessed energies of the 1960s to the debilitating frictions of the 1990s. Whether the natural technical difficulty involved in expounding and understanding academic history of technology actively circumscribes its teaching is in fact debatable. Although a factor in the restriction of history of technology, as in the similar curtailment of much economic history and a good deal of intellectual history, an intrinsic difficulty of the essential subject matter is not likely to be the principal (and certainly not the only) cause of the problems of the domain. At least as persuasive are those arguments which centre on the character of the discourse and the problem of imagery. It seems certain that, regardless of specific advancements within the field, it is seen by many British historians (including those historians in Britain who do not work exclusively on British history) as specialized in a trajectory which is moving away from the mainline of teaching, exposition and research – whatever the end of history, its present position is approaching the institutional and the cultural as writers attempt to encapsulate the mentalities of the past in the absence of all old 'centrics', Euro or otherwise. A purpose of this paper is to illustrate that such a characterization of the history of technology is more a caricature than a portrait. Put at its most direct, we now suggest that major interpretations of global or very general history rest on both the findings of historians of technology and interpretive perspectives which place technological imperatives and innovations at or near the centre of discourse. That is, major contributions to recent global histories which address the great issues of war and peace, rise and decline, inclusion and exclusion, mobility and stasis, leadership and followership, and so on, do not merely refer *en passant* to technological 'factors', but at times depend on a technological argument of some sort, whatever the saliency, wherever the origin. Interestingly, in such dependency we seem to find more about productive machines than about extractive techniques or civil structures, the reverse of that tendency within the field itself which is noted by Rupert Hall.

Secondly, and very much a logical extension from such a position, we might be able to argue both that the history of technology should be considered a natural component of historical training and that specialized courses in history of technology in general-history degrees should be structured to embrace not only the formal problems within this area of study (e.g., issues of innovation, of creativity and origin, of diffusion and transfer, of cognition and acceptance or rejection), but rather should quite formally address the more conventional problematics of history. These most certainly include such weighty matters as the causes and powers of imperialism (Headrick), the global exclusion of whole social systems or civilizations from successful social and economic development (Hill), or the inner characteristics of industrial cultures (Pollard), each of which inquiries have been investigated in a most fruitful and thoughtful manner by scholars presenting at least something of a 'history of technology' which ranges from the 'soft' to the 'hard' of Rupert Hall's distinctions.[1] In this manner, insidiously if not institutionally, history of technology may indeed 'establish itself in British universities' through intellectual conduction, transmitting light rather than heat.

A word on the 5244 pages that have been chosen from a wonderfully increasing universe of texts. We begin our real discussion with David Landes, Paul Kennedy, and Nathan Rosenberg and L.E. Birdzell, because, of the selected books, these are the three that are most involved with the exposition of some explicit technological view of historical change.[2] Thus, in rejecting natural-resource explanations of historical economic achievement, Rosenberg and Birdzell turn towards a conception of the historical relationship between technologies and institutions – although the invention of technologies may seem to be of obvious importance, the invention of institutions was an essential component of how the West grew rich: 'As an economic domain emerged in western Europe, it had to devise its own institutions, sometimes alone and sometimes by interaction with the political realm.'[3] So, when historians of the material world consider innovation and invention, they must follow Schumpeter or Weber in embracing a definition of technological change which includes new products, new territories and new organizations of production. That is, they must return to older notions of 'discovery' or 'inventor'.[4]

While it is very nice to have such interpretations in place and selling well, they are already doing their job of demonstrating the general interpretive role of the history of technology, although this does bear some investigation. There is a sense, however, in which the most interesting disclosures of technology in history might be found in texts which explicitly favour *other* ways of interpreting the historical world, of expanding the historical enterprise. So, on the face of it, the two volumes by Michael Mann seek to examine global history through comparative identification of the ideological, economic, military and political sources of social power, and in doing so they adopt a strong sociological stance.[5] Within this frame there are fascinating discussions of technology, particularly those that surround a general recognition that 'technological stagnation' occurred over long periods in ancient empires of both the East and the West.[6] Gunder Frank's most challenging contribution is to impel the reader to reconsider the common Eurocentric position that China and other great civilizations were gradually overtaken by an accumulating Western superiority in culture, technology and political organization. More than any other recent writer, Frank shifts the 'victory' of Europe into the nineteenth century, and explains it in terms of the fresh wealth earned in the bullion, slavery and new trades inherent in the thorough exploitation of the Americas, and the opportunity thus offered to turn on a global switch and 'buy into the wealth of Asia itself', the richest trade in the world.[7] Braudel asks us to consider entire 'civilizations' in a manner reminiscent of Toynbee, every one of which 'depends on economic, technological, biological and demographic circumstances'.[8] Such 'structures' are ancient, distinctive and essential, and they embrace 'religious beliefs, for instance, or a timeless peasantry, or attitudes to death, work, pleasure and family life'.[9] Clearly enough, the Braudel approach entwines technology within a complexity of underlying processes or structures, and thus technological history appears in explanations of stagnancy and reversal, in episodes of both adoption and refusal of the influence and artefacts of other, competing civilizations. Fernandez-Armesto, a master of style, impression and intuition, provides many tantalizing hints in a work which so often eludes any definition.[10] Technology is certainly seen as a superior asset of the West, one which carried with it a gale of destruction into non-European places, particularly perhaps in Africa, where 'traditional traders were condemned with astonishing speed to extermination or extinction'. However, the role of technological progress in establishing initial superiorities on the world stage is less clear-cut, for although twists of 'initiative' seem to serve the same function as Mann's social power in generating a decisive influence on the world by 'communicating ideas, creating or adapting technology and undertaking exploration, colonisation or aggression', it is also true that 'the relative performance of rival cultures depends on the self-perceptions of the peoples concerned'.[11] In global history, technology sits uneasily.

Yet each of these works requires the history of technology. To claim so is not reductive and certainly not any sort of debunking – I have selected these books because, although sometimes very annoying, they are all exceptional, they are all ambitious, they promise to be influential in the further development of history as a subject, they are selling at a rate which generates envy amongst us all.

In my exposition I can capture brevity only by batching these big books into two groups, and then subjecting them to a fundamental test: to what extent are stories that are designed to explore the material success of the West over the great civilization of East Asia, in fact dependent upon a history of technology?

TECHNOLOGY AND POLITICAL ECONOMY

All three of the works which deal most explicitly with technology also encompass discussion of the divergence between West and East. Rosenberg and Birdzell argue that it was the political divisions within Europe which created competitive environments, but in particular the financing of military competition 'was an important factor in overcoming the inherited distaste of the rural military aristocracy for the new merchant class'.[12] In contrast to such recent writers as Pomeranz, who stress the costliness of nationalism and its tendency to divert resources from more commercially productive pursuits, Rosenberg and Birdzell utilize their argument to explain the subsequent material tardiness of China despite its undoubted technological advancement. They thus indirectly invoke the Hume and Bagehot type of thesis concerning the 'cake of custom': for David Hume the retardation effects of established mores were weakened by interactions among freely trading systems, for Bagehot, endogenous processes of education, intellectual improvement and internal cultural leadership together served to crumble the cake of custom.[13] Because China severely reduced the extent of its interaction while maintaining the centralizing powers of the mandarinate, this meant that there could only be 'glacially slow advance, both in technology and in economic growth'. In contrast, in Europe the weak and late-developed civil services could not counteract the pervasive commercial and technological outcomes of national competition, 'they came to power too late to prevent the rise of capitalism, and their only recourse for expressing mandarin values was a gradual, Fabian exertion of authority over the aspects of capitalism not too mercurial to elude their grasp'.[14] Thus, in this sort of approach, statism may be expensive, as in block B of Chart 1, but this is offset through time by the activities and processes down the line at blocks D and E. In conclusion, Rosenberg and Birdzell present a model in which the role of the state was to set the competitive scene rather than to create environments or institutions of technological change directly. There is no need to argue for military spin-off, thus the single direction of arrow (a) represents commercial inducement, not technology-spillage or transfer, and the key to Europe's material advantage lies in the emergence of an autonomous merchant culture unimpeded by bureaucratic indulgences or regulations. The approach to technological change at block E is Rosenberg-Mokyr,[15] with progress arising cumulatively from a series of revolutionary breakthroughs followed by 'normal science' incrementalism 'as a response to economic or social needs'.[16] As with the model more generally, it is the absence in Europe of dynamic impedimenta that induces historical progress in blocks E and F and hence the improvements of material life. Thus economic needs unproblematically induced 'patient technical work', although the explanation for 'major inventions' remains undetermined, and that for more routinized chains of incrementalism lies in a combination of useful (in Rosenberg and Birdzell seemingly 'necessary but not sufficient') market and non-market institutions, as in block F and arrow (c) of Chart 1.[17] The story of Western science is shadowy in Rosenberg and Birdzell. Again, there is a clear enough contrast with China or other empires, for courts and hierarchies 'could not have created Western science'. But, although the depiction of Western scientific communities as intellectually competitive, relatively open and unimpaired by governments, and strengthened in authority by their use of experiment and informal peer review, does follow the boundaries of the general approach, the authors render this story contingent rather then explanative by claiming – in convention with a host of general historians – that prior to an unnamed miracle event of *c.*1875, 'science' and technique bore little of a relationship. By thus confusing or eliding science and reliable knowledge, Rosenberg and Birdzell again favour market institutions over cultural or other institutional forces as dynamic interpretive elements in their historical model. Thus the unsettled position of block G and arrow (d) in Chart 1.

At its outset, the Paul Kennedy approach closely parallels that of Rosenberg and Birdzell but does not refer to it. Statist competition within Europe engendered 'a primitive form of arms race among the city-states and then the larger kingdoms'. This in turn 'encouraged artisans and inventors to improve their wares, so as to obtain new orders', and this was a unique impetus to technological improvement in all arms and ancillary equipment and structures. In contrast, monopolies of central power elsewhere on earth meant a technological introversion, whether this be led by Chinese and Japanese scholars or Islamic janissaries or Russian army commanders. Thereby, in technology, Europe 'took a decisive lead over the other civilisations and power centres' a generalization best verified in the development of the long-range armed sailing ship.[18] To a far greater degree than Rosenberg and Birdzell, Kennedy captures the inducement effects of European expansion and colonial competition. When rivalry became transoceanic so too came improved cartography, new instrumentation, and better techniques of shipbuilding. Thus there arose the European breakaway surge: metallurgy, deep-mining, astronomy, medicine and physics, the printing-press spread of the products of the rational mind. Cumulatively, the 'effect of this explosion of knowledge was to buttress Europe's technological – and therefore military – superiority still further . . . it was going to be extremely difficult for other societies to ascend the ladder of world power when the more advanced European states occupied all the top rungs'.[19] In the absence of competitive political

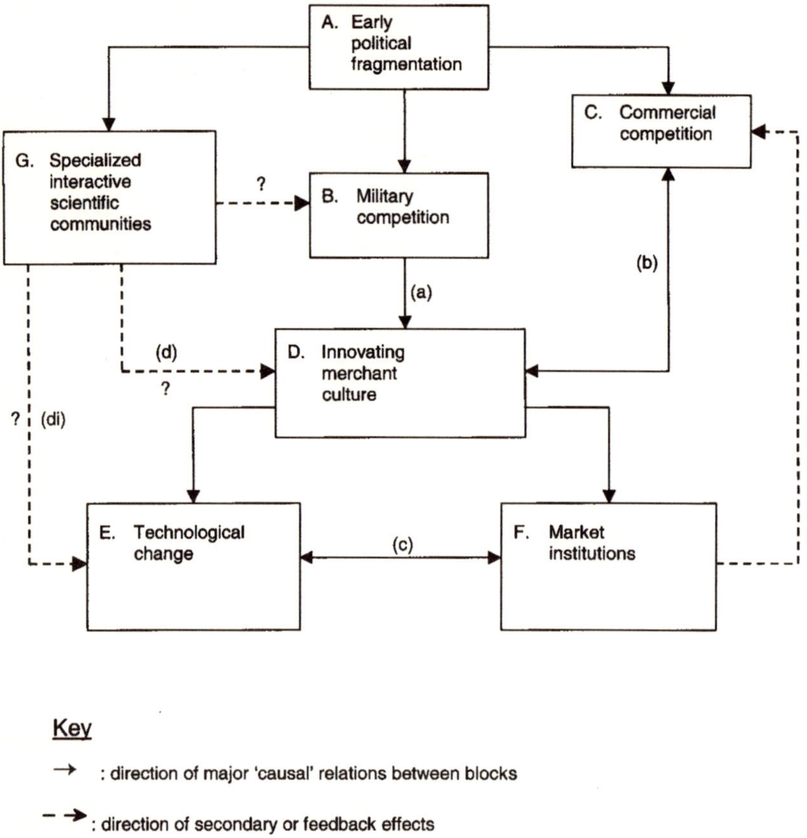

Chart 1 Growing rich – The Rosenberg-Birdzell schema.

plurality and markets, this difficulty could only be compounded. Again, the Kennedy approach reproduces the emphasis on a relative absence of barriers to economic growth in Europe when compared to even the best of the rest. Kennedy then elaborates the Smithian position that the military expenses of colonial or imperial expansion increase disproportionately for any one system as time goes by, and in a complex manner thereby determine the interchange of leaders and followers (the 'rise and fall' of his title) within the pale of the successful Western experience. So, in terms of Chart 1, Kennedy centres more on block B but also emphasizes the contrasting role of military competitiveness between the early stages of economic and technological growth and the later phases of capitalist maturity. Kennedy is also more adroit in considering the derivative processes captured at blocks E and F and arrow (c). Oceanic expansion and competition did not only induce incremental technological change (the flintlock rifle, the carronade), but by the eighteenth century promoted organizational innovations which transferred from leaders such as France or Britain to hard-pressed followers such as Imperial Russia. By this Kennedy means the 'monopolisation and bureaucratisation of military power by the state' through the creation of war ministries, barracks, hospitals, academies, ship-repair yards, 'with administrators to run them'. While Rosenberg might see such eighteenth-century bureaucratic developments as intruding on free innovation, and Pomeranz would emphasize their expense to the state (and thereby to the capital and risk environment of the nascent commercial or technological adventurer), Kennedy envisages such enhanced organizational activity of the state as essential to the winning of more modern wars, and thereby to the entrenchment of a pattern of both technological and institutional advancement and divergence from the old patterns of the Empires of the rest of the world. Only in the later nineteenth century did technological change seriously impinge upon the waging of war between the Western powers themselves, reversing the forces of Chart 1. It was the enormity of technological change which caused the great material divergence between Europe and Asia, and Kennedy broadly accepts the comparative magnitudes produced by Bairoch for 1750 and 1800 (see below and footnotes of Dodge Hall section), in which the per capita level of industrialization for all-Europe is not very far different from those of China, Japan, and even India, and where the differences in absolute sizes meant that China boasted a measurably larger share of world manufacturing output than all-Europe until around 1830, when the latter measurement moves very dramatically in favour of Europe and the American states.[20] The changes in world trade and output shares occurred not just because the West was growing so fast, but because Asian systems 'declined absolutely, that is, they de-industrialized, due to the penetration of their traditional markets by the far cheaper and better products of the Lancashire textile factories'. And alongside the disjunctions of new products came new diseases and new faiths. So the West and the rest diverged, with technology as both vehicle and passenger of a new world trajectory: 'The global dominance of the West, implicit since da Gama's day, now knew few limits.' As we shall see, much insidious interpretive weight is carried here by that small word 'implicit'. In particular, any consistency of approach demands that Kennedy is arguing that comparable levels of income, trade or industry per capita (of the Bairoch variety) do not mean very much when it comes to the academic interpretation of patterns of world history, of victory and defeat, of material rise and fall.

Of the books in this group, that by David Landes gives the most serious consideration to the older thesis that Europe's material superiority lay far back in the Judaeo-Christian past, when a unique combination of private property rights and belief in sovereign power circumscribed by God's authority and jurisdiction (the Church as God's surrogate on earth) 'made Europe very different from civilisations around'.[21] In contrast to the 'regulated and repressed' character of China, the despotisms of Europe had to contend with legal systems and divisions of power within sovereign areas, and the possibility that taxpaying merchant communities might migrate elsewhere. It was this series of limiting conditions which yielded the city commune and autonomy amongst merchants, those 'holes in the tissue of bondage' that in turn prompted migrations of knowledge and skills, emancipation, and novel privileges. Again, where in China lay authority might dictate the intellectual life of a homeostatic society, in Europe the separation of the religious and the secular allowed pockets of dissent and initiative, and such layers of fragmentation reduced the vulnerability to overall conquest that came with the 'brittleness' of encompassing imperial systems – bits of Europe might fall just as innovative fragments escaped new tyrannies and evolved as progressive elements. This was the basis for Europe's medieval economic revolution, associated with the wider exploitation of older innovations (cultivation by the wheeled plough and three-field rotation) and the emergence of the new – waterwheels with cranks and toothed gears, eyeglasses, the mechanical clock, movable-type printing, and gunpowder. Drawing on the work of Etienne Balazs and others,[22] Landes then argues that centralized statism and ideological repressions excluded such sustained developments in China or the Islamic world, this negating the long-term historical impacts of their earlier, undoubted technological virtuosity. It may be seen that the explanative apparatus of Landes is very similar indeed to that of Rosenberg and Birdzell as depicted in Chart 1. The possible difference lies in Landes' tendency to push the causative role of market forces further back in time because of his underlying belief that, at an early point, in Europe technological and institutional innovations 'worked and paid, and rulers and vested interests were limited in their ability to prevent or discourage innovation' as well as to suggest a more formative role for the applications of scientific rationality in Europe's material progression.

The account of technological change in Chapter 13 of Landes is one recognizable to historians of technology – advances in water power and steam power, coke smelting of iron, and the applications of powered machinery took place through chains of incremental innovation, interchangeability of basic processes (boring, printing, spinning) and growing precision through the use of improved instruments and materials. Indisputably, Landes identifies the 'critical, distinctively European sources of success' feeding into the Industrial Revolution as centred on intellectual and technological advances conditioned in an earlier build-up of intellectual autonomy, adversarial methods of knowledge-creation, verification through experiment and testing through debate, and the routinization of research and its diffusion. This combination was unique to Europe and sufficient to initiate the systematic, self-sustaining processes of industrial expansion. In contrast, 'cultural triumphalism combined with petty downward tyranny made China a reluctant improver and a bad learner', meaning that demonstrations of Western material prowess could not serve to effectively challenge existing Confucian mores or recipes, and for long the new artefacts of clocks or cannon remained court novelties in a social system where 'talent yielded to gentility'.[23]

TWISTS OF INITIATIVE AND SOURCES OF POWER

Much of Braudel's corpus follows from his generalization that since 'the very beginning everything has involved technical knowledge' and his belief that diffusion processes are not sufficient to explain pockets of particular creativity – although specific or discrete skills and products might travel quite freely, advancements such as the technology of seagoing navigation demanded an 'assimilation of several combined technical achievements' (e.g., the sternpost rudder, lap-jointed hull, compass, etc.) which required a control over a combination of skills that was unusual and that requires proper explanation.[24] In Braudel, that deeper explanation lies with the emergence of markets.[25] It is the market that induces innovative and entrepreneurial acts and harbours the early groups of innovators and entrepreneurs. Simple markets existed universally as places of periodic visit and exchange, characterized by the portable artisan's shop. Capitalism could develop in those areas where authorities were not able to dampen the rise of more complex markets, fairs, bourses and forms of credit, which came together within Europe to evolve in turn the 'private markets' of large merchants, free from the restrictive legislation of the traditional places of exchange, wherein merchant communities prospered in civic alliance with authorities who appreciated their ability to supply the city with essential goods and services. The failure to develop the bourse style of market as an intermediary step meant that China was condemned to a 'non development' of capitalism and to a related curtailing of the innovative behaviour associated with such zones of free operation, within which evolved large merchant monopolies and profits, global trading, and non-localized information systems. The superior technologies of capitalism arose out of such sites of 'superior knowledge, intelligence and culture'.[26] Where, as in China, an 'imperial administration blocked any attempts to create an economic hierarchy' only simpler exchange systems could exist in comfort, 'the shops and markets of city and small town'. Thus, although China may have demonstrated no overall material inferiority relative to the West, another type of 'inferiority lay in her economic structure, her market outlets and her merchant middle class, less well developed than that of Islam or the West', with a merchant class diverted from profits by vanity and a love of literature.[27] Thus in Chart 2, the strength of arrow (a) might have been in inverse proportion to the power of the state to revert to conservative control and fiscal policies.[28] Chart 2 also tries to illustrate the complexity of the position of science as reliable knowledge in the Braudel approach. In contrast to the writers surveyed above, Braudel sees a greater role for scientific knowledge in material progress prior to the late nineteenth century. The turning-point was Cartesian thinking: 'From Descartes we have the first systematic and modern critical appraisal of knowledge, a first heroic struggle against all intellectual or metaphysical illusions, all errors arising "from poetic intuition".' From then on all was 'economic impetus', the spur that China lacked.[29] In terms of a transition to industrial revolution, however, Braudel explicitly rejects technical change as a sufficient argument, appealing to Humean-Rostovian elements

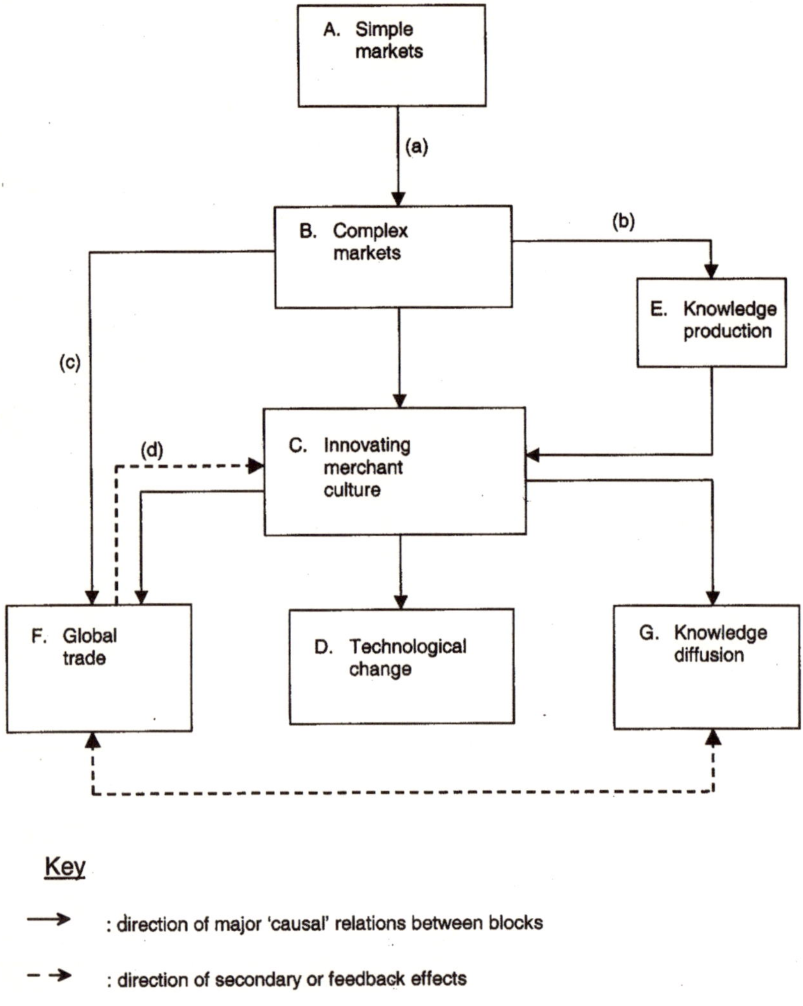

Chart 2 Merchants and knowledge production in Braudel.

of political stability, demographic change and agricultural improvement as the factors giving rise to the British siting of the new world.[30] Thus, as in Chart 2, markets and knowledge may stimulate fundamental technological change, but industrialization as a global phenomenon awaited changes in the conditions of demand, which is a reasonably conventional position.

Braudel's subsequent account of technological change in Europe is substantial and conventional.[31] So the Europe-China divergence is posed in terms of labour-surplus in the latter, wood shortage is placed as a reverse salient in Europe, militarism and nationalism in Europe are seen as inducements to technological change, liberalism and a relative lack of social barriers as Britain's special advantages. Boulton and Watt remain as special heroes. Chart 2 attempts to capture the essence of Braudel's interpretation of technological change in a highly generic and stylized manner.

Do Mann and Fernandez-Armesto alter this perspective radically? The latter rejects any historical demonstration of a 'correlation between industrialization and an identifiable set of social values' and follows the Polanyi position that commanding industrial ideologies did not reflect prior value orientations but were among the constructed results of industrialization, and also agrees with several recent global analysts that European industrial advancement was from a position of relative economic backwardness (e.g. Frank) rather than any ancient forwardness (e.g. Landes), this meaning that there was no high-level equilibrium trap 'to snare developing technologies'.[32] Thus, once more, demand is resorted to as the condition of technological progress. The resulting Western conquest of the globe reflected 'the industrial advantage ... empowered by industrial technology' in the form of steam boats, quinine pills and steel guns.

Mann begins more certainly, with technological changes being some function of the 'power organizations' within which they were situated.[33] Thus in a very interesting discussion of technological change in north-western Europe in the Middle Ages, Mann begins with the village and manor communities and their participation in networks of economic interaction involving institutions governing property rights, production relations and market exchanges, not dissimilar to Braudel. It was this kind of social organization that induced innovations in ploughing, in harnessing or in field rotation that together increased the yield from heavier soils, and thus the greater commercial exploitation of northern and western Europe around AD 1000. In global terms, this technological switch 'probably gave western Europe a decisive agricultural edge over Asia and particularly over Chinese intensive rice-cultivation techniques', again a distinction drawn by Braudel also.[34] If manor organization could increase the yield of the heavy soil grains, then the 'water mill could efficiently grind them'. From this developed long-distance trade in bulk subsistence commodities, more likely to be taxed than plundered, thus more likely to lead to state-mercantile alliances of the Braudel type than in places where the plundering of high-value and crafted goods was common. Wherever invention originally arose, its diffusion and incremental improvement through commerce and communications of this sort (long-distance trading, ocean sailing, movable-type printing) was precisely 'the European contribution to the world history of power' and the mechanism of the transition – 'capitalism and the national state formed a loose but coordinated and concentrated alliance, which was shortly to intensify and to conquer both heaven and earth'. The subsequent coverage of the industrialization of Europe follows lines sketched by Landes and is broadly conventional, with chains of incremental technological improvement stimulated by demand and unrelated to both 'big capital and complex science', but where 'innovation remains mysterious'.[35]

In terms of Chart 2, the Mann analysis would seemingly accept the role of the state in the movement from block A to block B, but would also emphasize the role of the state and other power organizations in all flows from block B, especially those through block C in the early period and through to block E in the period following British industrialization, when technological advancement became very much a matter of statist competitions, in a manner similar to Kennedy.

Frank's is the most explicitly polemic of the approaches, and the one that puts most weight on the nineteenth century as the period of transition from Asian dominance (following a long prelude of Asian superiority) to European dominance of the material world. Thus at the outset he brands Braudel's approach to the China-Europe contrast as 'historically inaccurate'.[36] Much of this overall position is exposed in the Dodge Hall section below, and here we note only some of the technological assumptions of the Frank schema.

Frank rejects the whole notion that European sociocultural superiority led to more materially progressive use of generic technologies there than in China or elsewhere during the years following 1400 or 1500, and explicitly attacks the Eurocentric assumptions of such standard sources as Singer *et al. History of Technology*, or the works of Lynn White jr, A.C. Crombie or, most recently, Floris Cohen, but remains seemingly agnostic concerning the great Needham corpus. Frank queries both the assumptions that European science led the world and that science had much to do with technology before the late nineteenth century. As at least one of these two assumptions is now seriously questioned by all global analysts and most historians of science, there is little novelty here. But Frank goes on to survey historical work on specific technologies – guns, ships, printing, textiles, metallurgy, transport – in an attempt to establish Asian technological parity or supremacy, not only prior to *c.*1500 (which many historians would accept without much demur) but up to 1700 or 1800 (altogether more problematic). Thus McNeill is harnessed to the technical proficiency of Ottoman military capabilities, Needham testifies to the quality of Chinese shipping and Sangwan to that of Indian shipbuilding, and we are reminded that movable metal type came from Korea, quality silk textiles from China, Persia and the subcontinent, while European supremacy in ironworking and in coal and other mining was only of importance after 1800.[37] The problem with the Frank approach is that it uses very selective itemizing and does tend to neglect any historical counter-detail after *c.*1700, moving to weaker statements of the sort that 'watermills were in use in China, India and Persia', cases not really in dispute and not apposite to questions of normal production and average efficiency, or to the potentials for further material advantage of particular sites and manners of technological development or adaptation. Secondly, Frank surprisingly falls into the very Eurocentric trap of focusing on the core technologies themselves, not on the manner of their improvement, commercial utilization, diffusion, explication within existing bodies of reliable knowledge and so on, even though these elements may indeed be among the key *explicanda* of actual material progress, the real links between artefacts and economic domination.

Frank is clear on the technological transition *c.*1800: 'technological advance takes place in Europe, but not in Asia, which had higher population growth but also more polarization in its distribution of income and scarcity of capital'. It was indeed global economic growth in the 1600s and 1700s which increased Asian population growth and 'polarized the distribution of income, and thereby constrained effective domestic demand for mass consumption goods', rather than the workings of long-term institutional or cultural factors within Asia/China itself. Technological lag occurred very late and as a function of global commercial rather than regional cultural forces. Thus the Chinese case conforms to Elvin's well-known high-level equilibrium trap, with the twist that the elements of such a trap were not ancient or even long term, but coalesced in the short term as a result of global commercial and demographic changes centred in the eighteenth century.[38] In contrast to Asia and other parts of the world, in Europe 'higher wages and higher demand, as well as the availability of capital including that flowing from abroad, now made investment in labour-saving technology both rational and possible'. The global transition is conjunctural and technological. This is a genuinely novel outcome of the Frank approach and a major historiographical contribution, although it embodies a fairly standard approach to demand, which once again is the factor separating Europe from China. Thus a more unequal distribution of income inhibits effective demand for basic manufactures that 'militates against technological innovation in labor-saving devices and investment in power-generating processes'.[39]

450 DODGE HALL – TECHNOLOGY IN THE LANDES-FRANK SHOWDOWN

On 2 December 1998 André Gunder Frank and David Landes met at Dodge Hall, Northeastern University, under the auspices of the World History Center.[40] Indeed, in front of an audience of some two hundred people, the two eminent academics chose not to dodge, but at times blundered usefully against each other in a manner which exposed many of the underlying propositions of the new global history. Frank construed Landes as believing 'that it is our precious Judaeo-Christian culture and values that fundamentally distinguish the west from the rest', identified this position at bottom as that of the liberal, dualistic Enlightenment and its nineteenth-century descendants, and claimed that it rested upon the false assumption that the rest lagged behind the West on all major, measurable and unmeasurable indicators, long prior to the industrial, military and commercial surge of Europe. In particular, Frank claimed that the Landes approach states that 'China is a culturally and intellectually homoeostatic society that could live with little change, that they had indifference to technology, technological and scientific torpor, lacked institutions for finding and learning, abhorred mercantile success and were not motivated by greed and passion. They showed deliberate introversion, isolation, risk aversion, irrationality, xenophobia, arrogance', and indeed an entrenched complexity of cultural traits opposed to those through which the West secured its own material victories. In this Landes approach, as simplified by Frank, technological retardation in China was ancient and an epiphenomenon of cultural singularity, or at least of social and institutional traits which were not in accordance with the necessary cultural inputs of Western growth and triumph. In reply, Landes agreed that his approach to the modern world assumed that 'this story, over the last thousand years has been set and driven by what was happening in Europe and some of its overseas extensions', that Europe had reached superiority in science 'and the application of science' by *c.*1500 from an earlier time of relative backwardness, so that 'the Europeans could plant themselves anywhere in the world within reach of naval cannon'. While Europe learnt from Islam, India and China, it also 'surpassed its teachers' so that in the Pacific theatre it possessed 'an important technological advantage and, in fact, that remains a key consideration in the penetration by Europe in the whole Asian and South Asian and Southeast Asian market'. In the Landes model, as encapsulated by Landes, the Judaeo-Christian tradition produced a particular attitude to the applications of 'science' to 'technology' which in turn yielded that 'European superiority of power' which in time secured the globe.

Clearly, Landes did not accept Adam Smith's generalization that around 1776 China was equal or superior to Europe on material measures.[41] In addition, Landes argued that the trajectory is what matters when systems clash – by the eighteenth century, Europe was on the way up, China as firmly on the way down, perhaps because 'the Chinese reacted counter-productively to their encounter in [sic] Europe'. For Landes, this failure of interaction was clearly describable in primarily technological terms – the rejection of clockmaking, eyeglasses, printing, and so on, together made the difference. This position remained unacceptable to Frank, simply because on his measures China prevailed as competitive with and possibly ahead of Europe into the nineteenth century, indeed 'until 1800, Asia was vastly ahead of Europe'. What is needed, therefore, is an argument explaining a real and large discontinuity or reversal, and this Frank finds in global commerce, while claiming that Landes has no such mechanism for sudden change in a world determined by long-drawn-out cultural traits and configurations. For Frank, culture remains 'derivative and institutions are derivative'. This appears to mean that culture and institutions are never fundamentally explanative or causal. The switch in world history associated with the exploitation of the Americas occurred, according to Frank, because a new inexhaustible market for European commodities 'gave occasion to a new division of labor and the improvements of art, which in the narrow circle of ancient commerce, could never have taken place for want of a market to take off the greater part of their produce'. Commerce not culture was the limiting factor of the great divide of the world.

At least as exhibited in the Landes-Frank clash, it seems that the incomplete and at times incoherent global history argument does suffer from a reluctance to examine technological trajectories and their inner lives and outer exigencies. In particular, the global history stories at this very general level do fail to distinguish between the processes of technological innovation, transfer, diffusion, adaptation and learning, and neglect the historical functions of information systems and other institutions and the relations of these to such generalized analytical components as 'culture'. Neglect here might be of great significance – thus Frank ultimately argues against the notion of European technological superiority on the grounds that all technologies may migrate, that no site has a monopoly of new ideas and techniques.[42] This fails to accord with detailed work in the history of technology and technology transfer, where it is found that technology transfer between varying socioeconomic systems seldom works, usually entails massive expenses, opportunity costs and organizational energies, and often creates enclaves of modernity amid acres of depression and victimization. Even Landes' chapters on technology and industry do not penetrate the interstices. For instance, none of the histories examined here makes much of analytical or historical distinctions between 'information' and 'knowledge', or between formal and tacit knowledge. But the institutions of technology are not naturally malleable, and historians assume plasticity at their cost.[43] Thus, it remains possible that in the eighteenth century, say, 'average welfare' in China was ahead of that in Europe.[44] But, given that a competitive interaction, military tussle and cognitive dissonance were now about to happen on a near-global scale, did the advanced commercial areas of China possess technological resources and information-diffusion mechanisms which were equivalent *in battle and at the point of contact* to those found in the very-most advanced parts of Europe? If so, then perhaps this is worth the documentation. If not, then much may be explored within the ensuing hugely complex process of *interaction itself*, and it is possible that Frankian commercial mechanisms will find some place alongside technological and institutional trajectories. Comparing what are in fact unmeasurable 'averages' between two great systems is inconsequential if those averages never compete, are never really representative of the active forces at the cutting-edge of victory and defeat. The resources of victory may lie mostly in advanced pockets within large systems, and such advanced pockets may be describable in technological terms.

TECHNOLOGY IN THE WORLD

> I set no value on strange and ingenious gadgets.
>
> Emperor Ch'ien-lung, 1793

> Steam and Barbarians, each in their own age, were the senseless agencies driving their respective civilizations away from inherited modes of order.
>
> Alfred North Whitehead, 1933

An interesting outcome of this comparison of texts is that global approaches which are of extreme novelty and interest, such as those of Braudel or Frank, resort to conventional accounts of technological change, in which inventions and innovations respond in timely manner to changes in the quantity or quality of demand, where artefacts may travel at low cost and with little engendering of cognitive dissonance, and where the importance of the cost-reducing (i.e. supply-side) effects of better or more apposite institutions (a form of Schumpeterian innovation that might embrace Mann's power organizations), of information and knowledge systems, of sites of dispute and experiment, are basically neglected. More allusive approaches, such as that of Fernandez-Armesto, seem to avoid such overt omissions, but perhaps this is only because they are indeed allusive. It is also of note that the really global and more interesting approaches to technology, and their use in wider historical explanations, appear mostly in pre-eighteenth-century stories, with the 'industrial revolution' being treated of in a very familiar, conventional and uncritical manner e.g., again, the absence of examination of actual sites of technological advancement, organs of improvement and diffusion. Finally, much of the global history descends to complete neglect of any technological story in the post-industrial revolution/British period of climacteric spread of modernized industrialization through Europe, to Japan and latterly on a wider front to the newly industrializing group of nations and, of course, to China itself, whose more recent success then appears as a rhetorical flourish. The assumptions concerning the technology of industrial revolution are carried into later years with the addition of some comments about 'science', although there is useful acknowledgement of Gerschenkron in Mann.[45]

There appear to be several reasons why general history resorts to the history of technology. Clearly, technology and institutions overlap empirically and conceptually – when is a factory a complex technology and when is it an institution? – and this often brings the former into sharp focus within the narrative. The technology-culture relationship is always complex and will remain conceptually perennial. In particular, recent approaches to material progress that have stressed the importance of blockages/barriers or their absence have perhaps inadvertently opened the way to much further work on institutions, mores, values and cultural modes as inhibitors or facilitators of historical technological progress, research that will indisputably require the contribution of expert historians of technology yet might be of prime interest to the most general of analysts. Of course, in the end technology is produced (invented, modified and moved) by both the imagination and the material needs of humankind, even if its location or progress at a point or site might result from ancillary cost-reducing and dynamic elements such as the immigration of skills or the reduction of the social space between investigators and producers. This allows general historians to insert the 'technological factor' at key moments as they offer interpretations of very particular material advances. The common divorcing of 'science' from 'technology' in a causal sequence may also serve to increase the utilization of technology as explicans (e.g. Frank pp. 185–92) for it has the strategic advantage of directing the 'history of technology field' away from the other hard area of 'science' history, and towards the domains of its embeddedness – in the cultural, the social, the political or the economic. However, historians of technology might surely beware jumping on any strategic bandwagon here – although technology contributed greatly to science, although technology followed from craft, and although science required its technicians, none of these positions in itself serves to negate or reverse the possible importance of science as organized, reliable knowledge.[46] Although I remain sceptical concerning the overall Landes position, his remark – that in some manner science and technology (reliable knowledge and material applications) both reflect and feedback upon a more underlying environment of inquiry, discourse and dissent – makes certain sense, in that the societal 'response to new knowledge of either kind is of a piece, and the society that closes its eyes to novelty from the one source has already been closing it to novelty from the other'.[47] In particular, several of the most important papers in this journal have illustrated in detail the precise contributions of useful and reliable knowledge to specific technological formulations and assemblages, nowhere better expressed than in the recent paper on Denis Papin by Alan Smith.[48] So have science and technology, in Gooday's terms elsewhere in the present volume, been 'overlapping domains' over a long period, and only integrated domains in the years since *c.*1870? Historically, in the East or in the West, science as information and procedure, as a resource of technique, has been important, but this is not equivalent to claiming a great intimacy of new 'breakthrough' techniques with 'eureka' scientific discoveries – some historians were looking in the wrong places, that is all. In particular, it might be productive to combine technological and intellectual history in examining the historical relations of knowledge, information and technique within the sites where change takes place.

Although it is not presently thought appropriate to argue such a case, it might be that technological history does best explain major discontinuities, and thus will always be central to the big debates in global history.[49] The clearest approach to this might be through the clever essay by Gerschenkron of 1968, in which operational definitions are given for constancy of direction, periodicity of events, endogenous change, length of causal regress, and stability of the rate of change.[50] With all the work on the social embeddedness of technology in its proper place, there yet exists no coherent denial of the sometimes discontinuous, disruptive and discrete character of technological breakthroughs and clusters. Such breakthroughs might indeed be historically explicable in terms of legal regimes and social institutions which themselves exhibit 'constancy of direction', but it is just as clear that the incremental and experimental absorption of new technologies into old or new industries and productive procedures may create a path-dependency which measures a sustained divergence between two or more large social or commercial systems, as one of these suffers the admittance of exogenous forces and an acceleration in the rate of change. Whether the broadly 'social factors' that condition technological advances are identical to those that hasten the acceptance, adoption, and adaptation of such advances remains problematic. It is certainly a theme neglected by the big books.

It might be noticed that at an earlier point I somewhat fudged the distinction between global history and 'very general history'. Although the best formal examples of the former do tend to rely on a partial telling of old technological stories, there are several examples of less-than-global histories which have addressed technological matters as fundamental in the construction of enormously useful accounts of large historical themes. They have not been considered as within the scope of the present brief account. This is unfortunate in that it may well be work at this level which does serve to breach existing domains or terrains of discourse. Thus in 1986, Otto Mayr published his wonderful *Authority, Liberty, and Automatic Machinery in Early Modern Europe*, a book which seriously entwined political, cultural and technological history in an illuminating history of the Continental clock and the English engine of self-regulation. As much as any more conventional account, this one threatened to unravel the causal elements of English exceptionality and ultimate commercial success.[51]

It is hoped that even such brief exposition does illustrate Buchanan's point that the history of technology is at once a 'distinct entity' and of 'inestimable value to the processes of education' just because it may be so readily and usefully incorporated into 'any sound assessment of human history'. Again, examination of recent global history does suggest, as with Hall, that 'hard' history of technology is even now 'essential for a full historical account of the development of western society', and all other societies besides, even if that real technique appears in leading historical texts sometimes as mimetic reportage rather than as thorough investigation. The room is there, both for the thorough investigation to take place elsewhere, and for it to expand to incorporate alternative global approaches of its own. In this the exponents of history of technology are certainly helped by the fact that they may unambiguously make the claim, agreeing alongside Buchanan, that 'technology is of its essence a social phenomenon', which means that they can get on and *do* a variety of social, cultural and institutional history of a most exciting sort, without becoming bogged down in Science Wars or their equivalents, without the pulling of rank, the drawing together of ranks, the production of papers of authority to proceed. It seems clear enough that general, and indeed universal, approaches to the historical enterprise depend upon the findings of the academic history of technology, often in an all but slavish fashion. One slight advantage of the mystique of the machine, which can be exploited by its historians, might indeed be that the work of specialists is so faithfully – if you will, uncritically – reproduced in the general historians' accounting of their universe.

If this all makes any sense, then it follows that I am not here advocating the teaching of 'history of technology' only as a necessary part of the fuller, liberal training of engineers, nor am I arguing for strong demarcations among the historians based on agreed notions of domain and discourse. Disaster lies in those very different ways. Rather, we affirm that history of technology is central to the historical enterprise at its most ambitious and as it is being written today in Britain and elsewhere, and that an acknowledgement of this is a recognition of the power of the specialist to rekindle old debates or inflame the new. General histories will not long survive their own theses if they may only draw upon a redundant, dormant or unfinished programme in the history of technology.

Notes and References

1. D.R. Headrick, *The Tools of Empire: Technology and European Imperialism in the Nineteenth Century* (New York, 1981); S. Hill, *The Tragedy of Technology* (London, 1988); S. Pollard, *Peaceful Conquest: The Industrialisation of Europe* (Oxford, 1981).

2. David Landes, *The Wealth and Poverty of Nations* (London, 1998); Paul Kennedy, *The Rise and Fall of the Great Powers* (New York, 1987); Nathan Rosenberg and L.E. Birdzell, Jr., *How the West Grew Rich. The Economic Transformation of the Industrial World* (London, 1986), henceforth respectively Landes 1998, Kennedy 1987, Rosenberg and Birdzell 1986.

3. Rosenberg and Birdzell 1986, 139.

4. Ian Inkster, 'Inertia and Technological Change: An Elementary Typology', in Pascal Bye and Daniel Hayton (eds.), *Industrial History and Technological Development in Europe* (Luxembourg, 1999), 343–8.

5. Michael Mann, *The Sources of Social Power. Volume I. A History of Power from the Beginning to A.D. 1760* (Cambridge, 1986), and *Volume II: The Rise of Classes and Nation States 1760–1914* (Cambridge, 1993), henceforth Mann 1986, Mann 1993.

6. Mann 1986, 280–95.

7. Gunder Frank, *ReOrient: Global Economy in the Asian Age* (Berkeley, 1998), 281, henceforth Frank 1998. On the global transfer of American bullion to exploitation of the thriving Asian trades see also Om Prakash, *Precious Metals and Commerce* (Aldershot, 1994). In this paper we take Frank at face value in his approach to world history through political economy. But it should be noted that since then and elsewhere he has suggested that world development may be treated in analogy as a three-legged stool, each leg of which is 'equally' composed of three elements, yielding nine in all, from ecology (*à la* the *Annales* school) to ideology. Economics is only one of these nine and seemingly separated from politics, which besmothers his overall perspective somewhat. This is interesting but sits uneasily with Frank's frequent dismissal of 'cultural' or 'institutional' elements throughout *ReOrient* itself. See Frank, 'Around the World in Eighty Years', especially page 15, http:/csf.Colorado.edu/wsystems/archive/papers/gunder/80days.html

8. Fernand Braudel, *A History of Civilizations*, trans. Richard Mayne (London, 1994), first edition (Paris, 1963), 18; and Braudel, *Civilization and Capitalism 15th–18th Century. Volume I, The Structures of Everyday Life*, transl. from the French by Sian Reynolds (London, 1981), first edition (Paris, 1975), henceforth Braudel 1963 and Braudel 1975.

9. Braudel 1963, 28.

10. Felipe Fernandez-Armesto, *Millennium: A History of Our Last Thousand Years* (London, 1995), henceforth Fernandez-Armesto 1995.

11. Fernandez-Armesto 1995, 5, 7, 422.

12. Rosenberg and Birdzell 1986, 137.

13. Kenneth Pomeranz, *The Great Divergence: Europe, China and the Making of the Modern World* (Princeton, 2000).

14. See David Hume, *Essays, Moral, Political and Literary* (Edinburgh, 1741–2); Walter Bagehot, *Physics and Politics* (London, 1873).

15. Rosenberg's original work on incrementalism was, of course, of fundamental importance to historians. See Nathan Rosenberg, *Perspectives on Technology* (Cambridge, 1976) and his *Inside the Black Box* (Cambridge, 1982); Joel Mokyr, *The Lever of Riches* (Oxford, 1990).

16. Rosenberg and Birdzell 1986, 263.

17. *Ibid.*, 264.
18. Kennedy 1987, 22–6.
19. *Ibid.*, 29–30.
20. *Ibid.*, 147–50.
21. Landes 1998, 35.
22. See in particular E. Balazs, *Chinese Civilization and Bureaucracy: Variations on a Theme* (New Haven, 1964), and *La Bureaucratie céleste: recherches sur l'économie et la société de la Chine traditionelle* (Paris, 1968). It is worth noting that Braudel also draws on Balazs in depicting the role of Confucianism in traditional China, e.g. see Braudel 1963, 179–89.
23. Landes 1998, 336 and 342.
24. Fernand Braudel, *Afterthoughts on Material Civilization and Capitalism*, trans. P. Ranum (Baltimore, 1977).
25. The argument is portrayed in great detail in Braudel, *Civilization Vol. II: The Wheels of Commerce* (London, 1982), where there are interesting qualifications, e.g. p. 134 on China and the limitations of any 'market economy' approach.
26. *Ibid.*, 33, 57.
27. Braudel 1963, 194–5. Here Braudel relates the lack in China of commercial investment, credit systems, fairs and bourses and good communications. Elsewhere Braudel isolates the surplus of cheap labour as an 'ultimate' block on mechanization if not on technological change more generally, e.g. Braudel 1975, p. 339.
28. *Wheels of Commerce* (*op. cit.*, 25), 136–7.
29. Braudel 1963, 369 and 372.
30. See above for Hume. For Rostow see Braudel, *Civilizations*, 383–88, W.W. Rostow, *The Stages of Economic Growth* (Cambridge, 1962). See also Ian Inkster, *Technology and Industrialisation. Historical Case Studies and International Perspectives*, Ashgate (Aldershot, 1998), Ch. 3.
31. See especially Braudel 1975, 338–430.
32. Fernandez-Armesto 1995, 368, 370 and 358–63. My reference here is to Karl Polanyi, *The Great Transformation* (Boston, 1957) (1st ed. New York, 1944, and London, 1945).
33. Mann 1986, 404–9.
34. See especially Braudel, *Afterthoughts* (Baltimore, 1977), which divides the world into wheat (=Europe and Livestock), rice (= Asia and intensive human-powered cultivation) and corn (= the Amerindians and easy growth, free time and monuments). See also P. Chaunu, *L'expansion européenne du XIII au XV Siècle* (Paris, 1969).
35. Mann 1986, Vol. 1, p. 446; Mann 1993, Vol. 2, p. 95.
36. Frank 1998, p. 20.
37. The references are well known to readers but see Saptal Sangwan, 'The Sinking Ships: Colonial Policy and the Decline of Indian Shipping 1735–1835', in Roy Macleod and Deepak Kumar (eds.), *Technology and the Raj*, (New Delhi, 1995), 137–52. Such exemplars may be presented all too casually. Thus in the case of Korean printing we know that the King of Korea (the second of the Yi dynasty) began an extensive scheme for the casting of movable metal type and ordered by decree in 1403 the casting of characters in copper, a programme that continued into the 1500s so that when supplies of copper were exhausted they put into the melting pot bells of ruined monasteries, vases and instruments belonging to private individuals. There is even a claim concerning Korean metal type as early as 1234. But the case strikes the importance of social and institutional context. Thus the 1403 decree read that '[I]n order to govern it is necessary to spread the knowledge of laws and books in such a way as to fill the minds and render upright the hearts of men. Thus only can order and peace be achieved.' It is clear that the motif, agency and siting of such an advance in Korea was in quite striking contrast to that in Europe, and such a contrast may be of truly analytical importance in sketching our interpretations of comparative material success. For the background to this case see Maurice Courant, 'Introduction to the Bibliographie Coréenne', *Transactions of the Korean Branch of the Royal Asiatic Society*, 1936, XXV, 1–101.
38. Mark Elvin, *The Pattern of the Chinese Past* (Stanford, 1973). For early notions of equilibrium most analysts depart from Schumpeter's original development theory, but for the more Malthusian approach with applications to Japan see R.R. Nelson, 'A Theory of the Low Level Equilibrium Trap', *American Economic Review*, 1956, where, as in all such cases, technology is held constant or as only embodied in new capital; thus the trap is made up of population growth and capital growth curves.

39. Frank 1998, 301, 304.

40. The following account and page references are from 'Frank-Landes Debate. ReOrient vs The Wealth and Poverty of Nations', 2 December 1998; http://www.whc.neu.edu

41. This claim about Smith on China is much bandied about. In fact, Smith emphasized the debilitating effect on the economy of China of the relative absence of foreign trade combined with the high risks faced by 'owners of small capitals' who were forever fated to be 'pillaged and plundered at any time by the inferior mandarins'; thus his claim of a 12 per cent prevailing rate of interest in China. He also said that 'China is a much richer country than any part of Europe' but that 'the real recompense of labour is higher in Europe than in China, the greater part of Europe being in an improving state, while China seems to be standing still'. Smith, indeed, posed a combined commercial, technological and institutional interpretation of the special character of China: 'By a more extensive navigation, the Chinese would naturally learn the art of using and constructing themselves all the different machines made use of in other countries, as well as the other improvements of art and industry which are generated in all the different parts of the world.' See *Wealth of Nations*, Everyman edition in 2 vols (London, 1910), Vol. I: 85, 173, 187–9; Vol. II, 173–4, 217.

42. Frank 1998, 204–5.

43. Ian Inkster, 'Technology Transfer in the Great Climacteric. Machinofacture and International Patenting in World Development circa 1850–1914', *History of Technology*, 1999, 21: 87–106.

44. The figures on per capita incomes suggest that China was outcompeting Europe prior to 1800. All such quantities are heroic, and possibly approach foolishness when estimating comparative productivities. The work often relied on is Paul Bairoch, *Economics and World History. Myths and Paradoxes* (Hemel Hempstead, 1993) together with the several important papers that led up to this volume, especially those in the *Journal of European Economic History*, 1976, Vol. 5, pp. 273–341, and 1982, Vol. 11, pp. 269–335. Frank 1998 also makes something of the claim that higher incomes per head were maintained in East Asia despite the greater population growth of the *c*1500–1800 period. We have seen that Kennedy 1987 uses these estimates; so too does Mann 1993, p. 14. See also Frank 1998, 171–4, for critical modifications that give more weight to Asia. For some clever and considered comparisons see Om Prakash, *Asia and the Pre-Modern World Economy*, International Institute for Asian Affairs (Leiden, 1995).

45. Mann 1993, p. 492. See Alexander Gerschenkron, *Economic Backwardness in Historical Perspective. A Book of Essays* (Cambridge, MA, 1962); Ian Inkster, 'Politicising the Gerschenkron Schema: Technology Transfer, Late Development and the State in Historical Perspective', *Journal of European Economic History* (forthcoming).

46. An attempt to address the big comparative and global issues through the lens of reliable knowledge was made at the conference 'Regimes for the Generation of Useful and Reliable Knowledge in Europe and Asia 1368–1815', Cumberland Lodge, Windsor, UK, 14–16 April 2000.

47. Landes 1998, p. 343. Whether this formulation helps in understanding the Europe–China divergence is another matter entirely.

48. Alan Smith, 'A New Way of Raising Water by Fire: Denis Papin's Treatise of 1707 and its Reception by Contemporaries', *History of Technology*, 1998, 20: 139–181. In the same volume see also the excellent paper by Marcus Popplow.

49. As Frank asked of Landes in the debate of 1998, how is it that 'your allegedly cultural continuities led to contingent disjunctures?', a query he repeated when he summarized that 'the challenge is to explain a contingent, sudden change on the basis of a permanent condition'. See the precise intelligence of the Introduction to Jack Goody, *The East in the West* (Cambridge, 1996).

50. Alexander Gerschenkron, *Continuity in History and Other Essays* (Cambridge, MA, 1968), esp. 21–35.

51. Otto Mayr, *Authority, Liberty and Automatic Machinery in Early Modern Europe* (Baltimore, 1986). This work seems to have been neglected by recent global historians.

History of Technology in Germany: A Success Story?*

HANS-JOACHIM BRAUN

I

On examining the state of the history of technology as an academic discipline, the situation in Germany looks to me to be comparatively favourable. Currently the subject is being taught at 22 Universities and Technical Universities and at three Polytechnical Institutes. Depending on the combination of the history of technology with other subjects – economic and/or social history, history of science, etc. – the number of professorships exceeds the above figures. And compared to other countries, for example the United Kingdom, the field seems to be thriving. This is particularly true if one takes countries in Southern and Eastern Europe into consideration. In Eastern Europe the difficult economic situation has led to a preference for subjects which are considered of immediate practical use for the country. The same goes for the so-called 'developing countries' and 'threshold countries'.

On the other hand in some Western countries, for example in Sweden or the Netherlands, history of technology has recently expanded. In the United States a considerable number of university institutes for the history of technology has been established from the 1960s onwards. In several of those institutes in the US, as well as in Sweden, Norway, and the Netherlands, history of technology is linked with 'Science and Technology Studies' (STS), an approach which has gained ground from the early 1980s. There is nothing wrong with this combination. On the contrary: by posing the question as to what lessons can be learnt from the history of technology for coming to grips with present-day problems of technology, STS can provide many useful insights and challenges. But one should be careful: science and technology studies generally restrict themselves to the twentieth century, and indeed in many cases there is a strong emphasis on post-1945 developments. It is obvious that the reach of the history of technology is wider and that in any case its possibilities as policy adviser are severely limited.

In this article I shall attempt to assess the present state of the academic discipline of history of technology in Germany. I shall start with a brief

* For Wolfhard Weber on his sixtieth birthday.

overview of the field from the 1960s onwards. This is necessary in order to elucidate the special conditions of the discipline in Germany. I shall then examine its development from around 1990 in more detail, discuss the effects of German reunification, deal with other factors which proved influential and conclude with the issue of a possible 'German style' of the historiography of technology. It goes without saying that this overview is only a subjective, preliminary assessment of the situation. Matters are still too much in flux to allow more substantial conclusions.

II

The history of technology as an academic discipline came into its own in the 1960s and 1970s.[1] Before that time there had been few places in Germany where historians of technology, generally with a background in engineering or science, dealt with the subject. But in the early 1960s political and academic interest in the history of technology became broader. In the United States history of technology was institutionalized in 1958 in the 'Society for the History of Technology' which soon started publishing its journal, *Technology and Culture*. Technological developments, such as nuclear technology, jet propulsion and rockets, which had been implemented during World War II, created an enormous impact on civil technology, and the 'Sputnik Shock' of 1957 demanded a thorough investigation into the origins of technological innovation. In Germany concepts of university reform, mainly adapted from the United States, stressed the function of 'interdisciplinary studies' in which the history of technology could perform the function of a bridge builder between the 'two cultures'. Economic growth in the Federal Republic, sometimes referred to as an 'economic miracle', highlighted the role of technological innovation. Historically orientated innovation studies could, academic policy-makers hoped, contribute to shedding some light on this question.

As it happened, however, the difference between high hopes and practical implementation was considerable. Still, between 1963 and 1969 Federal German policy formers of higher education established four professorships and a lecturership in the history of technology, sometimes in combination with the history of science or economic and social history.[2] Although this proved less than optimistic scholars had hoped for, it was sufficient to make an impact on historical research in Germany. During the 1970s the different Federal German states added more chairs in the discipline and scholars started to meet at annual symposia. In line with a new concept which regarded history primarily as a social science, some social historians set up the concept of a 'modern' approach to the history of technology.[3] The increasing interest of social historians in problems of the history of technology led, *inter alia*, to an emphasis on research into the relationship between technology and labour.[4] Not so much the consequences of technical development on conditions of labour stood in the forefront – this had been an important theme of social history for a long time – but research on changes in the production process. Scholars from the fields of economic and social history as well as from the history of science and technology shared an interest in innovation. They divided innovation into the stages of invention, innovation in some pioneer firms, and diffusion. Historians of science often concentrated on the invention process and on the relationship between science and technology in invention and early innovation, whereas historians with a background in social and economic history emphasized innovation and its diffusion. Although many general as well as social and economic historians stressed the importance of the history of technology for historical studies in general, the impact of the new discipline remained rather limited.

At historical conferences, it is true, matters of technology played some role, but historians of technology did not manage to place these at the centre of historical debate. Generally, this is still the situation today and applies not only to Germany but also to countries like the United States and France. In view of the well-known fact that the selection of fields of historical inquiry depends very much on themes which play a large role in contemporary debate, it is surprising that historians do not pay more attention to issues of technology which are obviously of great importance today. There are several reasons for this. One reason seems to be that the interest of many historians in technological matters is rather limited. If they had had a larger interest in technology, they would probably not have become historians. But what about the more 'technologically minded' students? As far as the history of technology as an academic discipline is concerned, not too much hope should be placed on them, either. If they study engineering, not many of them will have the time to deal with history of technology full-time. But the figure is considerably higher if the history of technology is a side subject in their curriculum.

In spite of the limiting factors, history of technology in German enjoyed some moderate expansion in the 1970s and 1980s. Implementing the concept of technology and labour with an emphasis on production processes, several *Länder* and cities in the Federal Republic of Germany founded industrial museums. These focused on industries like coal-mining, power machinery, automobile production or textiles, tried to re-create the conditions of the workplace and were of considerable importance for disseminating historical knowledge.[5] They recruited staff from university institutes of the history of technology and sometimes set up their own research departments, thus generating historical knowledge and not merely disseminating it. Apart from the venerable Deutsches Museum, founded in 1903, there were, *inter alia*, the decentralized Westphalian Industrial Museum with its headquarters in Dortmund (1979),[6] the Museum of Transport and Technology in Berlin, founded in 1982, which renamed itself the 'German Museum of Technology' in 1996, the Museum of Technology and Labour in Mannheim, established in 1982, which opened in 1985, and the Museum of Labour in Hamburg, established in 1989 and opened in 1997. Apart from local museums, for example the museum in Rüsselsheim (Hesse), which focused on automobile production, there were museums concentrating on special industrial branches such as the German Mining Museum in Bochum. Founded already in 1930, its mining archives opened in 1967 and a research institute on mining history was established a decade later.[7] The German Shipping Museum at Bremerhaven, established in 1971, concentrated on the history of shipbuilding, navigation and related matters.

The history of everyday life as well as environmental history were other areas of study which made an impact on the way newly established industrial museums were conceived. As with the emphasis on the production process, it was not so much technology *per se,* but the way technology affected the individual and society as a whole that became the focus of interest. This, of course, required new concepts of curatorial representation.[8]

From the late 1970s onwards issues of environmental history gained increasing importance in the academic history of technology.[9] In 1981 the Historical Commission of the Association of German Engineers held a symposium on environmental history, taking stock of what had been done and suggesting topics for further research.[10] In the late 1980s environmental history as an aspect of the history of technology and, indeed, of general history, was institutionalized by establishing an association devoted to this field. After the discipline in the Federal Republic of Germany had 'taken off' in the late 1960s and expanded in the 1970s it made its presence increasingly felt in an international context. The annual meetings of the Historical Association of German Engineers from the late 1970s onwards and other symposia had almost exclusively been national affairs. But from the early 1980s onwards German historians of technology played an increasingly active role in international associations and congresses, for example in the 'Society for the History of Technology' (SHOT) with its main base in North America, and in the 'International Committee for the History of Technology' (ICOHTEC) which is centred in Europe.[11] Sponsored by the Association of German Engineers and the German Research Foundation, ICOHTEC met for a symposium on 'Energy in History' in Lerbach near Cologne in 1984, and in 1989 German historians of technology were active in co-organizing the International Congress of History of Science in Hamburg and Munich with several sessions devoted to the history of technology.

In the German Democratic Republic history of technology was to a large extent part of the 'history of productive forces'.[12] On the basis of Marxist-Leninist philosophy the subject combined aspects of the history of technology with social and economic history. In the second half of the 1980s GDR historians published a three-volume history of the productive forces in Germany, covering the period 1800–1945.[13]

The Technical University of Dresden, the Freiberg Mining Academy (a Technical University), as well as the Technical University of Karl-Marx-Stadt (Chemnitz-Zwickau after reunification) were centres of the history of technology in East Germany. In the mid-1970s the industrial historian Rolf Sonnemann founded the 'Centre for the History of the Engineering Sciences' in Dresden, and, together with two colleagues, edited a 'History of Technology' in 1978 which took the literature of the West into consideration and proved useful as a general survey of the field.[14] But the major achievement of the 'Dresden School' was its research into the history of engineering sciences.[15] East German scholars presented much of their research findings at the 1986 ICOHTEC symposium in Dresden which was followed by a book in 1990.[16] As to the Freiberg Mining Academy it is not surprising that scholars there concentrated on the history of mining, as also on metallurgy. Already in 1978 ICOHTEC had held its ninth symposium on the history of mining and metallurgy in Freiberg. In the former Karl-Marx-Stadt (Chemnitz) history of techno-logy was taught in combination with the history of science.

III

For the history of technology as well as for many other academic disciplines German reunification in 1989 posed several problems. But there were also challenges and new opportunities. Some former disciplines in East Germany became obsolete, for example the obligatory courses in Marxist-Leninist philosophy of which, to an extent, the history of technology had been a part. Also, for obvious reasons, socialist political economy was no longer needed and the law subjects were radically transformed.

But what about the former history of technology as part of the history of the productive forces? Was reunification a blessing or a curse for the history of technology? Who were the winners, who the losers?

Obviously, after reunification the discipline could no longer be pursued in the traditional manner. Several East German historians of technology were made redundant. But reviewing the situation as a whole, 'colonization from the West' did not take place. This is true of Dresden: there was some disappointment, because the grand initial plans of the Saxon ministry of research and education had not been realized. Relating to the history of technology and regarding the quality of former research in Dresden, three professorships were envisaged: history of technology combined with the history of the technical sciences; industrial archaeology and industrial history; and the history of technology.[17] Mainly because of budgetary problems in the context of reunification only one professorship was finally re-established: the chair of the history of technology and engineering sciences was taken by a member of the former 'Dresden School'. As the Mining Academy (Technical University) of Freiberg was close to Dresden, the Saxon science policy-makers transformed the former chair in the history of productive forces and economic history into one of 'history of technology and industrial archaeology', thus trying to compensate for the failed attempt at establishing industrial archaeology at Dresden. The professorship for economic, cultural and regional history which had been envisaged for Freiberg was, again for budgetary reasons, not established. So, as far as figures are concerned, the numbers of scholars dealing with the history of technology in Dresden and Freiberg shrank after unification. It is still too early to assess if that loss in quantity has been made up by gains in quality.

In Chemnitz-Zwickau the professorship in the history of science and technology was transformed into one in the history of science, technology and technical education and given to a colleague who had gained his academic credentials in Dresden. The Technical University of Ilmenau had offered courses in the history of technology from 1983 onwards. In 1995, and to an extent due to various recommendations of the Association of German Engineers on engineering qualifications in the future, the state of Thuringia established a professorship in the history of technology and economic history which was also taken by a scholar of the 'Dresden School'. The Technical University of Brandenburg at Cottbus, newly founded in 1991, embarked on a reform of the engineering curriculum, stressing practical application of research findings and interdisciplinary co-operation. As in Freiberg, the professor in the history of technology came from West Germany. After a professorship in the history of civil engineering had been established in 1993 the chair of the history of technology followed in 1994. There was an emphasis on environmental history as well as on the industrial archaeology of the region, the latter in line with efforts in Freiberg and Chemnitz, but also with those at universities in West Germany like Bochum and Hamburg.

Federal and regional budgetary constraints, to an extent a consequence of the enormous financial expenditure of the 'old' Federal Republic in the context of reunification, also affected the western states of the Federal Republic. Whereas from the 1960s to the late 1980s there had been – moderate – expansion, consolidation was on the agenda in the 1990s. This meant that the times of newly founded chairs and institutes in the history of technology as well as of technical museums were to a large extent over. In fact, in some instances the term 'consolidation' served as a euphemism for contraction: in the case of the University of Bremen the chair of social and technological history, founded in 1970, was not renewed after the former chairholder had reached retirement age.

However, and seen in an international perspective, this consolidation phase took place at a relatively comfortable level. Research in different fields of the history of technology called for synthesis. Already in 1982 several German historians of technology had published a general history of technology designed for the 'intelligent layman' which reached a wide readership.[18] After the end of the 1980s a history of technology in Germany covering the nineteenth and twentieth centuries was published which stressed the advantages of 'small' as opposed to 'big' industries and emphasized regional development as well as aspects of environmental history. The five-volume *'Propyläen Technikgeschichte'* of the early 1990s featured the history of technology with an emphasis on Western Europe and North America,[19] whereas the ten-volume publication 'Culture and Technology' (1990–1995), sponsored by the 'Georg Agricola Society for the Advancement of the History of Science and Technology',[20] concentrated on the relationship between technology and other fields such as science, philosophy, religion, the state, the economy and the arts. Apart from these major works of synthesis German historians of technology absorbed recent international trends in their field and tried to make them applicable to studies of technological development in Germany: gender studies,[21] new approaches to cultural history which had gained considerable importance in France and the United States,[22] as also a stronger emphasis on 'everyday technology',[23] and on the use and consumption of technology as opposed to its design and production.[24] At institutional level an informal discussion group of mainly younger scholars interested in the history of technology was established in 1982[25] and the (German) Society for the History of Technology was founded in 1990.

IV

As mentioned above, consolidation at a relatively high level seems to be a fitting description of the current state of the history of technology in Germany. From the institutional point of view there are several opportunities open to scholars. But when confronted with some of the expectations of policy-makers who ask for advice on technology policy matters, historians of technology find it hard to justify their existence. Policy-makers do not always understand that historians of technology are ill-equipped to give this sort of advice. Regarding the opportunities of academic historians of technology in Germany much depends (as is true of other scholarly disciplines) on the energy and inspiration of individuals. If those qualities are there, funds for research are generally available.

Several scholarly associations in Germany are active in the history of technology; this year the Society of the History of Technology (SHOT) will hold a major international conference in Munich, where the 'Munich Centre for the History of Science and Technology' was established at the end of 1997. In the German media, a scholarly informed history of technology plays a distinctly larger role than two or three decades ago.

But there are still significant deficits. In the 1970s and the 1980s the discipline enjoyed some recognition in the historical sciences, although it certainly did not have a place at or near the centre of historical university curricula. At the meetings of the German Historical Society one or two sessions were generally devoted to the history of technology, but that was normally all. Although historians of technology in Germany have grown in numbers, their objective to play a more prominent role in the historical discipline has not been reached to a satisfactory degree. This is not a particularly German phenomenon but is also true of the United States where resources for the history of technology are more ample than in Germany.[26]

What are the reasons? As hinted at above, the general problem seems to reside in the fact that most historians seem to find it difficult to establish a relationship to matters technological. This is still the case, although the current role of technology in everyday life is visible everywhere. But there is reason for – at least moderate – hope. A shift of emphasis from the producer to the consumer, from production to use, may make a *rapprochement* between the history of technology and other historical disciplines easier, because such questions are of equal interest to historians of technology, of culture, and of society. Indeed, thinking along these lines and observing some recent trends in historiography, those formal boundaries are becoming obsolete: the issue to be investigated is at the centre and historians of different denominations are invited to bring their expertise to the task. On the other hand issues which had been in the forefront of research in the history of technology in the late 1960s or 1970s have not been taken from the agenda completely. To the contrary: we still have only a limited knowledge of the historical processes of invention and innovation, of the relationship between science and technology or of issues of technology transfer and technical education. In order to secure a larger role for technology in the discipline of history it would be vital to intensify efforts to allow a larger role for the history of technology in school history curricula. The more students of history are introduced to these matters the better are the chances of curricular reforms. But one should not be too optimistic. These are long-term processes in which much bureaucratic conservatism has to be overcome.

Another matter is the role of the history of technology in German engineering curricula for universities and technical universities.[27] Much is to be desired there, too. Although there is some agreement that 'non-engineering-subjects' should play a larger role in higher technical education, policy-makers generally favour courses in modern languages, business studies or patent law. Although the possibilities of practical application of historical knowledge are distinctly limited, it should be clear that informed decisions concerning future developments can only be made in the knowledge of how the present situation has come into being.

Is there a distinct 'German style' of research in the history of technology? From the point of view of content most German historians of technology who are knowledgeable about international developments in the field subscribe to methodological ideals and choices of subject matter which do not differ significantly from those of colleagues in, say, the US, Britain or France. Again much depends on personal interest, initiative, energy and open-mindedness. Probably reservations against 'big technological solutions' are stronger in German historiography of technology than in other countries. Very likely this has still to be seen in the context of the 'Third Reich' and its consequences in Germany.

From an institutional point of view there is a considerable number of scholarly societies which, in one way or another, deal with the history of technology. This number exceeds that in other industrial countries. Although the 'Gesellschaft für Technikgeschichte' ([German] Society for the History of Technology) is the most important of them, others do make their contribution, too. Obviously, there have been repeated proposals to concentrate efforts in one or, if need be, two societies. So far these ideas have not prevailed; indeed, a process of differentiation has set in with the objective of increasing the scope of the societies concerned. It will be interesting to watch future developments.

Notes and References

1. For a survey see W. Weber, 'History of Technology after 1945', *ICON, Journal of the International Committee for the History of Technology*, 1995, Vol. 1, 148–171 and L. Engelskirchen, 'Establishing History of Technology as an Academic Discipline in the Federal Republic of Germany after 1945', *ICON, Journal of the International Committee for the History of Technology*, 1998, Vol. 4, 146–156; W. Weber and L. Engelskirchen, *Streit um die Technikgeschichte in Deutschland 1945–1975* (Münster, 2000).

2. Engelskirchen (note 1), pp. 151–2.

3. A prominent example is K. Hausen and R. Rürup (eds.), *Moderne Technikgeschichte* (Cologne, 1975).

4. See *Geschichte und Gesellschaft*, 1978, Vol. 4, 143–290.

5. R. Wirtz, 'Industriekultur – Zur Konjunktur eines Konzeptes', in G.A. Stadter and A. Kuisle (eds.), *Technik zwischen Akzeptanz und Widerstand. Gesprächskreis Technikgeschichte 1982–1996* (Münster, 1999), 73–80.

6. H. Bönninghausen and R. Wirtz, 'Das Rheinische und Westfälische Industriemuseum. Zwei dezentrale Museen für Industrie- und Sozialgeschichte in Nordrhein-Westfalen', *Industriekultur, Denkmalpflege, Kulturlandschaft, Umwelt und Technikgeschichte*, 1995, 7–12.

7. W. Kroker, 'History of Technology at the German Mining Museum, Bochum', *History of Technology*, 1994, Vol. 16, 185–198.

8. W. Weber, 'Die Gründung technischer Museen in Deutschland im 20. Jahrhundert', *Museumskunde*, 1991, Vol. 56, 82–93; H.J. Klein, 'Vom Meisterwerk zum Werkmeister – Publikumsinteressen und Wahrnehmungsweisen bei der Darstellung von Technik und Industriekultur', *Museumskunde*, 1991, Vol. 54, 140–161.

9. See G. Bayerl, 'Materialien zur Geschichte der Umweltproblematik', *Technologie und Politik*, 1980, Vol. 16, 180–219. Already in 1967 Ilja Mieck had researched in this field. See I. Mieck, 'Aerem corrumpere non licet. Luftverunreinigung und Immissionsschutzgesetz in Preußen bis zur Gewerbeordnung 1869', *Technikgeschichte*, 1967, Vol. 34, 36–78. For the 1980s and 1990s see *inter alia* R.P. Sieferle, *Der unterirdische Wald. Energiekrise und Industrielle Revolution* (Munich, 1982); J. Radkau, 'Zur angeblichen Energiekrise des 18. Jahrhunderts: Revisionistische Betrachtungen über die 'Holznot', *Vierteljahrschrift für Sozial- und Wirtschaftsgeschichte*, 1986, Vol. 73, 1–37; F.J. Brüggemann und Th. Rommelsbacher (eds.), *Besiegte Natur. Geschichte der Umwelt im 19. und 20. Jahrhundert* (Munich, 1987); G. Bayerl, N. Fuchsloch, T. Meyer (eds.), *Umweltgeschichte. Methoden, Themen, Potentiale. Tagung des Hamburger Arbeitskreises für Umweltgeschichte, Hamburg 1994* (Münster, 1996); G. Bayerl and U. Troitzsch, *Quellen zur Geschichte der Umwelt von der Antike bis Heute* (Göttingen, Zürich, 1998).

10. U. Troitzsch, 'Historische Umweltforschung: Einleitende Bemerkungen über Forschungsstand und Forschungsaufgaben', *Technikgeschichte*, 1981, Vol. 48, 177–190.

11. H.J. Braun, 'Reforming ICOHTEC: International Committee for the History of Technology', *History of Technology*, 1994, Vol. 16, 212–215.

12. R. Sonnemann, 'Das Konzept der Geschichte der Produktivkräfte in der DDR-Geschichtswissenschaft', *Dresdener Beiträge zur Geschichte der Technikwissenschaften*, 1996, Vol. 24, 1–19.

13. R. Berthold *et al.* (eds.), *Produktivkräfte in Deutschland 1800–1945*, 3 vols (Berlin, 1985–1990).

14. R. Brentjes, S. Richter and R. Sonnemann (eds.), *Geschichte der Technik* (Leipzig, 1978).

15. Th. Hänseroth, K. Mauersberger and S.H. Richter (eds.), 'Geschichte der Technikwissenschaften an der TU Dresden', *Wissenschaftliche Zeitschrift der TU Dresden*, 1990, Vol. 39, No. 5, 1–6; Th. Hänseroth and K. Mauersberger, 'Das Dresdener Konzept zur Genese technikwissenschaftlicher Disziplinen – eine Bilanz', *Dresdener Beiträge zur Geschichte der Technikwissenschaften*, 1996, Vol. 24, 20–45; Th. Hänseroth (ed.), *Technik und Wissenschaft als produktive Kräfte in der Geschichte. Rolf Sonnemann zum 70. Geburtstag* (Dresden (Technical University), 1998). In the present context, the contributions by Hänseroth, Mauersberger, Troitzsch and Weber are of special interest.

16. G. Buchheim and R. Sonnemann (eds.), *Geschichte der Technikwissenschaften* (Leipzig, 1990).

17. For this, the following compilations by W. Weber, 'Naturwissenschaft- und Technikgeschichte in der Bundesrepublik Deutschland 1989–1992. Eine Übersicht über Forschung und Lehre an den Instituten', *Berichte zur Wissenschaftsgeschichte*, 1993, Vol. 16, 1–116; with index as book: Weinheim 1993, and 'Naturwissenschaftsgeschichte und Technikgeschichte in Deutschland 1993–1996. Eine Übersicht über Forschung und Lehre an den Institutionen', *Berichte zur Wissenschaftsgeschichte*, 1997, Vol. 20, pp. 1–130; with index as book: Weinheim 1997, are particularly useful.

18. U. Troitzsch and W. Weber (eds.) *Die Technik. Von den Anfängen bis zur Gegenwart* (Brunswick, 1982, 3rd ed. 1989).

19. W. König (ed.), *Propyläen Technikgeschichte*, 5 vols, (Berlin, 1990–1992).

20. A. Hermann (ed.), *Kultur und Technik*, 10 vols, (Düsseldorf, 1990–1995).

21. M. Osietzki, 'Männertechnik und Frauenwelt. Technikgeschichte aus der Perspektive des Geschlechterverhältnisses', *Technikgeschichte*, 1992, Vol. 59, 45–72; D. Schmidt and K. Zachmann, 'Der Ansatz der Geschlechtergeschichte in der Technikgeschichte oder: Wann die Technikgeschichte Geschlechtergeschichte braucht', *Blätter für Technikgeschichte*, 1995/96, Vol. 57/58, 87–98; N.E. Lerman, A.P. Mohun and R. Oldenziel (eds.), 'Gender Analysis and the History of Technology', Special Issue of *Technology and Culture*, 1997, Vol. 38(1).

22. M. Osietzki, 'Für eine neue Technikgeschichte', *Österreichische Zeitschrift für Geschichtswissenschaften*, 1992, Vol. 3, 293–318. On this and related approaches see A. Paulinyi, 'Wi(e)der eine neue Technikgeschichte (?)!', *Blätter für Technikgeschichte*, 1995/96, Vol. 57/58, 39–48.

23. B. Orland, 'Wie kann man den Alltagsbegriff für die Technikgeschichte nutzbar machen?', *Ferrum, Nachrichten aus der Eisenbibliothek*, 1999, Vol. 71, 4–10.

24. C. Pursell, 'Seeing the Invisible: New Perceptions in the History of Technology', *ICON, Journal of the International Committee for the History of Technology*, 1995, Vol. 1, 9–15. Also on recent trends H. Lackner and H. Luxbacher, 'Technikgeschichte in Deutschland und Österreich', *Österreichische Zeitschrift für Geschichtswissenschaften*, 1992, Vol. 3, 279–292; U. Troitzsch, 'Technikgeschichte', in H.-J. Goertz (ed.), *Geschichte. Ein Grundkurs* (Reinbek bei Hamburg, 1998), 379–393; H.J. Braun, 'Current Research in the History of Technology in Europe', *History of Technology*, 1999, Vol. 21, 167–188.

25. G.A. Stadler and A. Kuisle (eds.), *Technik zwischen Akzeptanz und Widerstand. Gesprächskreis Technikgeschichte 1982–1996* (Münster, 1999).

26. On this general issue see W. Weber, 'History of Technology in Teaching the Humanities and the Social Sciences', in Claude Debru (ed.), *History of Science and Technology in Education and Training in Europe* (Luxemburg, 1999), 125–128.

27. H.J. Braun, 'History of Technology in the Training of Engineers', *op. cit.* (26), 67–73.

Contents of Former Volumes

* indicates out of print

SEVENTH ANNUAL VOLUME, 1982*

EIGHTH ANNUAL VOLUME, 1983*

NINTH ANNUAL VOLUME, 1984*

TENTH ANNUAL VOLUME, 1985*

ELEVENTH ANNUAL VOLUME, 1986*

TWELFTH ANNUAL VOLUME, 1990

THIRTEENTH ANNUAL VOLUME, 1991*

FOURTEENTH ANNUAL VOLUME, 1992

FIFTEENTH ANNUAL VOLUME, 1993

SIXTEENTH ANNUAL VOLUME, 1994

SEVENTEENTH ANNUAL VOLUME, 1995

EIGHTEENTH ANNUAL VOLUME, 1996

NINETEENTH ANNUAL VOLUME, 1997

TWENTIETH ANNUAL VOLUME, 1998

TWENTY-FIRST ANNUAL VOLUME, 1999